Studies in Computational Intelligence

Volume 737

Series editor

Janusz Kacprzyk, Polish Academy of Sciences, Warsaw, Poland
e-mail: kacprzyk@ibspan.waw.pl

About this Series

The series "Studies in Computational Intelligence" (SCI) publishes new developments and advances in the various areas of computational intelligence—quickly and with a high quality. The intent is to cover the theory, applications, and design methods of computational intelligence, as embedded in the fields of engineering, computer science, physics and life sciences, as well as the methodologies behind them. The series contains monographs, lecture notes and edited volumes in computational intelligence spanning the areas of neural networks, connectionist systems, genetic algorithms, evolutionary computation, artificial intelligence, cellular automata, self-organizing systems, soft computing, fuzzy systems, and hybrid intelligent systems. Of particular value to both the contributors and the readership are the short publication timeframe and the worldwide distribution, which enable both wide and rapid dissemination of research output.

More information about this series at http://www.springer.com/series/7092

Mirjana Ivanović · Costin Bădică
Jürgen Dix · Zoran Jovanović
Michele Malgeri · Miloš Savić
Editors

Intelligent Distributed Computing XI

Springer

Editors
Mirjana Ivanović
Department of Mathematics and Informatics, Faculty of Sciences
University of Novi Sad
Novi Sad
Serbia

Costin Bădică
Faculty of Automatics, Computer Science and Electronics
University of Craiova
Craiova
Romania

Jürgen Dix
Institut für Informatik
Technische Universität Clausthal
Clausthal-Zellerfeld
Germany

Zoran Jovanović
School of Electrical Engineering
University of Belgrade
Belgrade
Serbia

Michele Malgeri
Dipartimento di Ingegneria Elettrica, Elettronica e Informatica
Universita degli Studi di Catania
Catania
Italy

Miloš Savić
Department of Mathematics and Informatics, Faculty of Sciences
University of Novi Sad
Novi Sad
Serbia

ISSN 1860-949X ISSN 1860-9503 (electronic)
Studies in Computational Intelligence
ISBN 978-3-319-88230-7 ISBN 978-3-319-66379-1 (eBook)
https://doi.org/10.1007/978-3-319-66379-1

Printed on acid-free paper

This Springer imprint is published by Springer Nature
The registered company is Springer International Publishing AG
The registered company address is: Gewerbestrasse 11, 6330 Cham, Switzerland

Preface

Intelligent distributed computing emerged as the result of the fusion and cross-fertilization of ideas in Intelligent Computing and Distributed Computing. Its roots come from artificial intelligence in the 1970s, when the idea of cooperating agents came to life. Its outcome is the development of a new generation of intelligent distributed systems, by combining methods and technology from classical artificial intelligence, computational intelligence, and multi-agent systems taking into account, also, security concerns and emerging IoT applications.

This volume contains the proceedings of the 11th International Symposium on Intelligent Distributed Computing, IDC'2017. The symposium was hosted by the School of Electrical Engineering from the University of Belgrade, in Belgrade, Serbia, from 11 to 13 October 2017. IDC'2017 continues the tradition of the IDC Symposium Series that started 11 years ago as an initiative of two research groups:

(i) Systems Research Institute, Polish Academy of Sciences, Warsaw, Poland, and
(ii) Software Engineering Department, University of Craiova, Craiova, Romania.

The IDC Symposia welcome submissions of original papers on all aspects of intelligent distributed computing ranging from concepts and theoretical developments to advanced technologies and innovative applications. The symposia aim to bring together researchers and practitioners involved in all aspects of intelligent distributed computing. IDC is interested in works that are relevant for both Distributed Computing and Intelligent Computing, with scientific merit in these areas.

The IDC'2017 event comprised the main conference organized in eight sessions: (1) Distributed Algorithms and Optimization, (2) Reasoning and Decision Making in Distributed Environments, (3) Multi-agent Systems, (4) Data Analysis, Mining, and Integration, (5) Machine Learning, (6) Internet of Things and Cloud Computing, (7) Service-based Distributed Systems, and (8) WASA 2017 (7th Workshop on Applications of Software Agents). The proceedings book contains contributions 22 regular, and 6 short papers selected from a total of 52 received submissions from 30 countries (counting the country of each coauthor for each paper submitted). Each submission was carefully reviewed by at least three members of the Program

Committee. Acceptance and publication were judged based on the relevance to the symposium topics, clarity of presentation, originality and accuracy of results, and proposed solutions. The acceptance rates were 46.15%, counting only regular papers, and 61.54% when including also short papers (four accepted papers were withdrawn during the finalization process). The 28 contributions published in this book address many topics related to theory and applications of intelligent distributed computing including: cloud computing, P2P networks, agent-based distributed simulation, ambient agents, smart and context-driven environments, Internet of Things, network security, mobile computing, unmanned vehicles, augmented physical reality, swarm computing, team and social computing, constraints and optimization, and information fusion.

We would like to thank Janusz Kacprzyk, editor of Studies in Computational Intelligence series and member of the Steering Committee, for his continuous support and encouragement for the development of the IDC Symposium Series. Also, we would like to thank the IDC'2017 Program Committee members for their work in promoting the event and refereeing submissions. A special thank you to all colleagues who submitted their work to this event.

We are thankful for the talks delivered by our invited speakers Eva Onaindia (Valencia, Spain), Bela Stantic (Brisbane, Australia), and Karl Tuyls (Liverpool, United Kingdom): Thank you very much for these interesting lectures.

Finally, we acknowledge and appreciate the efforts of the main organizers from the Department of Mathematics and Informatics, Faculty of Sciences, University of Novi Sad, Serbia, for organizing this event. A special thanks also go to the co-organizers from the School of Electrical Engineering, University of Belgrade, Serbia, for hosting the event and such a beautiful location.

Novi Sad, Serbia — Mirjana Ivanović
Craiova, Romania — Costin Bădică
Clausthal-Zellerfeld, Germany — Jürgen Dix
Belgrade, Serbia — Zoran Jovanović
Catania, Italy — Michele Malgeri
Novi Sad, Serbia — Miloš Savić
July 2017

Contents

Part I
Distributed Algorithms and Optimization

A Performance Analysis of Self-★ Evolutionary Algorithms on Networks with Correlated Failures

Rafael Nogueras and Carlos Cotta

Abstract We consider the deployment of island-based evolutionary algorithms (EAs) on unstable networks whose nodes exhibit correlated failures. We use the sandpile model in order to induce such complex, correlated failures in the system. A performance analysis is conducted, comparing the results obtained in both correlated and non-correlated scenarios for increasingly large volatility rates. It is observed that simple island-based EAs have a significant performance degradation in the correlated scenario with respect to its uncorrelated counterpart. However, the use of self-★ properties (self-scaling and self-sampling in this case) allows the EA to increase its resilience in this harder scenario, leading to a much more gentle degradation profile.

Keywords Evolutionary algorithms · Self-★ properties · Ephemeral computing · Sandpile model

1 Introduction

The use of parallel environments is of paramount interest for tackling intensive computational tasks. In particular, evolutionary algorithms (EAs) have a long success story in this kind of environments, dating back to the 1980s. In this sense, there has been during the last years an important focus on the use of EAs in emergent computational scenarios that depart from classical dedicated networks so common in the past. Among these we can cite cloud computing [13], P2P networks [21], or volunteer computing [5], just to name a few. The dynamic nature of the underlying computational substrate is one of the most distinguished features of some of these new scenarios—consider for example a P2P network in which nodes enter or leave the system subject to some uncontrollable dynamics caused by user interventions, network disruptions, eventual crashes, etc. The term *churn* is used to denote this phe-

R. Nogueras · C. Cotta (✉)
ETSI Informática, Campus de Teatinos, Universidad de Málaga,
29071 Málaga, Spain
e-mail: ccottap@lcc.uma.es

M. Ivanović et al. (eds.), *Intelligent Distributed Computing XI*,
Studies in Computational Intelligence 737, https://doi.org/10.1007/978-3-319-66379-1_1

nomenon [17]. Under some circumstances, a potential solution to this issue might be to hide these computational fluctuations under an intermediate layer, providing a virtual stable environment to algorithms running on it. Nonetheless, this can constitute a formidable challenge, mainly in situations in which the underlying substrate is composed of nodes with low computing power just providing brief, ephemeral bursts of computation (think of, e.g., a large collection of low-end networked devices—cell phones, smart wearables, etc.—contributing their idle time) [6]. The alternative is making the algorithm aware of the dynamic nature of the environment, endowing it with the means for reacting and self-adapting to the volatility of the computational substrate. EAs are in this regard well-suited to this endeavor, since they are resilient techniques that have been shown to be able to withstand—at least to some extent—the sudden loss of part of the population [11], and can be readily endowed with self-★ properties [2] so as to exert self-control on their functioning.

Recent work has precisely studied the use of self-★ properties such as self-scaling [15] and self-healing [14] in this context, providing some evidence on the contribution of these techniques to the robustness of the algorithm when run on unstable computational environments. Quite interestingly, these previous studies have however only considered simple network models in which the dynamics of each node is independent of the rest of the network, that is, the availability of a computing node does not depend on the availability of other nodes. A more general situation would encompass correlated availability patterns, that is, the dynamics of each node might be affected by the dynamics of other nodes, see e.g., [10]. Overall, the presence of correlated failures puts to test the robustness and resilience of the EA, and hence studying it can provide a wider perspective on the usefulness of self-★ techniques to cope with computational instability.

2 Methodology

We consider an island-based EA running on a simulated unstable environment. Each island runs on a computational node of the system, whose availability fluctuates along time. When a computational node goes down, its contents are lost. Similarly, when a computational node is reactivated, the island running on it must be created anew using some particular procedure. In the following subsections we shall describe in more detail the model of the computational scenario and the mechanisms used by the EA to cope with instability.

2.1 Network Model

Let us consider a network composed on n_ι nodes interconnected following a certain topology. More precisely, we consider a regular lattice with von Neumann connectivity (virtual topology used for the purposes of migration in the island model) overlaid on a scale-free network (underlying topology for the purposes of failure correlation) as it is often the case in P2P networks, e.g., [12]. In the latter, node degrees are distributed following a power-law and hence there will be a few hubs with large connectivity and increasingly more nodes with a smaller number of neighbors. To generate this kind of networks we use the Barabási-Albert model [1], whereby the network is grown from a clique of $m + 1$ nodes by adding a node at a time, connecting it to m of the nodes previously added (selected with probability proportional to their degree—the so-called, preferential attachment mechanism) where m is a parameter of the model.

As stated before, these nodes are volatile, and may abandon the system and re-enter it at a later time, eventually repeating the process over and over again. To model this instability we consider two scenarios: (i) independent or non-correlated failures and (ii) correlated failures. The first one is the simplest model. Therein, the dynamics of each node is independent of other nodes. Each of them can switch from active to inactive or vice versa independently of other nodes with some probability $p(t)$ that only depends on the time it has been in its current state. Following previous work, as well as the commonly observed behavior of e.g., P2P systems [17], $p(t)$ follows a Weibull distribution. This distribution is controlled by two parameters β and η. The first one is the scale parameter and captures the spread of the distribution. The larger this parameter, the less frequent failure events are. The second one is the shape parameter and captures the effect that time has on failure events: for $\eta > 1$ (resp. $\eta < 1$), the longer the time elapsed, the more (resp. less) likely a failure event will be. If η was exactly 1, failures would be time-independent and hence exponentially distributed.

As to the correlated scenario, it features node failures that will be influenced by neighboring nodes. Consider for example the case of sensor networks in which nodes with a large number of active neighbors have their energy depleted faster due to the increased energy toll for communications, or the case of networks that carry load and in which the failure of a node makes other ones absorb the load of the latter, eventually resulting in additional overload failures [10]. This can be modeled in different ways, e.g., [4, 20]. In this work we have considered the sandpile model in order to induce cascading failures [8]. Much like in the previous case, we consider micro-failure events happening on each node with a certain probability $p(t)$. Now, each node i will have an associated threshold value θ_i, indicating the number of micro-failure events required for it to go down. When the number of such micro-failures effectively equals this threshold, the node is disconnected from the system, and each of the active neighbors of this node receives an additional micro-failure event.[1] In case any of these neighbors now accumulated a number of micro-failures equal to its own threshold, it would go down as well, propagating in turn another micro-failure to its active neighbors, and so on (hence the possibility of cascading failures). Figure 1 shows an example: after node b (which was in a critical state, i.e., one event short of going down) fails, neighboring nodes a and e (which were also in

[1]It must be noted that these so-called micro-failures are not intended to represent any real phenomenon, but are just used as a means to introduce failure interdependencies.

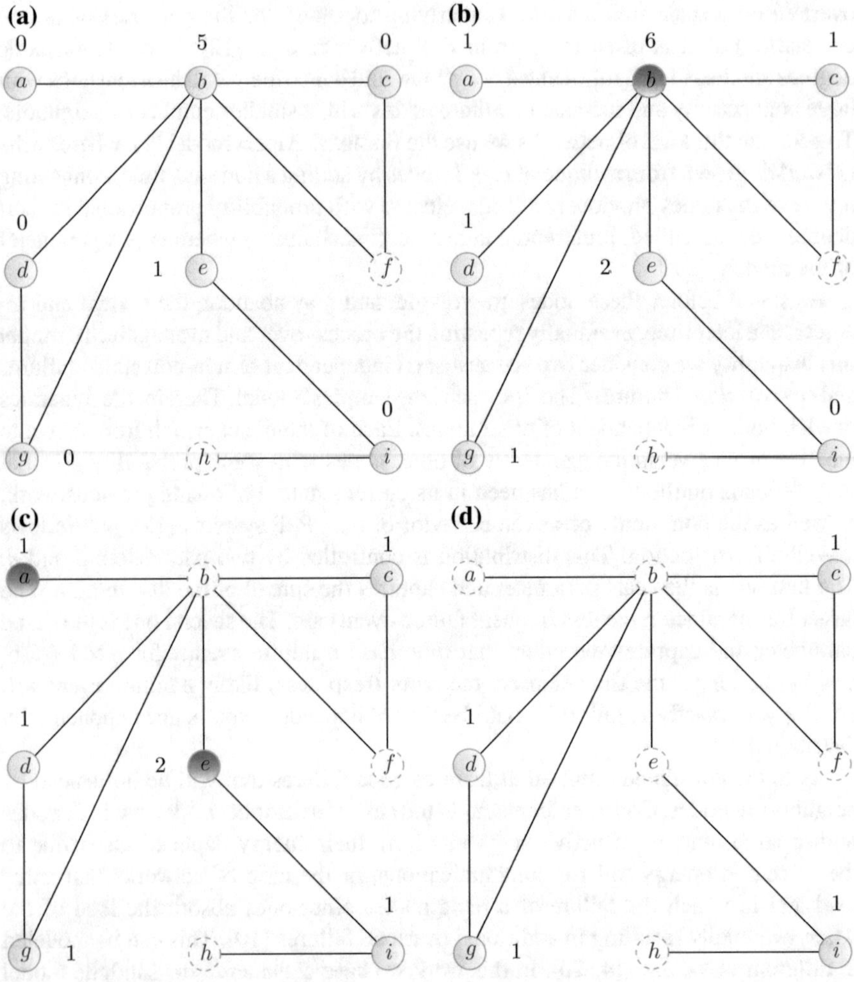

Fig. 1 Example of failure propagation in the sandpile model. Active (resp. inactive) nodes are depicted with *solid* (resp. *dashed*) *borders*. The numbers next to each active node indicate the cumulated number of failure events. The threshold θ_i for each node equals here its degree. **a** Initial state, **b** Failure on node b, **c** Failure propagation to nodes a and e. **d** Final state

such a critical state) fail as well. We have considered for simplicity that the threshold θ_i of each node i is constant and equal to the number of neighbors (active or inactive) of the node. As to reactivation, just like in the non-correlated case a single event is required.

2.2 Algorithmic Model

Two variants of the island-based EA are considered: (i) a basic one (termed noB) in which every island has a fixed size and random reinitialization is used whenever a new node enters the system, and (ii) a self-★ EA (termed LBQ) that uses self-scaling and self-sampling to re-size each island individually in response to fluctuations in the number of active neighbors and in the population sizes of these. In both cases, the islands run a basic steady-state EA and stochastically perform migration (with probability p_{mig}) of a single individual to neighboring islands. In each migration event the migrant is randomly selected from the current population and the receiving island inserts it in its population by replacing the worst individual.

Regarding the self-★ properties considered, self-scaling attempts to attain a rather stable global population size across the islands active in each moment. To this end, each island periodically monitors the state of its neighbors to determine: whether they are active or not, their population sizes and the number of active neighbors they have in turn. When a neighboring island is detected to have just gone down, the island increases its own population size in order to compensate the loss of the former. This is done by calculating the fraction of the population size of the deactivated island corresponding to the number of active neighbors it had. On the other hand, if all neighboring islands are active then they exchange individuals in order to balance their population sizes. See [15] for details. Note that this is a completely autonomous and decentralized policy and therefore each node cannot comprehend the global state of the network, for instance, a node does not account for the simultaneous failure of nodes that are themselves neighbors, hence the interest of studying the robustness of the EA in the correlated scenario.

As to self-sampling, it amounts to maintaining within each island a probabilistic model of its current population in order to sample it whenever the population needs to grow. This has the advantage of introducing diversity due to the stochastic sampling, keeping as well the momentum of the search since the newly created individuals are coherent with the current state of the population (unlike the case of using random individuals to this end). In this work we have considered the use of a tree-like bivariate probabilistic model such as that used in the COMIT estimation of distribution algorithm [3]—see also [14] for details.

3 Experimental Results

We consider $n_ι = 64$ islands whose initial size is $\mu = 32$ individuals and a total number of evaluations *maxevals* = 250 000. Each island runs a steady-state EA with one-point crossover, bit-flip mutation, binary tournament selection and replacement of the worst parent. We use crossover probability $p_X = 1.0$, mutation probability $p_M = 1/\ell$, where ℓ is the genotype length, and migration probability $p_{mig} = 1/160$. Regarding the network parameters, we use $m = 2$ in the Barabási-Albert model in order to define the topology; as for node deactivation/reactivation, we use the shape parameter $\eta = 1.5$ (larger than 1 and hence implying an increasing hazard rate with time), and scale parameters $\beta = -1/\log(p)$ for $p = 1 - (kn_\iota)^{-1}, k \in \{1, 2, 5, 10, 20\}$. To interpret these parameters, note that they would correspond to an average of one micro-failure event every k cycles if the failure rate was constant. This provides different scenarios ranging from low volatility ($k = 20$) to very high volatility ($k = 1$). To gauge the results, we also perform experiments with $k = \infty$ (situation corresponding to a stable network without failures). Also, in order to have a more meaningful comparison between both scenarios (accommodating the fact that several micro-failure events are required in order to take down a node in the correlated case but only one is needed in the non-correlated case), in the non-correlated scenario we adjust k values as $k' = k\tilde{\theta}$, where $\tilde{\theta}$ is the average of all θ_i values in the correlated scenario (which in this case is also the average degree of the network). Note at any rate that the main focus of the experimentation is the relative behavior of the algorithms considered in either scenario rather than a comparative between scenarios in absolute terms.

As stated in Sect. 2.2, we consider two algorithmic variants: noB (a standard island-based EA with fixed island sizes and random reinitialization of islands upon reactivation) and LBQ (the island-based EA endowed with self-sampling and self-scaling). The experimental benchmark comprises three test functions, namely Deb's trap function [7] (TRAP, concatenating 32 four-bit traps), Watson et al.'s Hierarchical-if-and-only-if function [19] (HIFF, using 128 bits) and Goldberg et al.'s Massively Multimodal Deceptive Problem [9] (MMDP, using 24 six-bit blocks). We perform 25 simulations for each algorithm, problem, volatility scenario and failure model.

Figure 2 shows a summary of the results (detailed numerical data for each problem, algorithm, and network model are provided in Tables 1 and 2). Let us firstly

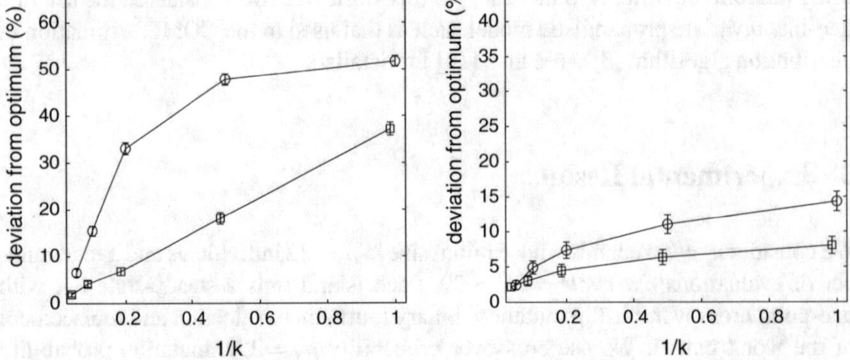

Fig. 2 Average deviation from the optimal solution across all problems for each algorithmic variant and network failure model. **a** noB. **b** LBQ

Table 1 Results (averaged for 25 runs) of the different EAs on the three problems considered under the network model with non-correlated failures. The median ($\tilde{x}$), mean ($\bar{x}$) and standard error of the mean ($\sigma_{\bar{x}}$) are indicated

Strategy	k	TRAP		H-IFF		MMDP	
		$\tilde{x}$	$\bar{x} \pm \sigma_{\bar{x}}$	$\tilde{x}$	$\bar{x} \pm \sigma_{\bar{x}}$	$\tilde{x}$	$\bar{x} \pm \sigma_{\bar{x}}$
–	∞	0.00	0.00 ± 0.00	0.00	5.33 ± 1.49	1.50	1.50 ± 0.17
noB	20	0.00	0.10 ± 0.07	0.00	3.78 ± 1.27	1.50	1.84 ± 0.20
	10	0.00	0.25 ± 0.10	11.11	9.44 ± 1.42	3.00	2.73 ± 0.26
	5	1.25	1.20 ± 0.23	16.67	14.47 ± 1.53	4.49	4.74 ± 0.31
	2	10.00	9.20 ± 0.61	32.64	31.97 ± 0.96	13.15	13.21 ± 0.34
	1	30.00	29.88 ± 0.80	53.65	53.35 ± 0.58	28.96	28.25 ± 0.57
LBQ	20	0.00	0.05 ± 0.05	0.00	6.22 ± 1.37	0.00	0.30 ± 0.12
	10	0.00	0.00 ± 0.00	11.11	9.11 ± 1.66	0.00	0.06 ± 0.06
	5	0.00	0.10 ± 0.07	16.67	13.00 ± 1.66	0.00	0.30 ± 0.12
	2	0.00	0.35 ± 0.11	19.44	18.22 ± 1.39	0.00	0.48 ± 0.14
	1	0.00	0.90 ± 0.22	22.22	22.28 ± 0.87	0.00	0.96 ± 0.23

focus on noB (Fig. 2a). As expected, the performance of the algorithm degrades as node volatility increases (that is, as we move to the right along the X axis). It is nevertheless interesting to note how the degradation profile of noB is more marked in the correlated scenario. More frequent and simultaneous node failures have a clear toll on performance. If we now consider the case of LBQ, two major observations stand out: on one hand, the performance of LBQ is notably better than that of noB for the same volatility rate. This had been already observed in the non-correlated case (albeit for multimemetic algorithms—this behavior is hence extended for plain EAs as well) and is now confirmed in the correlated scenario, indicating than the self-★ properties seem to keep providing robustness to the algorithm in this case too. As a matter of fact—and this leads to the second observation—the degradation of performance in the correlated case is much less marked for LBQ than it was for noB. More precisely, if we conduct a ranksum test on the results obtained by each algorithm on each problem and network scenario we observe that the performance of noB significantly (at level $\alpha = 0.01$) degrades in the correlated scenario with respect to the non-correlated one for all churn rates, whereas LBQ is only significantly degraded for moderate and

Table 2 Results (averaged for 25 runs) of the different EAs on the three problems considered under the network model with correlated failures. The median ($\tilde{x}$), mean ($\bar{x}$) and standard error of the mean ($\sigma_{\bar{x}}$) are indicated

Strategy	k	TRAP		H-IFF		MMDP	
		$\tilde{x}$	$\bar{x} \pm \sigma_{\bar{x}}$	$\tilde{x}$	$\bar{x} \pm \sigma_{\bar{x}}$	$\tilde{x}$	$\bar{x} \pm \sigma_{\bar{x}}$
–	∞	0.00	0.00 ± 0.00	0.00	5.33 ± 1.49	1.50	1.50 ± 0.17
noB	20	1.25	1.47 ± 0.21	16.67	13.18 ± 1.68	4.49	4.93 ± 0.41
	10	6.87	7.15 ± 0.41	25.87	26.97 ± 1.05	11.98	12.19 ± 0.43
	5	26.25	26.15 ± 0.85	47.40	47.67 ± 0.63	25.97	25.35 ± 0.48
	2	46.88	46.33 ± 0.58	61.46	61.19 ± 0.29	35.46	35.87 ± 0.40
	1	51.25	51.27 ± 0.50	63.72	63.80 ± 0.21	39.95	40.08 ± 0.36
LBQ	20	0.00	0.05 ± 0.05	11.11	7.22 ± 1.49	0.00	0.06 ± 0.06
	10	0.00	0.10 ± 0.07	16.67	14.06 ± 1.57	0.00	0.60 ± 0.23
	5	0.00	0.70 ± 0.18	19.44	20.61 ± 1.19	0.00	0.78 ± 0.21
	2	2.50	2.10 ± 0.21	27.78	27.08 ± 0.83	4.49	3.95 ± 0.37
	1	5.00	5.68 ± 0.41	31.94	30.53 ± 1.04	7.49	6.83 ± 0.42

high churn rates ($k \leqslant 5$ for TRAP and HIFF and $k \leqslant 2$ for MMDP). This is not to say that LBQ is not adversely affected by the new scenario (in the non-correlated case the performance of LBQ was only significantly degraded with respect to the stable $k = \infty$ case for $k \leqslant 5$ in HIFF and $k \leqslant 2$ in TRAP, whereas in the correlated scenario there are statistically significant differences for $k \leqslant 2$ in MMDP, $k \leqslant 5$ in TRAP and $k \leqslant 10$ in HIFF) but this degradation is mostly in the most volatile cases (unlike noB, whose performance is degraded with respect to $k = \infty$ in the correlated case for all churn rates in all three problems) and not so large in magnitude as for noB. A result consistent with this can also be seen in Fig. 3, in which the genetic diversity of the population (measured using Shannon's entropy) is depicted for each algorithm and scenario (the data corresponds to the TRAP function, but the behavior is qualitatively similar in the remaining problems). Notice how noB faces increasingly large difficulties to converge as the volatility goes up, and how these difficulties are noticeable even for low-volatility settings in the correlated scenario. LBQ can however maintain a better focus on the search, and seems mostly affected in the most volatile settings of the correlated scenario.

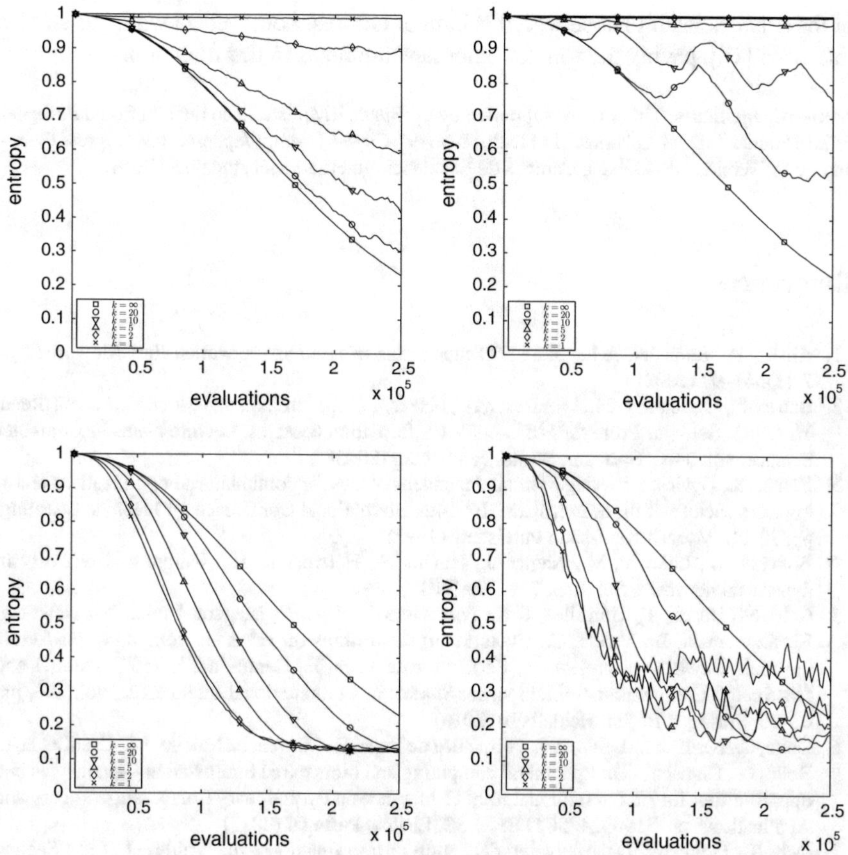

Fig. 3 Genetic diversity for the TRAP function. The *top row* corresponds to noB and the *bottom row* to LBQ; the *left column* corresponds to non-correlated failures, and the *right row* to correlated failures

4 Conclusions

The presence of correlated failures constitutes a major threat to the robustness of computing networks. We have analyzed in this work how this phenomenon may affect the performance of an island-based EA, and observed a marked degradation of the results in absence of appropriate policies to deal with this harder scenario. Endowing the EA with self-★ properties can however increase its resilience and make it able to withstand from low up to moderately high volatility.

There are several avenues for further work. First of all, other topologies could be tried. Work is under way here. Also, different models of correlated failures could be tried, using either dynamic thresholds (work is in progress in this area [16]) or other alternative models [4, 20]. In the longer term, a related problem is the optimization of the network itself to cope with this kind of failures. Some recent work has tackled this issue [18], paving the way for other developments in this direction.

Acknowledgements This work is supported by the Spanish Ministerio de Economía and European FEDER under Project EphemeCH (TIN2014-56494-C4-1-P)—http://ephemech.wordpress.com— and by Universidad de Málaga, Campus de Excelencia Internacional Andalucía Tech.

References

1. Albert, R., Barabási, A.L.: Statistical mechanics of complex networks. Rev. Modern Phys. **74**(1), 47–97 (2002)
2. Babaoğlu, Ö., Jelasity, M., Montresor, A., Fetzer, C., Leonardi, S., van Moorsel, A., van Steen, M. (eds.): Self-star Properties in Complex Information Systems. Lecture Notes in Computer Science, vol. 3460. Springer, Berlin, Heidelberg (2005)
3. Baluja, S., Davies, S.: Using optimal dependency-trees for combinatorial optimization: learning the structure of the search space. In: 14th International Conference on Machine Learning, pp. 30–38. Morgan Kaufmann Publishers (1997)
4. Böttcher, L., Luković, M., Nagler, J., Havlin, S., Herrmann, H.J.: Failure and recovery in dynamical networks. Sci. Rep. **7**, 41729 (2017)
5. Cole, N., Desell, T., González, D.L., Fernández de Vega, F., Magdon-Ismail, M., Newberg, H., Szymanski, B., Varela, C.: Evolutionary algorithms on volunteer computing platforms: the milkyway@home project. In: Fernández de Vega, F., Cantú-Paz, E. (eds.) Parallel and Distributed Computational Intelligence. Studies in Computational Intelligence, vol. 269, pp. 63–90. Springer, Berlin, Heidelberg (2010)
6. Cotta, C., Fernández-Leiva, A.J., Fernández de Vega, F., Chávez, F., Merelo, J.J., Castillo, P.A., Bello, G., Camacho, D.: Ephemeral computing and bioinspired optimization—challenges and opportunities. In: 7th International Joint Conference on Evolutionary Computation Theory and Applications, pp. 319–324. SCITEPRESS, Lisboa, Portugal (2015)
7. Deb, K., Goldberg, D.: Analyzing deception in trap functions. In: Whitley, L. (ed.) Second Workshop on Foundations of Genetic Algorithms, pp. 93–108. Morgan Kaufmann Publishers, Vail, Colorado, USA (1993)
8. Dorogovtsev, S.N., Goltsev, A.V., Mendes, J.F.F.: Critical phenomena in complex networks. Rev. Mod. Phys. **80**, 1275–1335 (2008)
9. Goldberg, D., Deb, K., Horn, J.: Massive multimodality, deception and genetic algorithms. In: Männer, R., Manderick, B. (eds.) Parallel Problem Solving from Nature—PPSN II, pp. 37–48. Elsevier Science Inc., New York, NY, USA (1992)
10. Kong, Z., Yeh, E.M.: Correlated and cascading node failures in random geometric networks: a percolation view. In: 2012 Fourth International Conference on Ubiquitous and Future Networks (ICUFN), pp. 520–525. IEEE, Phuket, Thailand, July 2012
11. Lombraña González, D., Jiménez Laredo, J., Fernández de Vega, F., Merelo Guervós, J.J.: Characterizing fault-tolerance in evolutionary algorithms. In: Fernández de Vega, F., et al. (eds.) Parallel Architectures and Bioinspired Algorithms. Studies in Computational Intelligence, vol. 415, pp. 77–99. Springer, Berlin, Heidelberg (2012)
12. Matei, R., Iamnitchi, A., Foster, P.: Mapping the Gnutella network. IEEE Internet Comput. **6**(1), 50–57 (2002)
13. Meri, K., Arenas, M., Mora, A., Merelo, J.J., Castillo, P., García-Sánchez, P., Laredo, J.: Cloud-based evolutionary algorithms: an algorithmic study. Nat. Comput. **12**(2), 135–147 (2013)
14. Nogueras, R., Cotta, C.: Self-healing strategies for memetic algorithms in unstable and ephemeral computational environments. Nat. Comput. **16**(2), 189–200 (2017)
15. Nogueras, R., Cotta, C.: Studying self-balancing strategies in island-based multimemetic algorithms. J. Comput. Appl. Math. **293**, 180–191 (2016)

16. Nogueras, R., Cotta, C.: Evaluating island-based EAs on unstable networks with complex failure patterns. In: Proceedings of GECCO' 17 Companion (late breaking abstract). Berlin, Germany (2017), 2 pp.
17. Stutzbach, D., Rejaie, R.: Understanding churn in peer-to-peer networks. In: 6th ACM SIGCOMM Conference on Internet Measurement—IMC 2006, pp. 189–202. ACM Press, New York, NY, USA (2006)
18. Tang, X., Liu, J., Hao, X.: Mitigate cascading failures on networks using a memetic algorithm. Sci. Rep. **6**, 38713 (2016)
19. Watson, R., Hornby, G., Pollack, J.: Modeling building-block interdependency. In: Eiben, A., et al. (eds.) Parallel Problem Solving from Nature—PPSN V. Lecture Notes in Computer Science, vol. 1498, pp. 97–106. Springer, Berlin, Heidelberg (1998)
20. Watts, D.J.: A simple model of global cascades on random networks. Proc. Natl. Acad. Sci. **99**(9), 5766–5771 (2002)
21. Wickramasinghe, W., Steen, M.V., Eiben, A.E.: Peer-to-peer evolutionary algorithms with adaptive autonomous selection. In: Thierens, D., et al. (eds.) Genetic and Evolutionary Computation—GECCO 2007, pp. 1460–1467. ACM Press, New York, NY, USA (2007)

Spatially Structured Evolutionary Algorithms: Graph Degree, Population Size and Convergence Speed

Carlos M. Fernandes, Juan L.J. Laredo and Agostinho C. Rosa

Abstract An evolutionary algorithm (EA) is said to be spatially structured when its individuals are arranged in an incomplete graph and interact only with their neighbors. Previous studies argue that spatially structured EAs are less likely to converge prematurely to local optima. Furthermore, they have been initially designed for distributed computing and it is often claimed that their parallelization is simpler than the equivalent non-structured algorithm. However, most of the empirical studies on spatially structured EAs use a predefined and fixed population size, whereas the full potential of this or any other any kind of EA can only be explored if the population size is properly set. This paper investigates optimal population sizes of spatially structured EAs (cellular EAs, in particular) and the relationship between that size, convergence speed and the degree of the structuring network. EAs structured by regular graphs with different degrees have been tested on different types of fitness landscapes. We conclude that in most cases graphs with low degree require smaller populations to converge consistently to global optima. However, if the population size is properly set, EAs structured by graphs with higher degrees not only converge to global optima with high probability, but also converge faster.

Keywords Evolutionary computation · Spatially structured genetic algorithms · Optimal population size · Distributed EAs

---

C.M. Fernandes (✉) · A.C. Rosa
LARSyS: Laboratory for Robotics and Systems in Engineering and Science, University of Lisbon, Lisbon, Portugal
e-mail: cfernandes@laseeb.org

A.C. Rosa
e-mail: acrosa@laseeb.org

C.M. Fernandes
Department of Computer Architecture, University of Granada, Granada, Spain

J.L.J. Laredo
LITIS, University of Le Havre, Le Havre, France
e-mail: juanlu.jimenez@gmail.com

M. Ivanović et al. (eds.), *Intelligent Distributed Computing XI*,
Studies in Computational Intelligence 737, https://doi.org/10.1007/978-3-319-66379-1_2

1 Introduction

Evolutionary Algorithms (EAs) [2] are a class of metaheuristics based on the theory of evolution. Initially, an EA generates a population of solutions. Then, a set of those solutions is selected according to their fitness and recombined for generating new individuals. The new population replaces the whole or part of the parents' population and the process repeats until a stopping criterion is met. This simple procedure increases the average fitness of the population and, eventually, finds a local or global solution to the problem.

Standard EAs use what is known as panmictic populations: each individual can interact (recombine) with every other individual. However, parallel and distributed implementations of EAs may benefit from alternative, restricted forms of recombination. In recent years, spatially structured EAs [13], which restrain the interaction according to a population structure, are gaining increasing attention. The structure specifies a network of acquaintances for individuals to interact, that is, mating or selection is restricted to neighborhoods within the network structure. As argued in [2], non-panmictic EAs, such as cellular [1] or distributed EAs [3, 8], provide a better sampling of the search space and improve the performance of the equivalent panmictic EA.

This paper focuses on the particular case of spatially structured EAs called cellular EAs (cEAs). The efficiency of cEAs has been systematically demonstrated [1, 2, 13] and is attributed to their ability to maintain fitness and genetic diversity [2]. Since individuals only interact with a restricted number of other individuals, information diffuses slower through the network. This means that the balance between exploration and exploitation of panmictic EAs (under the same selection and recombination strategies) is severely altered: exploration is more intense, while exploitation takes place only in local neighborhoods. This results in higher takeover times: the diffusion of good individuals is slower. Consequently, the convergence is also slower, but the algorithm is less likely to converge to local optima.

There are several studies that investigate selection pressure, convergence speed and takeover times of cellular EAs [1, 2, 4, 6, 7]. However, to the extent of our knowledge, the relationship between population size, convergence speed, accuracy and the degree of the underlying graph has not been studied yet. Since population size is a key factor not only in the convergence speed of EAs, but also for efficient parallel implementations, we propose to investigate the optimal population size of structured EAs on regular graphs with different degree. For that purpose, we use the bisection method for assessing optimal population size in different fitness landscapes. Under these settings, we are able to determine which graph maximizes the performance of the algorithm in each type of landscape, as well as the smallest population that guarantees a high probability of convergence to the global optimum. With such knowledge, we can improve our comprehension of the mechanisms behind efficient cEAs, while optimizing the computational resources required for real-world implementations of cEAs.

The remainder of the paper is structured as follows: Sect. 2 gives a background review on cellular EAs; Sect. 3 describes the methodology used in this study; Sect. 4 presents and discusses the results; Sect. 5 concludes the paper and outlines future lines of work.

2 Background Review and Motivation

Genotypic representation, operators, selection schemes and population size are typical EAs moduli that require design choices. Population size, in particular, must be set to a minimal size that guarantees a sufficient supply of raw building blocks. If the population is too small, the algorithm loses diversity prematurely and converges to local optima. Conversely, if the population is excessively large, the convergence speed of the algorithm may be affected. The population must grasp a proper balance between genetic diversity and convergence speed, and methods have been devised for determining the minimal size that assures a high probability of convergence to global optima [12]. These methods can be applied to any kind of EA, including cEAs.

The initial objective of spatially structured EAs was to develop a framework for studying massive parallelization. However, the need to provide traditional EAs with a proper balance between exploration and exploitation motivated several lines of research that explore the potentiality of different population structures in maintaining genetic diversity [13]. The primary focus of the field has been on static regular lattices: every individual has a fixed number of potential interaction partners. Giacobini et al. [7] present mathematical models for the selection pressure of cEAs on regular lattices. The experiments confirmed the theoretical results. The validation of the model has been made on 32×32 grids (1024 individuals), but the authors identified a breakdown of the usual logistic approximation for low-dimensional lattices.

Alba and Dorronsoro [2] dynamically change the ratio that defines the neighborhood of interaction in cEAs. Since the ratio may affect selection pressure, the authors analyze its influence on the balance between exploration and exploitation. However, the base-structure of the cEA (i.e. a grid lattice) is maintained throughout the run and the population size is set to fixed value for all problems and configurations of the algorithm.

Standard cEAs have some drawbacks: synchronicity (in most cases) and a strong dependence on the problem since the genetic diversity promoted by a prefixed topology is uncorrelated with the problem structure. In order to overcome these limitations, complex population structures have been also studied, sometimes using recent developments in network theory [10]. Giacobini et al. [6] studied takeover times in random and small-world structures. Again, the population size is set to a fixed value in every experiment. Whitacre et al. [14] focus on two important conditions missing in EA populations: a self-organized definition of locality and interaction epistasis. With that purpose in mind, they propose a dynamic structure and conclude that these two features, when combined, provide behaviors not observed in the canonical EAs or traditional spatially structured EAs. The most noticeable change in the behavior is an unprecedented capacity for sustainable coexistence of genetically distinct individuals within a single population. The population size varies on the range [50, 400], but the authors not give a reason for choosing this interval. Fernandes et al. [5] proposed dynamic and partially connected ring topologies for cEAs. The structures improve the rate of convergence to global optima when compared to cEAs with standard topologies on quasi-deceptive, deceptive and NP-hard problems. In this case, the authors conducted optimal population size tests, demonstrating that the proposed topologies require smaller populations when compared to traditional cEAs.

Our purpose is to investigate how population size of cEAs correlates with the structure and the fitness landscape. Since takeover times decrease with graph degree, it is expected that structures with higher degrees require larger populations. However, since good solutions diffuse more quickly when the individuals have more neighbors, it is possible that larger populations required by higher degree graphs converge faster than smaller populations in less connected structures.

3 Methodology

In order to investigate the optimal population size of different types of graphs, we have implemented cEAs with increasing degree. Most of the studies on spatially structured EAs on regular graphs use 1-D or 2-D grids—see [1, 2, 9]. In fact, a grid topology does not restrict the study [11]. However, we have chosen a more general basic structure, exemplified in Fig. 1.

Starting from a ring structure (k = 2) the degree is doubled by linking each individual to its neighbors' neighbors, creating regular graphs with $k = \{2, 4, 8, 16, 32 \ldots\}$. Additionally, EAs with $k = n - 1$ (i.e., with panmictic populations), where n is the population size, have been tested.

This study is restricted to synchronous cEAs, i.e., offspring are placed in the secondary population and replacement is made when the size n' of the offspring

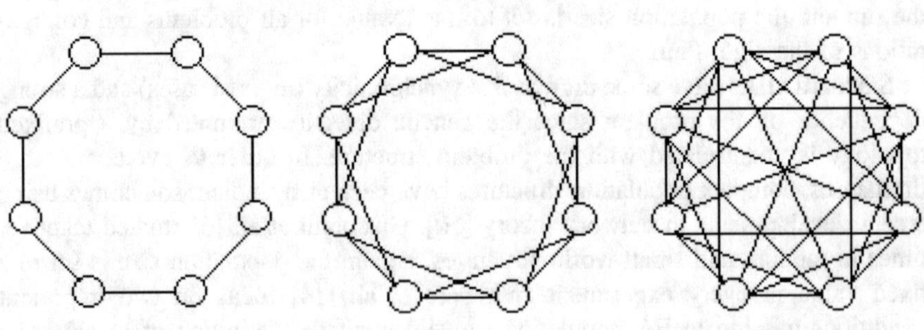

Fig. 1 Regular graphs for population size $n = 8$

population is equal to the size n of the parents population. The selection scheme is the binary tournament, two-point crossover is the recombination method and bit-flip is the mutation type. In each iteration, each individual (parent1) is recombined with one of its

Algorithm 1: cellular EA

1. For each individual $i \leftarrow 1$ to n:
   1.1. Initialize individual i
   1.2. Evaluate individual i: $f(\overrightarrow{x_i})$
2. For each individual $i \leftarrow 1$ to n:
   2.1. Compute neighborhood
   2.2. Parent 1 is individual i
   2.3. Parent 2 selected with binary tournament from the set of parent1 neighbours
   2.4 Crossover (parent 1, parent 2)
   2.5. Select randomly one of the offspring: offspring i
   2.6. Mutation (offspring i)
   2.7. Evaluate offspring i: $f(\overrightarrow{x'_i})$
   2.8. Insert offspring i in temporary population P_t
3. For each individual $i \leftarrow 1$ to n:
   3.1. Replace individual i by offspring I if $f(\overrightarrow{x'_i}) > f(\overrightarrow{x_i})$ (maximization problems)
4. If the stopping criterion is not met, go to 2

Algorithm 2: Bisection method

1. Start with small n
2. Double n until EA convergence criteria is met
3. (min,max)=(n/2,n)
4. repeat until (max-min)/min < T
   n =(min+max)/2
   if n leads to convergence criteria
   then max = n else min = n
5. Compute the statistics for the problem size using population size = max

neighbors (parent2). From the set of two children generated by crossover, one is randomly chosen and replaces parent1 if its fitness is higher. The pseudo-code of the cEA is in Algorithm 1.

Finding an appropriate population size for a given problem is critical for the performance of any EA. To determine the optimal population size of the cEAs we have used a selectorecombinative version of the algorithms (i.e., without mutation) and the bisection method [12]. Please note the bi-section method is performed in EAs without mutation. The objective is to determine the minimal population size that guarantees a sufficient supply of building blocks for the search process to converge to the global optimum without needing mutation. Then, it is expected that smaller populations can be used effectively when mutation probability is set to a non-zero value.

The bisection method is a simple yet effective technique used to determine the optimal population size of selectorecombinative EAs and it is described by Algorithm 2. For this particular case the threshold T was set to 0.1 and initial population size was set to 200. Every configuration was run 30 times before updating and the convergence criteria is met if 29 of those 30 runs converge towards the global optimum. The algorithms were tested with $p_c = 1.0$. Mutation probability was set to 0. After determining the optimal population size, that configuration was executed for 50 times and the number of evaluations necessary to reach the optimum was averaged over the successful runs.

4 Experimental Setup and Results

The algorithms were tested with onemax, 2-trap, 3-trap, 4-trap and MMDP. A trap function is a piecewise-linear function defined on *unitation* (the number of ones in a binary string) that has two distinct regions in the search space, one leading to the global optimum and the other leading to a local optimum. Depending on its parameters, trap functions may be deceptive or not. The trap functions in these experiments are defined by:

$$F(\vec{x}) = \begin{cases} k, & \textit{if } u(\vec{x}) = k \\ k - 1 - u(\vec{x}), & \textit{otherwise} \end{cases} \tag{1}$$

where u($\vec{x}$) is the unitation function and k is the problem size (and also the fitness of the global optimum). With these definitions, order-3 traps are in the region between deceptive and non-deceptive, while order-2 are non-deceptive and order-4 are fully deceptive. For the experiments, order-2, -3 and -4 trap functions were constructed by juxtaposing, respectively, 150, 75 and 60 subproblems, generating 300- (2-trap), 150- (3-trap) and 120-bit (4-trap) problems. The fitness values of the best solutions are, respectively, 300, 150 and 120.

The onemax problem is the 1-trap function and consists in maximizing the number of ones in a binary string. The size of the string in the onemax problem used for this study is $l = 400$, corresponding to an optimal fitness of 400.

The MMDP is an NP-hard, deceptive and multimodal. It consists of k 6-bits subproblems with two global optima and a deceptive attractor in the middle of the fitness landscape. Each subproblem fitness values depend on the unitiation function. Table 1 shows the contribution of each subproblem to the fitness value of a string. For the experiments, 120-bit strings were used. Optimal solutions have fitness values of 20. Table 2 summarizes the test set.

Table 1 MMDP. Contribution of each subproblem configuration to the fitness value

u($\vec{x}$)	0	1	2	3	4	5	6
$F(\vec{x})$	1.000000	0.000000	0.360384	0.640576	0.360384	0.000000	1.000000

Table 2 Functions: type, string size and best solution fitness

Function	Type	String size	Best fitness
onemax	Non-deceptive	400	400
2-trap	Non-deceptive	300	300
3-trap	Nearly deceptive	150	150
4-trap	Deceptive	120	120
MMDP	Deceptive	120	20

Table 3 Optimal population size

	onemax	2-trap	3-trap	4-trap	MMDP
$k = 2$	400	500	350	400	500
$k = 4$	450	500	350	450	500
$k = 8$	500	550	350	500	500
$k = 16$	550	700	400	550	500
$k = 32$	800	750	500	550	550
$k = 64$	1000	1000	650	650	650
$k = 128$	1200	1300	700	750	750
$k = n - 1$	2200	2275	1100	1200	800

First experiments determine the optimal population size of cEAs with $k = \{2, 4, 8, 16, 32, 64, 128, n - 1\}$, where n is population size. Results are in Table 3. As expected, optimal size increases with the degree of the underlying structures.

Table 4 shows the average number of evaluations required to reach the global optimum when the population size is set to the values found previously and shown

Table 4 Convergence speed: average number of evaluations and standard deviation

	onemax	2-trap	3-trap	4-trap	MMDP
$k = 2$	319,986.21	348,483.33	133,712.07	146,560.00	168,266.67
	±17,253.56	±22,773.05	±16,085.69	±18,423.25	±29,590.81
$k = 4$	219,930.00	222,433.33	86,205.00	104,167.24	108,866.67
	±12,784.26	±26,610.77	±10,812.42	±12,786.66	±14,151.84
$k = 8$	157,233.33	147,836.21	54,961.67	73,206.90	66,100.00
	±7,747.89	±1,1261.46	±5,178.70	±10,743.37	±8,515.10
$k = 16$	**114,210.34**	129,173.33	41,701.67	51,645.00	44,724.14
	±5,271.88	±7,495.93	±4,602.52	±4,915.85	±5,630.89
$k = 32$	122,560.00	**99,795.00**	38,683.33	40,425.00	35,806.90
	±5,366.15	**±5,847.96**	±3,100.29	±4,175.58	±2,988.51
$k = 64$	117,933.33	104,068.97	40,913.33	**37,812.07**	33,979.31
	±4,448.42	±4,008.30	±2,966.80	**±2,395.00**	±3,254.16
$k = 128$	120,331.03	110,196.67	**37,727.59**	38,700.00	33,725.00
	±3,887.07	±4,601.69	**±1,947.32**	±3,275.69	±2,904.63
$k = n - 1$	181,462.07	160,463.33	51,920.00	49,646.67	**33,296.55**
	±5,010.59	±3,995.89	±3,792.95	±3,989.5	**±2,142.51**

in Table 3. The optimal convergence speed (in bold) is attained with highly connected graphs. The panmictic population ($k = n - 1$) does not attain the best convergence speed values in every function but it is significantly better than lower degree graphs ($k = 2$ and $k = 4$) in every fitness landscapes. (In this study, Mann-Whitney U tests were performed at the 5% level of significance to determine if two distributions of numerical results are significantly different). Furthermore, $k = n - 1$ attains the best performance in the MMDP problem.

The following experiment was performed to stress out the importance of determining the optimal population size of EA for a particular fitness landscape.

Table 5 Fitness (median, best and worst values) and success rates (SR)

		onemax $n = 200$	2-trap $n = 250$	3-trap $n = 175$	4-trap $n = 200$	MMDP $n = 250$
$k = 2$	Median	400	300	150	120	20
	Best	400	300	150	120	20
	Worst	400	300	149	118	19.64
	SR	30	30	28	24	23
$k = 4$	Median	400	300	150	120	20
	Best	400	300	150	120	20
	Worst	400	300	148	118	19.64
	SR	30	30	25	23	22
$k = 8$	Median	400	300	150	119	20
	Best	400	300	150	120	20
	Worst	400	300	148	118	19.64
	SR	30	30	20	12	23
$k = 16$	Median	400	300	149	119	20
	Best	400	300	150	120	20
	Worst	400	300	146	116	19.64
	SR	30	30	7	7	22
$k = 32$	Median	400	300	147	117.5	20
	Best	400	300	150	120	20
	Worst	400	300	144	115	19.28
	SR	30	30	1	1	20
$k = 64$	Median	400	300	146	116	19.64
	Best	400	300	149	120	20
	Worst	400	300	142	113	19.28
	SR	30	30	0	1	6
$k = 128$	Median	400	300	145	116	19.64
	Best	400	300	148	120	20
	Worst	400	300	140	112	18.20
	SR	30	30	0	1	16
$k = n - 1$	Median	400	300	146	116	19.64
	Best	400	300	149	119	20
	Worst	400	300	142	113	18.92
	SR	30	30	0	0	9

Table 6 Convergence speed: average number of evaluations and standard deviation

	onemax	2-trap	3-trap	4-trap	MMDP
$k = 2$	95,900.00	235,716.67	99,497.22	110,708.33	124,673.91
	±3,626.24	±23,382.20	±15799.90	±21,653.64	±58,068.80
$k = 4$	72,460.00	148,950.00	65,856.00	67,452.17	151,840.91
	±2,769.36	±16,527.33	±12,553.7	±10,967.43	±200,212.74
$k = 8$	55,493.33	117,900.00	58,403.95	43,233.33	114,608.70
	±2,183.79	±49,055.90	±89,744.99	±5,718.13	±166,207.75
$k = 16$	43,646.67	217,041.67	25,300.00	26,914.29	116,761.36
	±2,058.24	±371,359.30	±2,710.82	±4,989.13	±230,706.28
$k = 32$	35,793.33	423,041.67	17,675.00	18,600.00	162,950.00
	±1,960.64	±203,743.01	–	–	±230,706.28
$k = 64$	32,866.67	615,841.67	–	15,000.00	86,708.33
	±1,590.78	±228,150.17	–	–	±119,648.17
$k = 128$	32,333.33	608,983.33	–	13800.00	287,468.75
	±2,074.03	±239,684.09	–	–	±304,905.86
$k = n - 1$	31,600.00	611,441.67	–	–	127,277.78
	±1,702.74	±232,371.70	–	–	±153,606.36

For each problem, the population size n of each cEA was set to $n_{min}/2$, where n_{min} is the population size in Table 3. Mutation probability is set to $p_m = 1/l$, where l is the string size, and crossover probability is $p_c = 1.0$. The algorithms were all run for 1,000,000 function evaluations or until reaching the global optimum. Results are averaged over 30 runs and shown in Tables 5 and 6.

Table 5 shows the median, best and worst fitness attained by each configuration in each problem, as well as the success rates (the number of runs in which the algorithm found the global optimum). The global optimum of onemax and 2-trap functions is found in every run by every cEA. These are simple and unimodal problems without local optima. Therefore, provided with variation and mutation operators and given enough time, an EA will eventually find the optimum. Comparison of performance can therefore be made using convergence speed. Table 5 shows that the convergence speed in the onemax problem increases with k. Optimal speed in 2-trap is attained with $k = 8$. However, better convergence speed is attained for 2-trap using larger populations—see cEAs with $k = 32$ and $k = 64$ in Table 4.

For the 3-trap problems, better results are clearly attained by the cEAs with optimal population size, except for $k = 2$, which attains a good success rate with lower convergence speed. For 4-traps, the accuracy is clearly degraded when population size is set to $n = 200$. The same goes for MMDP problem.

These results show that deceptive functions require a careful tuning of the population size. Furthermore, the numerical results in Tables 3, 4, 5 and 6 demonstrate that setting the population size to suboptimal values may mislead the conclusions on the performance of cEAs, mainly in deceptive and multimodal problems. These problems require a proper balance between the initial supply of building blocks, the selection pressure and the variation operator. The results in this paper call into question the efficiency of cEAs in deceptive and multimodal problems—please remember that the most efficient EA in the tested MMDP problem is the panmictic EA. However, further tests are required in order to confirm the hypothesis.

5 Conclusions

This paper investigates the relationship between population size, convergence speed and graph degree of cEAs with populations structured by regular graphs. In order to determine the minimal population size that guarantees convergence to global optimum with high probability, the bisection method has been applied to cEAs with different degree. The numerical results show that graphs with lower degree require smaller populations. However, the larger populations required by graphs with higher degree converge faster to global optima. These results suggest that when the population is properly set, higher degree or even panmictic populations are more efficient than cEAs in ring structures or low degree graphs. Furthermore, conclusions on the performance of the different structures are entirely different and almost certainly misleading if the population size is set to the same value for all configurations.

The study has been restricted to regular and static graphs. In the future, we intend to apply the same experimental procedure to random, small-world and dynamic structures. The takeover times of the different graphs with different population size will be investigated as well as the behavior and performance of the different structures with different string sizes (scalability tests). Finally, the implications of a proper setting of the population size in parallel and distributed cEAs will also be studied.

Acknowledgements First author wishes to thank FCT, *Ministério da Ciência e Tecnologia*, his Research Fellowship SFRH/BPD/111065/2015). This work was supported by FCT PROJECT [UID/EEA/50009/2013].

References

1. Alba, E., Tomassini, M.: Parallelism and evolutionary algorithms. IEEE Trans. Evol. Comput. **6**(5), 443–462 (2002)
2. Alba, E., Dorronsoro, B.: The exploration/exploitation tradeoff in dynamic cellular genetic algorithms. IEEE Trans. Evol. Comput. **9**, 126–142 (2005)
3. Bäck, T.: Evolutionary Algorithms in Theory and Practice. Oxford University Press, Oxford (1996)

4. Cantú-Paz, E.: Migration policies, selection pressure, and parallel EAs. Journal of Heuristics **7** (4), 311–334 (2001)
5. Fernandes, C.M., Laredo, J.L.J., Merelo, J.J., Cotta, C., Rosa, A.C.: Dynamic and Partially Connected Ring Topologies for Evolutionary Algorithms with Structured Populations, EvoApplications 2014: Applications of Evolutionary Computation, pp. 665–677 (2014)
6. Giacobini, M., Tomassini, M., Tettamanzi, A.: Takeover time curves in random and small-world structured populations. In: Proceedings of the 7th GECCO, pp. 1333–1340 (2005)
7. Giacobini, M., Tomassini, M., Tettamanzi, A.G.B., Alba, E.: Selection intensity in cellular evolutionary algorithms for regular lattices. IEEE Trans. Evol. Comput. **9**, 489–505 (2005)
8. Laredo, J.L.J., Bouvry, P., González, D.L., Fernandéz de la Vega, F., Arenas, M.G., Merelo, J.J., Fernandes, C.M.: Designing robust volunteer-based evolutionary algorithms. Genet. Program Evol. Mach. **15**(3), 221–244 (2014)
9. Payne, J.L, Eppstein, M.J.: Emergent mating topologies in spatially structured genetic algorithms. In: Proceedings of 8th GECCO, pp. 207–214 (2006)
10. Réka, A., Barabási, A.-L.: Statistical mechanics of complex networks. Rev. Mod. Phys. **74**, 47–94 (2000)
11. Sarma J., De Jong, K.: An analysis of the effect of the neighborhood size and shape on local selection algorithms. In: Proceedings of International Conference on Parallel Problem Solving from Nature IV, LNCS 1141, pp. 236–244. Springer (1996)
12. Sastry, K.: Evaluation-relaxation schemes for genetic and evolutionary algorithms. M.Sc. thesis, University of Illinois, Urbana, IL, USA (2001)
13. Tomassini, M.: Spatially Structured Evolutionary Algorithms. Springer, Heidelberg (2005)
14. Whitacre, J.M., Sarker, R.A., Pham, Q.: The self-organization of interaction networks for nature-inspired optimization. IEEE Trans. Evol. Comput. **12**, 220–230 (2008)

Heuristic of Anticipation for Fair Scheduling and Resource Allocation in Grid VOs

Victor Toporkov, Anna Toporkova and Dmitry Yemelyanov

Abstract In this work, a job-flow scheduling approach for Grid virtual organizations (VOs) is proposed and studied. Users and resource providers preferences, VOs internal policies, resources geographical distribution along with local private utilization impose specific requirements for efficient scheduling according to different, usually contradictive, criteria. With increasing resources utilization level the available resources set and corresponding decision space are reduced. In order to improve overall scheduling efficiency, we propose an anticipation scheduling heuristic. It includes a target (anticipated) pattern solution definition and a special replication procedure for efficient and feasible resources allocation. A proposed anticipation algorithm is compared against conservative backfilling variations using such criteria as average jobs response time (start and finish times) as well as users and VO economic criteria (execution time and cost).

Keywords Scheduling · Grid · Utilization · Heuristic · Job batch · Virtual organization · Cycle scheduling scheme · Anticipation · Replication · Backfilling

1 Introduction and Related Works

In distributed environments with non-dedicated resources such as utility Grids the computational nodes are usually partly utilized by local high-priority jobs coming from resource owners. Thus, the resources available for use are represented with a set of slots—time intervals during which the individual computational nodes are

V. Toporkov (✉) · D. Yemelyanov
National Research University "MPEI", ul. Krasnokazarmennaya, 14,
Moscow 111250, Russia
e-mail: ToporkovVV@mpei.ru

D. Yemelyanov
e-mail: YemelyanovDM@mpei.ru

A. Toporkova National Research University Higher School of Economics,
ul. Myasnitskaya, 20, Moscow 101000, Russia
e-mail: AToporkova@hse.ru

M. Ivanović et al. (eds.), *Intelligent Distributed Computing XI*,
Studies in Computational Intelligence 737, https://doi.org/10.1007/978-3-319-66379-1_3

capable to execute parts of independent users' parallel jobs. These slots generally have different start and finish times and a performance difference. The presence of a set of slots impedes the problem of coordinated selection of the resources necessary to execute the job-flow from computational environment users. Resource fragmentation also results in a decrease of the total computing environment utilization level [1, 2].

Two established trends may be outlined among diverse approaches to distributed computing. The first one is based on the available resources utilization and application level scheduling [3]. As a rule, this approach does not imply any global resource sharing or allocation policy. Another trend is related to the formation of user's virtual organizations (VO) and job-flow scheduling [4, 5]. In this case a metascheduler is an intermediate chain between the users and local resource management and job batch processing systems.

Uniform rules of resource sharing and consumption, in particular based on economic models, make it possible to improve the job-flow level scheduling and resource distribution efficiency. VO policy may offer optimized scheduling to satisfy both users' and VO common preferences. The VO scheduling problems may be formulated as follows: to optimize users' criteria or utility function for selected jobs [6, 7], to keep resource overall load balance [8, 9], to have job run in strict order or maintain job priorities [10], to optimize overall scheduling performance by some custom criteria [11, 12], etc.

VO formation and performance largely depends on mutually beneficial collaboration between all the related stakeholders. Thus, VO policies in general should respect all members and the most important aspect of rules suggested by VO is their fairness.

A number of works understand fairness as it is defined in the theories of cooperative games and mechanism design, such as fair job-flow distribution [8], fair quotas [13, 14] or fair user jobs prioritization [10]. The cyclic scheduling scheme (CSS) [15] implements a fair scheduling optimization mechanism which ensures stakeholders interests to some predefined extent. Thus, we elaborate a problem of parallel jobs scheduling in heterogeneous computing environment with non-dedicated resources considering users individual preferences and goals.

The downside of a majority centralized metascheduling approaches is that they lose their efficiency and optimization features in distributed environments with a limited resources supply. For example, in [2] a traditional backfilling algorithm provides better scheduling outcome when compared to different optimization approaches in resource domain with a minimal performance configuration. The general root cause is that in fact the same scarce set of resources (being efficient or not) have to be used for a job-flow execution or otherwise some jobs might hang in the queue. And under such conditions user jobs priority and ordering greatly influence the scheduling results.

A main contribution of this paper is a heuristic anticipation job-flow scheduling approach which retains optimization features and efficiency even in distributed computing environments with limited resources. The rest of the paper is organized as follows. Section 2 presents a general CSS fair scheduling concept. The proposed anticipation scheduling technique is presented in Sect. 3. Section 4 contains simulation experiment setup and results of comparison with conservative backfilling variations. Finally, Sect. 5 summarizes the paper.

2 Cyclic Alternative-Based Fair Scheduling Model and Limited Resources

Scheduling of a job-flow using CSS is performed in time cycles known as scheduling intervals, by job batches [15]. The actual scheduling procedure during each cycle consists of two main steps. The first step involves a search for alternative execution scenarios for each job or simply alternatives [16]. During the second step the dynamic programming methods [15] are used to choose an optimal alternatives' combination (one alternative is selected for each job) with respect to the given VO and user criteria. This combination represents the final schedule based on current data on resources load and possible alternative executions.

An example for a user scheduling criterion may be a minimization of overall job running time, a minimization of overall running cost, etc. This criterion describes user's preferences for that specific job execution.

Alongside with time (T) and cost (C) properties each job execution alternative has a user utility (U) value: user evaluation against the scheduling criterion. We consider a relative approach to represent a user utility $U \in [0\%, 100\%]$. Each alternative gets its utility in relation to the "best" and the "worst" optimization criterion values user could expect according to the job's priority. And the more some alternative corresponds to user's preferences the smaller is the value of $U \to 0\%$.

For a fair scheduling model the second step VO optimization problem could be in form of: $C \to \max, \lim U$, i.e. maximize total job-flow execution cost, while respecting user's preferences to some extent.

First step of CSS requires allocation of a multiple alternatives *nonintersecting* in terms of slots for each job. Otherwise irresolvable collisions for resources may occur if different jobs will share the same time-slots. Sequential alternatives search and resources reservation procedures help to prevent such scenario. However in an extreme case when resources are limited or overutilized only at most one alternative execution could be reserved for each job. In this case alternatives-based scheduling result will be no different from FIFO resources allocation procedure without any optimizations [2].

3 Heuristic Anticipation Scheduling

3.1 General Scheme

In order to improve scheduling efficiency for job batch the following heuristic is proposed. It consists of three main steps.

1. First, a set of all possible execution alternatives is found for each job not considering time slots intersections and without any resources reservation.
2. Second, CSS scheduling procedure is performed to select alternatives combination (one alternative for each job of the batch) optimal according to VO fair-share policy. The resulting alternatives combination most likely corresponds to an infeasible scheduling solution as possible time slots intersection will cause collisions on resources allocation stage.
3. Third, a feasible resources allocation is performed by replicating alternatives selected in step **2**.

After these three steps are performed the resulting solution is both feasible and efficient as it reflects scheduling pattern obtained from a near-optimal reference solution from step **2**. The following subsections will discuss these scheduling steps in more details.

3.2 Finding a Near Optimal Infeasible Scheduling Solution

CSS scheduling results are strongly depend on diversity of alternatives sets obtained for batch jobs. As we need to find alternatives for an apriori infeasible reference solution a reasonable diverse set of possible execution alternatives will do.

We used a modification of Algorithm searching for Extreme Performance windows (AEP) [16] to allocate a diverse set of execution alternatives for each job. Originally AEP scans through a whole list of available time slots and retrieves one alternative execution optimal according to the user custom criterion. During this scan, AEP estimates every possible and sufficient slots combination against user criterion and selects the one with the best criterion value. In order to retrieve all possible execution alternatives we save all distinct intermediate AEP search results to a dedicated list of possible alternatives.

After sets of possible execution alternatives are independently allocated for each job a CSS scheduling optimization procedure selects an optimal alternatives combination according to VO and users criteria [15]. More details on alternatives combination selection procedure were provided in Sect. 2.

3.3 Replication Scheduling and Resources Allocation

The resulting near-optimal scheduling solution in most cases is infeasible as selected alternatives may share the same time slots and thus cause resource collisions. However we propose to use it as a reference solution and replicate into a feasible resources allocation.

For the replication purpose a new *Execution Similarity* criterion is introduced. It helps AEP [16] to find a window with minimum *distance* to a reference pattern alternative. Generally we define a *distance* between two different alternatives (windows) as a relative difference or *error* between their significant criteria values. For example if reference alternative has C_{ref} total cost, and some candidate alternative cost is C_{can}, then the relative cost error E_C is calculated as $E_C = \frac{|C_{\text{ref}} - C_{\text{can}}|}{C_{\text{ref}}}$. If one need to consider several criteria the *distance D* between two alternatives may be calculated as a geometric distance in a parameters space: $D_g = \sqrt{E_C^2 + E_T^2 + \cdots + E_U^2}$.

AEP with *Execution Similarity* scans through a whole list of available time slots, for every possible slots combination calculates its distance from a reference alternative and selects the one with the minimum distance to a reference.

For a feasible job batch resources allocation AEP consequentially allocates for each job a single execution window with a minimum *distance* to a reference alternative. Time slots allocated for the i-th job are reserved and excluded from the slot list when AEP search algorithm is performed for the following jobs $i + 1, i + 2, \ldots$. Thus, this procedure prevents any conflicts for resources and provides scheduling solution which in some sense reflects near-optimal reference solution.

AEP and its modifications have a quadratic computational complexity with respect to the number of available computing nodes [16]. It is performed twice for each job (during alternatives search and replication steps), so the overall complexity linearly depends on the job-flow capacity. At the same time dynamic scheduling scheme [15] from the step **2** is pseudo-polynomial and additionally depends on a total budget allocated for the job-flow execution.

3.4 Replication Reference Setup

Anticipated near-optimal scheduling solution provides a heuristic insight on how each job should be executed with a reference to other users criteria, VO optimization policy and a current computing domain composition and utilization level. Basically this solution suggests what kind of resources should be allocated for each job in terms of performance and cost. Thus, available resources can be consistently distributed between the user jobs according to their performance or cost optimization targets.

At the same time the anticipated solution can't provide any meaningful reference on jobs' start and finish times. As anticipation procedure independently allocates a set of possible execution alternatives for each job, it does not consider resources reservation and utilization by other jobs. Thus, resulting anticipated jobs' start and finish times are randomly distributed on a whole scheduling interval with a bias towards the interval's start. In this way anticipation scheduling scheme can't provide neither adequate estimation on jobs' starting times, nor the common jobs' execution order.

In order to improve the anticipated reference solution we use backfilling algorithm to provide practical values for jobs start and finish times. Backfilling is able to minimize the whole job-flow execution makespan as well as to generally follow the initial jobs relative queue order [1]. These features make backfilling scheduling solution a good reference target for the anticipation scheduling scheme. Thus, for the replication step we set infeasible CSS solution as a reference for jobs execution runtime and cost, and backfilling solution—for jobs start and finish times.

Additionally we introduce a finish time approximation coefficient K_t to relate the anticipated finish times to backfilling reference solution. For example when $K_t = 1$ we use exact jobs finish times provided by backfilling as a reference for a replication step. $K_t = 0.5$ means that we strive to execute the job-flow twice as faster compared to backfilling. So just by changing K_t we are able stretch resulting anticipation solution on a desired time interval and at the same time preserve a general jobs execution order provided by backfilling.

4 Simulation Study

4.1 Simulation Environment Setup

An experiment was prepared as follows using a custom distributed environment simulator [2, 15–17]. For our purpose, it implements a heterogeneous resource domain model: nodes have different usage costs and performance levels. A space-shared resources allocation policy simulates a local queuing system (like in GridSim or CloudSim [18]) and, thus, each node can process only one task at any given simulation time. The execution cost of each task depends on its running time which is proportional to the dedicated nodes performance level. The execution of a single job requires parallel execution of all its tasks.

Virtual organization and computing environment properties are the following. The resource pool includes 25 heterogeneous computational nodes. A base cost of a node is an exponential function of its performance value, so any two nodes of the same resource type and performance have the same base cost. Effective node cost during the scheduling interval is then calculated by adding a variable distributed normally as ± 0.6 of a base cost, simulating discounts or extra charges up to 60%. The initial 5–10% resource load with owner jobs is distributed hyper-geometrically over the whole scheduling interval.

The job batch properties are specified as follows. Jobs number in a batch is 75. Nodes quantity needed for a job is an integer number distributed evenly on $[2, 5]$.

Node reservation time is an integer number distributed evenly on [100, 600]. Job budget varies in the way that some of jobs can pay as much as 160% of base cost whereas some may require a discount. Every request contains a specification of a custom criterion which is one of the following: *job execution runtime* or *overall execution cost*.

During each experiment a VO resource domain and a job batch were generated and the following scheduling algorithms were simulated and studied.

First we ran a conservative backfilling algorithm BF_s to obtain an exemplary job-flow scheduling solution. Conservative backfilling consequently starts each job as soon as possible on condition it does not delay execution of higher priority jobs. Next, we ran a conservative backfilling modification BF_f, which instead of minimizing jobs' start times, performs jobs' finish time minimization with the same restriction to delay high priority jobs. For this purpose we used AEP algorithm with a finish time minimization criterion to find and allocate suitable resources for each job.

Finally we performed anticipation scheduling procedure *ANT* with a $C \rightarrow \max$, $\lim U_a = 10\%$ policy and different approximation coefficient values $K_t \in \{0, 0.1, 0.5, 1, 1.1, 1.5, \}$ (see Sect. 3.4).

4.2 Simulation Results

More then 2000 scheduling cycles were simulated to obtain average job-flow scheduling results for BF_s, BF_f and *ANT*. Figures 1 and 2 show average job-flow starting and finishing times as a function of K_t parameter.

First it should be noted that BF_f algorithm at average provided 2% earlier jobs start times and 7% earlier finish times compared to a simple BF_s implementation. Thus, considered backfilling modification BF_f provides an even higher scheduling standard for an anticipation scheme. At the same time *ANT* provided earlier jobs

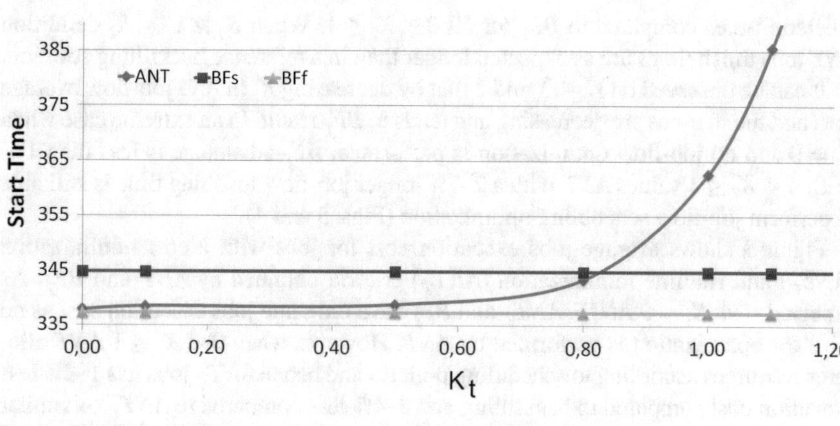

Fig. 1 Simulation results: average job execution start time

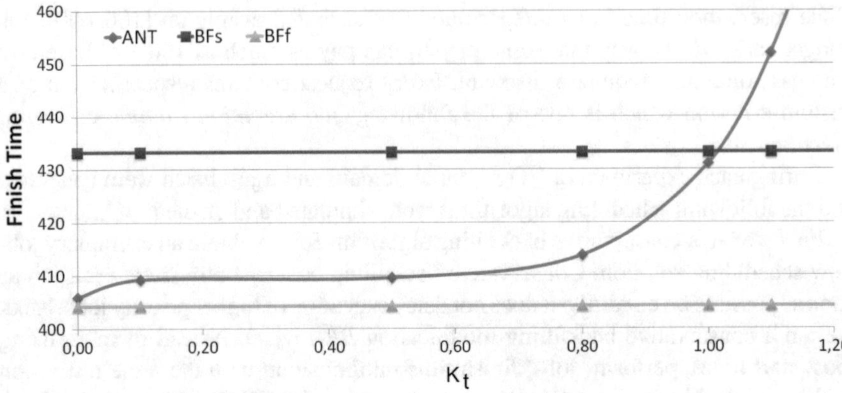

Fig. 2 Simulation results: average job execution finish time

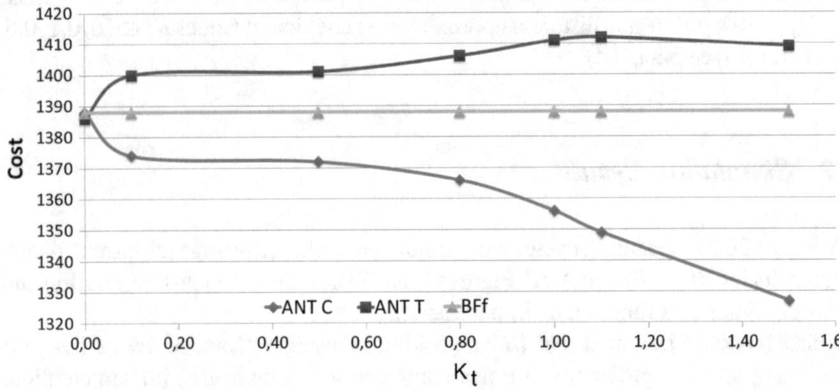

Fig. 3 Simulation results: average job execution cost

finishing times compared to BF_s for all $0 < K_t \leq 1$. When $K_t > 1$ by K_t definition *ANT* jobs finish times are as expected longer then in a reference backfilling solution.

It can be observed on Figs. 1 and 2 that by decreasing K_t in *ANT* job-flow average start and finish times are decreasing and tends to BF_f result. In an extreme case when $K_t = 0$ and no job-flow optimization is performed, BF_f advantage is less then 1%. With $0 < K_t \leq 1$ values *ANT* with a 2–7% longer job-flow finishing time is still able to perform job-flow scheduling optimization (Figs. 3 and 4).

Figure 3 shows average jobs execution cost for jobs with a cost minimization (ANT_C) and runtime minimization (ANT_T) criteria obtained by *ANT* and BF_f. As expected with $K_t = 0$ ANT_C, ANT_T and BF_f have the same jobs execution cost as no job-flow optimization is performed by *ANT*. However when $0 < K_t \leq 1$ *ANT* allocates resources according to scheduling policies and hence ANT_C jobs has 1–2% less execution cost compared to backfilling and 2–4% less compared to ANT_T. A similar picture is presented on Fig. 4 for an average jobs' execution runtime. With a rela-

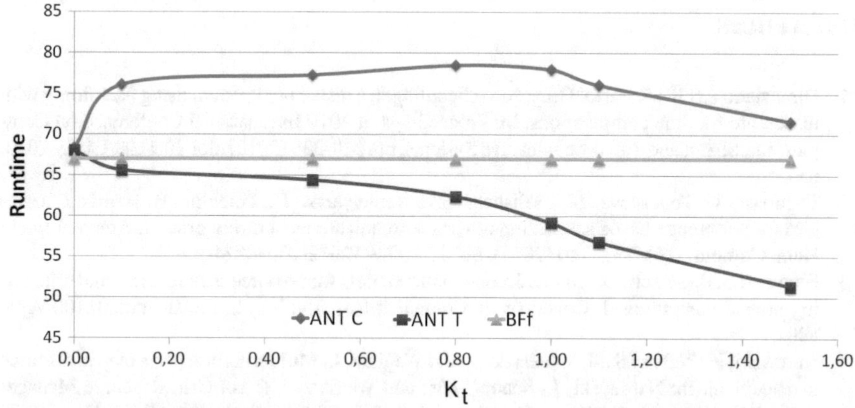

Fig. 4 Simulation results: average job execution runtime

tively small values $K_t < 0.8$ ANT_T provides up to 6% shorter jobs runtime compared to backfilling and 20% shorter compared to ANT_C jobs. With increasing K_t ANT_T advantage over backfilling increases and reaches 22% when $K_t = 1.5$.

Summarizing the results, anticipation scheduling algorithm is able to perform an efficient and fair resources allocation and provide competitive job-flow execution completion time. This achieved by a replication procedure which uses backfilling and CSS scheduling results combination as a reference target solution.

5 Conclusions and Future Work

In this paper we study the problem of a fair job batch scheduling with a relatively limited resources supply. We introduce a heuristic scheduling scheme which uses combination of a fair share scheduling policy with a common backfilling algorithm as a reference to allocate a feasible accessible solution.

Computer simulation was performed to study anticipation scheduling scheme and to evaluate its efficiency. The obtained results show that the new heuristic approach provides flexible solutions for different fair scheduling scenarios while job-flow execution time is only 2–7% longer compared to backfilling.

Future work will be focused on replication algorithm study and its possible application to fulfil complex user preferences expressed in a resource request.

Acknowledgements This work was partially supported by the Council on Grants of the President of the Russian Federation for State Support of Young Scientists and Leading Scientific Schools (grants YPhD-2297.2017.9 and SS-6577.2016.9), RFBR (grants 15-07-02259 and 15-07-03401) and by the Ministry on Education and Science of the Russian Federation (project no. 2.9606.2017/8.9).

References

1. Dimitriadou, S.K., Karatza, H.D.: Job scheduling in a distributed system using backfilling with inaccurate runtime computations. In: Proceedings of 2010 International Conference on Complex, Intelligent and Software Intensive Systems, pp. 329-336 (2010). doi:10.1109/CISIS.2010.65
2. Toporkov, V., Toporkova, A., Tselishchev, A., Yemelyanov, D., Potekhin, P.: Heuristic strategies for preference-based scheduling in virtual organizations of utility grids. J. Ambient Intell. Hum. Comput. **6**(6), 733–740 (2015). doi:10.1007/s12652-015-0274-y
3. Buyya, R., Abramson, D., Giddy, J.: Economic models for resource management and scheduling in grid computing. J. Concurrency Comput. **14**(5), 1507–1542 (2002). doi:10.1002/cpe.690
4. Kurowski, K., Nabrzyski, J., Oleksiak, A. and Weglarz, J.: Multicriteria aspects of grid resource management. In: Nabrzyski, J., Schopf, J.M. and Weglarz, J. (eds.) Grid Resource Management. State of the Art and Future Trends, pp. 271–293 (2003). doi:10.1007/978-1-4615-0509-9_18
5. Rodero, I., Villegas, D., Bobro, N., Liu, Y., Fong, L., Sadjadi, S.M.: Enabling interoperability among grid meta-schedulers. J. Grid Comput. **11**(2), 311–336 (2013). doi:10.1007/s10723-013-9252-9
6. Ernemann, C., Hamscher, V., Yahyapour, R.: Economic scheduling in grid computing. In: Revised Papers from the 8th International Workshop on Job Scheduling Strategies for Parallel Processing, vol. 2537, pp. 128–152. Springer, Berlin, Heidelberg (2002). doi:10.1007/3-540-36180-4_8
7. Rzadca, K., Trystram, D., Wierzbicki, A.: Fair game-theoretic resource management in dedicated Grids. In: IEEE International Symposium on Cluster Computing and the Grid (CCGRID 2007), pp. 343–350 (2007). doi:10.1109/ccgrid.2007.52
8. Penmatsa, S., Chronopoulos, A.T.: Cost minimization in utility computing systems. Concurrency Comput.: Pract. Experience **16**(1), 287–307 (2014). doi:10.1002/cpe.2984
9. Vasile, M., Pop, F., Tutueanu, R., Cristea, V., Kolodziej, J.: Resource-aware hybrid scheduling algorithm in heterogeneous distributed computing. J. Future Gener. Comput. Syst. **51**, 61–71 (2015). doi:10.1016/j.future.2014.11.019
10. Mutz, A., Wolski, R. and Brevik, J.: Eliciting honest value information in a batch-queue environment. In: 8th IEEE/ACM International Conference on Grid Computing, pp. 291–297, IEEE Computer Society (2007). doi:10.1109/grid.2007.4354145
11. Blanco, H., Guirado, F., Lrida, J.L., Albornoz, V.M.: MIP model scheduling for multi-clusters. Proc. Euro-Par **2012**, 196–206 (2012). doi:10.1007/978-3-642-36949-0_22
12. Takefusa, A., Nakada, H., Kudoh, T., Tanaka, Y.: An advance reservation-based co-allocation algorithm for distributed computers and network bandwidth on QoS-guaranteed grids. In: 15th International Workshop JSSPP 2010, vol. 6253, pp. 16–34 (2010). doi:10.1007/978-3-642-16505-4_2
13. Carroll, T., Grosu, D.: Divisible load scheduling: an approach using coalitional games. In: Proceedings of the Sixth International Symposium on Parallel and Distributed Computing (ISPDC 07), pp. 36–36 (2007). doi:10.1109/ispdc.2007.16
14. Kim, K., Buyya, R.: Fair resource sharing in hierarchical virtual organizations for global grids. In: Proceedings of the 8th IEEE/ACM International Conference on Grid Computing, pp. 50–57 (2007). doi:10.1109/grid.2007.4354115
15. Toporkov, V., Yemelyanov, D., Bobchenkov, A., Tselishchev, A.: Scheduling in Grid Based on VO Stakeholders Preferences and Criteria. Advances in Intelligent Systems and Computing, vol. 470, pp. 505–515. Springer International Publishing Switzerland (2016). doi:10.1007/978-3-319-39639-2_44
16. Toporkov, V., Toporkova, A., Tselishchev, A., Yemelyanov, D.: Slot selection algorithms in distributed computing. J. Supercomput. **69**(1), 53–60 (2014). doi:10.1007/s11227-014-1210-1

17. Toporkov, V., Tselishchev, A., Yemelyanov, D., Bobchenkov, A.: Composite scheduling strategies in distributed computing with non-dedicated resources. Proc. Comput. Sci. **9**, 176–185 (2012). doi:10.1016/j.procs.2012.04.019
18. Calheiros, R.N., Ranjan, R., Beloglazov, A., De Rose, C.A.F., Buyya, R.: CloudSim: a toolkit for modeling and simulation of cloud computing environments and evaluation of resource provisioning algorithms. J. Softw.: Pract. Experience **41**(1), 23–50 (2011). doi:10.1002/spe.995

On the Applications of Dijkstra's Shortest Path Algorithm in Software Defined Networks

Tihana Galinac Grbac and Nikola Domazet

Abstract The software defined networking has opened new opportunities for offering network resources to end users "as a service". For these purposes a number of technologies have been proposed and implemented to enable easy definition and management of network resources dynamically. In this paper we provide an overview of software defined network and technologies used for identifying network topology. We present three approaches based on Dijkstra's Shortest Path Algorithm and evaluate their performance in an experimental study.

Keywords Software defined network · Network topology · Dijkstra's shortest path algorithm · Dynamic network definition and management

1 Introduction

Maintenance and ownership of network devices involved into distributed computing tasks become more and more expensive. Variety of implementations of standard functions in network hardware is weakly interoperable, requires specialized competences and requires time consuming maintenance and expensive ownership. Demands for network resources will continuously increasing in near future, introducing various communicating users, devices and media [3–5]. Traditional network protocols and devices becomes increasingly ineffective, networks are overdimensioned for the maximal traffic and weakly exploited.

Software Defined Network (SDN) is new paradigm that enable network to dynamically adapt to current traffic needs. It introduces new abstraction by separating network device layer and network control layer. This is achieved with new standard

T. Galinac Grbac (✉)
Faculty of Engineering, University of Rijeka, Vukovarska 58, HR-51000 Rijeka, Croatia
e-mail: tihana.galinac@riteh.hr
URL: http://www.riteh.uniri.hr/~tgalinac/

N. Domazet
Ericsson Nikola Tesla, Krapinska 45, HR-10000 Zagreb, Croatia
e-mail: nikola.domazet@ericsson.com

M. Ivanović et al. (eds.), *Intelligent Distributed Computing XI*,
Studies in Computational Intelligence 737, https://doi.org/10.1007/978-3-319-66379-1_4

interface, Open Flow protocol [3–5] which enables introduction of new autonomous network management functions at control layer, independent of hardware vendor. Networks may reconfigure dynamically per flow basis, adapt its topology to accommodate dynamic computing needs, while optimizing network resource usage.

In this paper, as an illustration of SDN benefit, we created and test an network management applications that use the specialized network layer algorithms and dynamically reconfigure distributed computing traffic flows to optimize network cost. One example protocol used for topology discovery in traditional networks is Link layer discovery protocol (LLDP), whose various vendor's implementations provide the same service and functionality, but are not mutually usable. The main principle of their operation is to train LLDP devices periodically by sending LLDP packets to all outputs and the received information is stored internally within LLDP device thus building knowledge of the surrounding network. LLDP packets are not forwarded after receiving, so devices receive information only from directly related neighbors. In the SDN network paradigm, network devices are simplified and lose LLDP capabilities. However, the protocol can be implemented via the SDN controller. All received LLDP messages are sent to the controller that adequately processes and stores them.

Here, in this paper, we present the case of using LLDP functionality for management applications written in the Python script language that use OpenFlow protocol version 1.3. We used Ryu implementation of SDN controller and the topology is simulated via the Mininet program. At the network control layer the network topology can simply be presented using the graph, shown in Fig. 2, which enables runtime finding the most efficient paths through the network topology. The demonstrative example presented in this paper is based on the work [2] which use Dijkstra shortest path algorithm to optimize network use on per flow basis. Our contribution is introduction of algorithms with weights to the edges of the graph where higher weight means longer execution time at this edge.

After introduction, in Sect. 2 we provide details on topology algorithms. Then in Sect. 3 we provide implementation details and in Sect. 4 we provide results from experimental evaluation. In Sect. 5 we conclude the paper.

2 Topology Algorithms

The network topology is naturally represented as a graph, in which vertices represent SDN devices, and edges the direct communication links between them. It is usually viewed as a weighted graph, with edge weights representing some properties of the communication link, such as traffic load, time, cost, etc.

The topology algorithms considered in this work aim to find shortest paths through the network. The length of the path is usually measured as the sum of weights of all edges in the path. The most common algorithm for finding shortest paths in weighted graphs is Dijkstra's algorithm [1].

However, we follow here [2] and also use their extension of Dijkstra's shortest path algorithm. The extended Dijkstra's algorithm considers, not only the weights of edges, but also the weights of vertices. This is necessary in SDN network, because the SDN device (i.e., vertex in the graph) is processing the packets, and its performance depends on its load, capacity, processing speed, etc. The length of a path is then defined as the sum of weights of all edges and all vertices in the path. With this modification, starting from a given vertex as a root, the extended Dijkstra's algorithm determines the shortest path tree in the graph, essentially in the same way as the usual Dijkstra's algorithm. Difference is in the step which checks if the new path is shorter, the length of paths is calculated as the sum of weights of vertices and edges (not only edges).

The applications of the (extended) Dijkstra's algorithm to the SDN network require a way of assigning weights to edges and vertices. This can be done in several ways. In this work we propose a new way of assigning weights in the network graph and compare the network performance with definitions of weights from the literature.

To describe the algorithm precisely, we introduce some notation. Let $G = (V, E)$ be the graph representing the network topology, where V is the set of vertices, i.e., SDN devices, and E the set of edges, i.e., the direct communication links between SDN devices. It is a simple connected graph. Let $w : E \to \mathbb{R}_{>0}$ and $w' : V \to \mathbb{R}_{>0}$ be the weight function of edges and vertices, respectively.

In the unweighted case, we denote the weight functions by w_0 and w'_0. They are constant functions set to one. That is, $w_0(e) = 1$ for all edges $e \in E$ and $w'_0(v) = 1$ for all vertices $v \in V$.

In the first weighted case, as in [2], the weight functions measure the relative load of devices and links. In this case the weight functions are denoted w_1 and w'_1. Let $C : E \to \mathbb{R}_{>0}$ and $C' : V \to \mathbb{R}_{>0}$ be the capacity functions for edges and vertices, respectively. Then, $C(e)$, for an edge $e \in E$, is the bandwidth of the communication link represented by e, that is, the maximal number of bits per second that can pass through e. Similarly, $C'(v)$, for a vertex $v \in V$, is the processing capacity of the SDN device represented by v, that is, the maximal number of bits that can be processed by v in one second. For a traffic flow f let $\gamma(f)$ and $\gamma'(f)$ denote the traffic load of f on edges and vertices, respectively, measured in bits per second. Then the weight functions are defined as

$$w_1(e) = \frac{\sum_{f \in F_e} \gamma(f)}{C(e)}, \quad \text{for } e \in E,$$

$$w'_1(v) = \frac{\sum_{f \in F_v} \gamma'(f)}{C'(v)}, \quad \text{for } v \in V,$$

where F_e and F_v is the set of all traffic flows passing through the edge e and the vertex v, respectively, at the considered moment in time. The idea behind this approach is to determine paths with lower traffic load.

The second approach is based on the dynamic measurement of real time required for the packet to pass through a communication link. This is established by sending sample packets through links in the network, one at the time, not to influence the traffic load, and measuring the arrival time. In this way, the controller in SDN network gets the information about the average passing time of packets in the network. These times are the edge weights in this case. More precisely, let $e \in E$ be the edge connecting vertices v_1 and v_2. When the controller sends a sample packet through e, it measures the time t_{send} and t_{rec} when the packet is sent and received. However, one must also take into account the time required for the communication between controller and the SDN devices at vertices v_1 and v_2. If t_1 and t_2 are response times of vertices v_1 and v_2, the actual time required for the passage of a packet from outer source is obtained as

$$w_2(e) = t_{\text{send}} - t_{\text{rec}} - \frac{t_1 + t_2}{2}.$$

This value is the weight of the edge e in this approach.

As an example of differences between the two weighted approaches for finding the shortest path in SDN network, consider the network represented by the graph in Fig. 1. The edges are labelled by their weights $w_1(e)$ measuring their traffic load percentage [%], and $w_2(e)$ measuring the actual packet passage time in milliseconds [ms]. The outcome of the Dijkstra's algorithm applied to the two weighted graphs is shown in Fig. 2. We can see that the two approaches for assigning weights come up with different shortest paths.

Fig. 1 Example of a weighted graph representing network topology

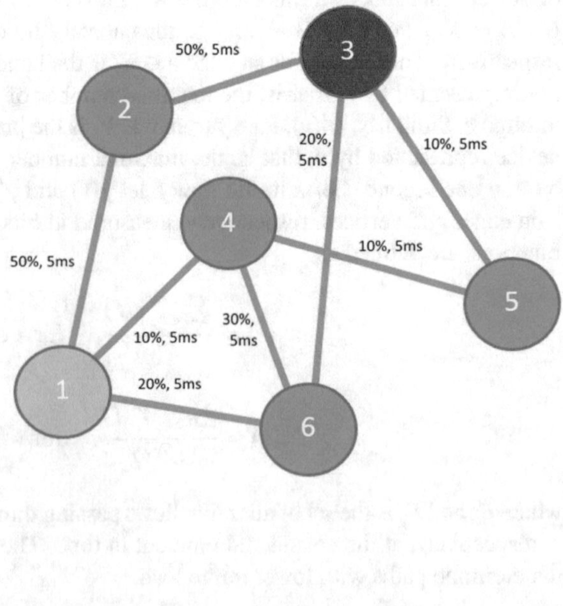

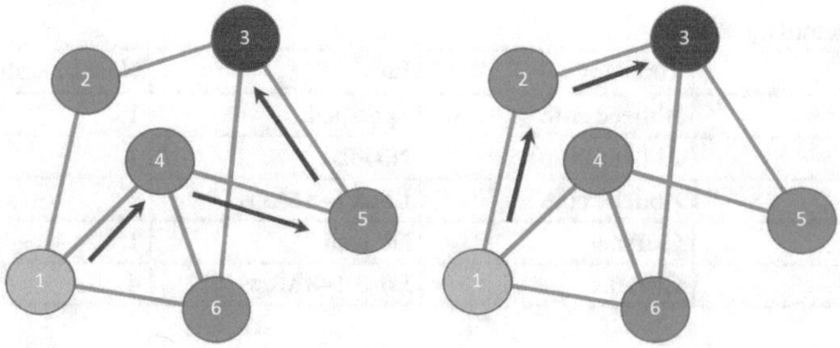

Fig. 2 Outcome of Dijkstra's algorithms for the network graph weighted by loads (*left*) and weighted by response times (*right*)

3 Implementation of Dijkstra Algorithm

Here we present our implementation of SDN application implementing topology algorithms presented in Sect. 2 based on Ryu implementation [6] written in Python programming language. For experimentation we used Ryu pre-installed VM image files [7]. The virtual machine drives Linux distribution of Ubuntu version 14.04.1 LTS. Within VM's is a built-in PyCharm Community operating environment with installed Ryu plugins implementing most of the OpenFlow-related functionality that can be easily used for writing Python management applications.

Our implementation considers three different approaches to calculate bridge weights by using the Dijkstra algorithm. The first and simplest control application uses the Dijkstra algorithm so that all links assign weight one. This gives us a management algorithm that finds the path through topology with minimum jumps. The second application uses the modification where weights are represented by link loads. The third control application uses for weights the real time measurement of the response time of each link in the network topology.

4 Experimental Evaluation

The verification was conducted on simulations of real networks by using Mininet, which simulates the real SDN network with ehd users and SDN network. The two simulated real networks considered in this paper are Abilen Core network (11 SDN devices and 14 communication links) and Claranet (15 SDN devices and 18 communication links). For both networks the bandwidth of all edges was set to $C(e) = 10\,\text{Mbps}$, and the passage time depends on the physical length of the edge. We conducted five tests of the Dijkstra's algorithms, three on Abilen Core network and two on Claranet, see Table 1.

The specified load in Test 1 is such that the two weighted Dijkstra algorithms would not come with the same shortest path at two parts of the network. The Tests

Table 1 Summary of tests

Test id.	Topology	Load	No. of iterations
Test 1	Abilene core	Specified	1
Test 2	Abilene core	No load	1
Test 3	Abilene core	Load 1–8 Mbps	4
Test 4	Claranet	No load	1
Test 5	Claranet	Load 1–8 Mbps	4

2 and 4 with no load look at the difference between algorithms when there is no traffic in the network. The Tests 3 and 5 chooses random load in interval between 1 and 8 Mbps. In that case we made four iterations of each test, because the change of random loads may produce different results.

The results of tests show that the Dijkstra's algorithm with no weights could be efficient only if the network traffic is very low, or all the links are approximately equally loaded. The Dijkstra algorithm with load weights turns out to be quite unstable. For low traffic or equally distributed traffic, the results are similar to no weight algorithm, but with unequal distribution of loads its efficiency growth. However, it is very sensitive to specific cases of network load and response time, such as the example given in Sect. 2. The Dijkstra algorithm with weights given in terms of response time turns out to be the best solution, as it is always more efficient than the other two algorithms. In the tests with no load or with equally distributed load, it is only slightly more efficient, but when the loads are random, it is often much faster than the other two. The difference is in some cases an order of magnitude. In particular, this happens in Test 1 in which the loads are specified in such a way that the two weighted approaches differ.

5 Conclusion and Future Work

To overcome distributed computing challenges of future networks a software defined network (SDN) technology is proposed. Here in this paper we present three approaches to traffic flow routing based on Dijkstra shortest path algorithm, compare their effectiveness and present how SDN technology may be used to dynamically manage traffic flows. We show implementation details and discuss benefits. Our preliminary results shows that dynamic decisions on flow routing are more efficient if they are based on temporal load measurements. Our future experimental tests would consider larger topologies and dynamic network conditions. Dynamic weights introduce potential instability and oscillations.

Acknowledgements This work has been supported in part by Croatian Science Foundation's project EVOSOFT UIP-2014-09-7945 and by the Univ. of Rijeka Res. Grant 13.09.2.2.16.

References

1. Dijkstra, E.: A note on two problems in connexion with graphs. Numerische mathematik **1**(1), 269–271 (1959)
2. Jiang, J.R., Huang, H.W., Liao, J.H., Chen, S.Y.: Extending Dijkstra's shortest path algorithm for software defined networking. In: The 16th Asia-Pacific Network Operations and Management Symposium, pp. 1–4, Sept 2014
3. Matsubara, D., Egawa, T., Nishinaga, N., Kafle, V.P., Shin, M.K., Galis, A.: Toward future networks: a viewpoint from ITU-T. IEEE Commun. Mag. **51**(3), 112–118 (2013)
4. Nunes, B.A.A., Mendonca, M., Nguyen, X.N., Obraczka, K., Turletti, T.: A survey of software-defined networking: past, present, and future of programmable networks. IEEE Commun. Surv. Tutorials **16**(3), 1617–1634 (2014)
5. Osseiran, A., Monserrat, J.F., Marsch, P.: 5G Mobile and Wireless Communications Technology, 1st edn. Cambridge University Press, New York, NY, USA (2016)
6. Ryu Project Team: RYU SDN Framework. https://www.osrggithubio/ryu-book/en/html/. Accessed 15 May 2017
7. Ryu Project Team: Ryu pre-installed VM image files. https://www.sourceforgenet/projects/ryu/files/vmimages/OpenFlowTutorial/. Accessed 15 May 2017

Part II
Reasoning and Decision Making in Distributed Environments

Towards a Paraconsistent Approach to Actions in Distributed Information-Rich Environments

Łukasz Białek, Barbara Dunin-Kęplicz and Andrzej Szałas

Abstract The paper introduces ACTLOG, a rule-based language capable of specifying actions paraconsistently. ACTLOG is an extension of $4QL^{Bel}$, a rule-based language for reasoning with paraconsistent and paracomplete belief bases and belief structures. Actions considered in the paper act on belief bases rather than states represented as sets of ground literals. Each belief base stores multiple world representations which can be though of as a representation of possible states. In this context ACTLOG's action may be then seen as a method of transforming one belief base into another. In contrast to other approaches, ACTLOG permits to execute actions even if the underlying belief base state is partial or inconsistent. Finally, the framework introduced in this paper is tractable.

Keywords Action languages · Paraconsistent reasoning · Paracomplete reasoning · Belief structures

1 Actions in Information-Rich Environments

Reasoning about actions and change is an important ingredient of many AI systems. Throughout the years a vast variety of advanced solutions has been introduced, developed, verified and used in this field. Although these solutions evoked a broad and intensive research on related problems (see, e.g., [19] and references there), the issue of inconsistent information has rarely been addressed in this context. In fact, due to

Ł. Białek · B. Dunin-Kęplicz · A. Szałas (✉)
Institute of Informatics, University of Warsaw, Warsaw, Poland
e-mail: andrzej.szalas@mimuw.edu.pl

B. Dunin-Kęplicz
e-mail: keplicz@mimuw.edu.pl

Ł. Białek
e-mail: bialek@mimuw.edu.pl

A. Szałas
Department of Computer and Information Science, Linköping University, Linköping, Sweden

M. Ivanović et al. (eds.), *Intelligent Distributed Computing XI*,
Studies in Computational Intelligence 737, https://doi.org/10.1007/978-3-319-66379-1_5

the distribution and heterogeneity of multiple information sources of different quality and credibility, *inconsistent* and *incomplete information* (further abbreviated in this paper by I3) is common and natural phenomenon in contemporary information-rich environments. In our approach, rather than fighting with missing or inconsistent information, we treat them as first-class citizens and provide tools for robust actions' specifications. The importance of addressing inconsistencies in a robust manner is emphasized, e.g., in [14] (see also [15]), where *inconsistency robustness* is phrased as:

> information system performance in the face of continually pervasive inconsistencies—a shift from the previously dominant paradigms of inconsistency denial and inconsistency elimination attempting to sweep them under the rug.

For related discussion see also, e.g., [2], in particular an overview in [1] where, among others, the authors point out that:

> inconsistency is useful in directing reasoning, and instigating the natural processes of argumentation, information seeking, multi-agent interaction, knowledge acquisition and refinement, adaptation, and learning.

The ultimate goal of our research is to develop a planning system rich enough to cope with I3 in realistic models of environments. Thus we will focus on a novel paraconsistent approach to actions' specification, keeping in mind a perspective of planning. The classical planning system, STRIPS, has been introduced already in 1971 [13]. Even though STRIPS strongly influenced many planning systems, it did not offer means for dealing with I3. Later, paraconsistent approaches have been proposed (see, e.g., [10, 21]), where non-standard states during planning process have been addressed. However, in majority of contemporary planners inconsistent/missing knowledge is projected into the two-valued classical framework. Typically this has been achieved by resolving I3 with the use of non-monotonic techniques or other heuristics. In our approach, also rooted in STRIPS, I3 is present in all phases of planning including actions' and planning specification as well as planning algorithms. That is, I3 is incorporated in the planning process, rather than being a subject of disambiguation at some point.

Since the very start of paraconsistent knowledge representation and planning, beliefs have been typically modeled via various combinations of multi-modal logics [8, 11], non-monotonic logics [19], probabilistic reasoning [24] or fuzzy reasoning [25], just to mention some of them. However, most of the solutions turn out to be unsatisfactory due to either the lack of tools for handling I3 or to high complexity. Importantly, a shift in perspective has been suggested in [5, 6], where rather than reasoning in modal logics or other formalisms of high complexity, a tractable approach based on querying paraconsistent belief bases has been indicated. In order to achieve the required expressiveness and modeling convenience, two additional truth values: $\mathfrak{i}$, standing for *inconsistent* and $\mathfrak{u}$, standing for *unknown* have been adopted.

The implementation tool our solution is based on is $4QL^{Bel}$ [3], extending the 4QL rule language [16–18]. While 4QL already allows one to flexibly resolve/disambiguate I3 at any level of reasoning, $4QL^{Bel}$ includes means for specifying paraconsistent beliefs and belief structures and referring to them in rules.

The goal of this research is to define a formal language ACTLOG for specifying actions in information-rich environments, thus allowing for:

- concise rule-based specification of actions and their effects;
- flexibility in evaluation of formulas in distributed paraconsistent belief bases;
- tractability of computing actions' preconditions and resulting belief bases;
- practical expressiveness meaning that all actions (and only such) with effects computable in deterministic polynomial time can be specified in ACTLOG.

The paper is structured as follows. First, in Sect. 2, we recall the $4QL^{Bel}$ rule-language, originally defined in [3] and used in ACTLOG. Next, in Sect. 3, we introduce the ACTLOG action specification language. Section 4 provides results concerning tractability of the approach. Section 5 contains examples illustrating the approach. Finally, Sect. 6 concludes the paper.

2 Querying Belief Bases

In this section we recall only the most important definitions provided in [3, 7]. For detailed description of belief bases and belief structures, see [3, 5–7] and of belief bases in the context of the $4QL^{Bel}$ language, see [3]. We will use truth and information ordering of all languages from 4QL family [16, 18, 23], shown in Fig. 1.

We assume that *Const* is a fixed finite set of constants, *Var* is a fixed set of variables and *Rel* is a fixed set of relation symbols. If S is a set then $\text{FIN}(S)$ denotes the set of all finite subsets of S. By $\mathbb{C} \stackrel{\text{def}}{=} \text{FIN}(\mathcal{G}(Const))$ we denote the set of all finite sets of ground literals over the set of constants *Const*.

Definition 1 By a *belief base over a set of constants Const* we understand any finite set Δ of finite sets of ground literals over *Const*, i.e. any finite set $\Delta \subseteq \mathbb{C}$. $\triangleleft$

According to Definition 1, a belief base consists of sets of ground literals. Each set in a belief base represents a possibly incomplete and/or inconsistent view of the world. For example, a belief base can consist of three sets: one containing facts based on measurements received from a ground robot's sensor platform, the second containing facts interpreting video streams from a drone's camera while the third representing views provided by ground operators.

Fig. 1 Orderings on truth values

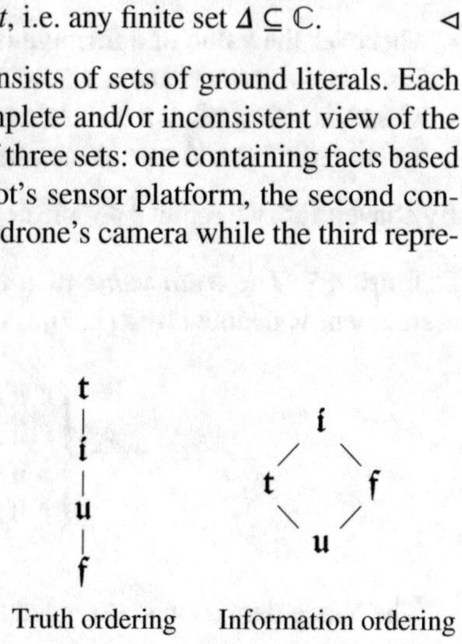

Table 1 Syntax of the base logic

$\langle Formula\rangle$ ::= $\langle Literal\rangle \mid \neg\langle Formula\rangle \mid \langle Formula\rangle \wedge \langle Formula\rangle \mid$
$\langle Formula\rangle \vee \langle Formula\rangle \mid \forall X\,\langle Formula\rangle \mid \exists X\,\langle Formula\rangle \mid$
$\langle Formula\rangle \in \langle TruthValues\rangle \mid$
$f(\Delta).\langle Formula\rangle \mid \mathrm{Bel}_\Delta(\langle Formula\rangle)$

where:

- $\langle Literal\rangle$ represents the set of literals;
- $\langle TruthValues\rangle$ represents nonempty subsets of $\{\mathfrak{t}, \mathfrak{f}, \mathfrak{i}, \mathfrak{u}\}$;
- Δ is a belief base,
- f is a mapping transforming a belief base into a (single) finite set of ground literals; if f is not specified, by default $f(\Delta) \stackrel{\text{def}}{=} \bigcup \Delta$, i.e., $\Delta.\alpha \stackrel{\text{def}}{=} (\bigcup \Delta).\alpha$.

Syntax of the base logic is given in Table 1. In addition to the classical syntax, we allow formulas of the form:

- $\mathrm{Bel}_\Delta()$, expressing beliefs related to belief bases (indicated by Δ);
- $f(\Delta).()$, allowing one to evaluate $\mathrm{Bel}()$-free formulas in belief bases: here f is a mapping transforming a belief base into a single set of ground literals, e.g., $f(\Delta)$ may be $\bigcup_{D\in\Delta} D$ or $\bigcap_{D\in\Delta} D$ (further denoted by $\bigcup \Delta$, $\bigcap \Delta$, respectively).

In general, f occurring in $f(\Delta).()$ is an arbitrary information fusion method, intended as a means to combine information included in sets of ground literals of Δ.

The semantics of disjunction is given by the maximum wrt truth ordering (see Fig. 1) and of conjunction is given by the minimum wrt truth ordering.
The semantics of negation is defined by:

$$\neg\mathfrak{t} \stackrel{\text{def}}{=} \mathfrak{f},\ \neg\mathfrak{f} \stackrel{\text{def}}{=} \mathfrak{t},\ \neg\mathfrak{i} \stackrel{\text{def}}{=} \mathfrak{i},\ \neg\mathfrak{u} \stackrel{\text{def}}{=} \mathfrak{u}.$$

That is,

- whenever the value of a formula is $\mathfrak{u}$, it is unknown whether a formula is true or false, so the same is unknown for its negation, too;
- whenever a formula is $\mathfrak{i}$, it is claimed true and false at the same time, so its negation is claimed false and true at the same time so is $\mathfrak{i}$, too.

By convention, we remove double negations, that is, $\neg\neg\alpha$ is always identified with α.

Definition 2 The *truth value* of a literal ℓ wrt a set of ground literals L and an assignment v, denoted by $\ell(L, v)$, is defined as follows:

$$\ell(L,v) \stackrel{\text{def}}{=} \begin{cases} \mathfrak{t} \text{ if } \ell(v)\in L \text{ and } (\neg\ell(v))\notin L; \\ \mathfrak{i} \text{ if } \ell(v)\in L \text{ and } (\neg\ell(v))\in L; \\ \mathfrak{u} \text{ if } \ell(v)\notin L \text{ and } (\neg\ell(v))\notin L; \\ \mathfrak{f} \text{ if } \ell(v)\notin L \text{ and } (\neg\ell(v))\in L. \end{cases}$$

◁

The above definition is extended to other formulas in Tables 2 and 3.

Table 2 Semantics of first-order formulas (with $\in$)

- if α is a literal then $\alpha(L, v)$ is defined in Definition 2;
- $(\neg\alpha)(L, v) \stackrel{\text{def}}{=} \neg(\alpha(L, v))$;
- $(\alpha \wedge \beta)(L, v) \stackrel{\text{def}}{=} \min\{\alpha(L, v), \beta(L, v)\}$;
- $(\alpha \vee \beta)(L, v) \stackrel{\text{def}}{=} \max\{\alpha(L, v), \beta(L, v)\}$;
- $(\forall X\alpha(X))(L, v) \stackrel{\text{def}}{=} \min_{a \in Const} \{(\alpha(X/a)(L, v)\}$;
- $(\exists X\alpha(X))(L, v) \stackrel{\text{def}}{=} \max_{a \in Const} \{(\alpha(X/a)(L, v)\}$;
- $(\alpha \in T)(L, v) \stackrel{\text{def}}{=} \begin{cases} \mathfrak{t} \text{ when } \alpha(L, v) \in T \\ \mathfrak{f} \text{ otherwise,} \end{cases}$

where:

- L is a set of ground literals;
- min, max are respectively minimum and maximum wrt truth ordering;
- $\alpha(X/a)$ denotes the formula obtained from α by substituting all free occurrences of variable X by constant a.

For a belief base Δ, we define $\alpha(\Delta, v) \stackrel{\text{def}}{=} \alpha(\bigcup \Delta, v)$.

Table 3 Semantics of the Bel() and $f()$.() operators

- $(\text{Bel}_\Delta(t))(v) \stackrel{\text{def}}{=} t$, for $t \in \{\mathfrak{t}, \mathfrak{f}, \mathfrak{i}, \mathfrak{u}\}$;
- $\big(\text{Bel}_\Delta(\alpha)\big)(v) \stackrel{\text{def}}{=} Lub\{\alpha(D, v) \mid D \in \Delta\}$;
- $(f(\Delta).\alpha)(v) \stackrel{\text{def}}{=} \alpha(f(\Delta), v)$,

where:

- Δ is a belief base;
- $v : Var \longrightarrow Const$ is an assignment of constants to variables;
- α is a first-order formula (for nested Bel() s, one starts with the innermost one.)
- Lub denotes the least upper bound wrt the information ordering (see Fig. 1).

As shown in [3], $4QL^{Bel}$ offers means to express queries to belief bases and more generally, to belief structures. That is, rules in $4QL^{Bel}$ have the form:

$$\langle Literal \rangle : - \langle Formula \rangle, \tag{1}$$

where $\langle Formula \rangle$ is an arbitrary formula as defined in Table 1. Like in 4QL, in $4QL^{Bel}$ we allow to encapsulate rules in modules and refer to them using traditional notation for remote calls: $m.A$, where m is a module name and A is a formula. The meaning of $m.A$ is the answer to query expressed by A evaluated in m.[1]

Definition 3 A $4QL^{Bel}$ *program* is a set of $4QL^{Bel}$ rules. ◁

The semantics for $4QL^{Bel}$ programs is given by well-supported models. A *model* M of a $4QL^{Bel}$ program P is a set of ground literals (not necessarily minimal) such that for every rule (1) in P,

[1] Acyclicity of references among modules is required (needed for tractability of computing queries).

- whenever ⟨*Formula*⟩ is **t** in M, the conclusion ⟨*Literal*⟩ is in M, and
- whenever ⟨*Formula*⟩ is $\mathfrak{i}$ in M, the conclusion ⟨*Literal*⟩ as well as its negation ¬ ⟨*Literal*⟩ are in M.

Intuitively, a model is *well-supported* when all literals of M are justified by reasoning starting from facts of P and using rules of P. Note that well-supportedness does not entail minimality. This is an intended feature of our approach: in many contexts minimality is not desired [4, 12, 16, 20, 22].

3 The ActLog Language

Now we are ready to extend $4QL^{Bel}$ towards specifying actions. All back-end operations like managing logical operations on the facts or reasoning will be handled by $4QL^{Bel}$.

Let us start from defining an action specification. The syntax is presented as Action 1, where:

- `name` is the action name and $\bar{x}$ are its parameters;
- $\alpha(\bar{x})$ is an arbitrary formula of $4QL^{Bel}$, called the *precondition* of action `name`;
- $\beta^+(\bar{x}), \beta^-(\bar{x})$ are $4QL^{Bel}$ programs, representing effects of action `name` by providing sets of literals to be added ($\beta^+(\bar{x})$) and to be removed ($\beta^-(\bar{x})$)
- it is assumed that α, β^+ and β^- contain no free variables other than those in $\bar{x}$.

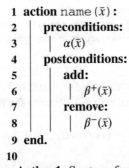

action `name` $(\bar{x})$ **:**
 preconditions:
  $\alpha(\bar{x})$
 postconditions:
  add:
   $\beta^+(\bar{x})$
  remove:
   $\beta^-(\bar{x})$
end.

Action 1: Syntax of actions in ActLog.

The definition reflects the general idea of action definition. As a novelty, our action is a belief bases transformer: a state of the environment, expressed as a first belief base, is transformed by an action into the resulting belief base. Next, the use $4QL^{Bel}$ to represent actions' effects ensures their concise representation. Finally, due to tractability results for $4QL^{Bel}$ [3], effects of actions can be computed in a tractable manner.

It is also important to notice that rules in action specification may use operators like, e.g., $\mathrm{Bel}_\Delta()$, referring to belief bases. This allows one to deal with distributed belief bases. In our paper such bases are known from the context, so we sometimes omit the subscript indicating a belief base.

Definition 4 If tuple of variables $\langle x \rangle$ and constants $\bar{a}$ have the same length, we call them *compatible*. Given a set of ground literals C, specification expressed as Action 1 and a tuple of constants $\bar{a}$ compatible with $\bar{x}$, the action $\mathtt{name}(\bar{a})$ is *executable* on C when its precondition $\alpha(\{C\}, v) = \mathbf{t}$, where v assigns constants $\bar{a}$ to variables $\bar{v}$, respectively.

An action is *executable* on a belief base Δ, if it is executable on some $C \in \Delta$. ◁

Note that in preconditions of actions (formula α of Action 1) one can use:

- any formula of the form defined in Table 1, in particular involving the $\mathrm{Bel}()$ operator as well as the operator '$\in T$', where $T \subseteq \{\mathbf{t}, \mathbf{f}, \mathbf{i}, \mathbf{u}\}$;
- the operator '$\in T$' allowing one can react to inconsistency and lack of knowledge.

Therefore an action can be executed when the state is inconsistent and/or some/all literals are unknown. Running actions in such circumstances is a unique feature of ActLog.

When action $\mathtt{name}(\bar{a})$ is executed, it transforms its input belief base Δ into the resulting belief base Δ' as shown in Algorithm 1. The belief base Δ' represents *effects of* action $\mathtt{name}(\bar{a})$ on Δ.

Input: action $\mathtt{name}(\bar{a})$, specified as Action 1, where $\bar{a}$ is a tuple of constants;
belief base Δ;
Result: belief base Δ' representing effects of input action execution on belief base Δ;

set $\Delta' = \emptyset$
foreach $C \in \Delta$ **do**
  if action $\mathtt{name}(\bar{a})$ is executable on C **then**
    set $M^+ = \emptyset$; **set** $M^- = \emptyset$
    compute the well-supported model of $\beta^+(\bar{a}) \cup C$ adding to M^+ each literal obtained as a consequence of a rule of $\beta^+(\bar{a})$;
    compute the well-supported model of $\beta^-(\bar{a}) \cup C$ adding to M^- each literal obtained as a consequence of a rule of $\beta^-(\bar{a})$;
    add the set $(C \cup M^+) \setminus M^-$ **to** Δ'
  else
    add the set C **to** Δ'
  end
end

Algorithm 1: Computing effects of actions.

4 Tractability of the Approach

For any ActLog specification of an action $\mathtt{name}(\bar{x})$, by $\#D$ we denote the sum of sizes of all domains in the specification and by $\#M$ we denote the number of $4\mathrm{QL}^{\mathrm{Bel}}$ modules occurring in the specification. For belief base Δ, by $\#\Delta$ we denote the number of all literals appearing in Δ. Note that $\#\Delta$ is polynomial in the size of $\#D$ (the size of relations is constant). In real-world applications, $\#M$ as well as $\#D$ are manageable by the hardware/database systems used, so is $\#\Delta$.

The following theorems can be proved analogously to corresponding results for 4QL [16–18] and $4QL^{Bel}$ [3].

Theorem 1 *Let Δ be a belief base. For every* ActLog *specification of action* `name`$(\bar{x})$ *and a tuple of constants $\bar{a}$, compatible with $\bar{x}$ precondition α as well as its effects can be computed in deterministic polynomial time in* $\max\{\#D, \#P, \#\Delta\}$. ◁

Theorem 2 ActLog *captures deterministic polynomial time over linearly ordered domains. That is, every action with polynomially computable preconditions and effects can be expressed in* ActLog. ◁

5 Examples of ActLog Specifications

Let us illustrate our approach by sample action specifications related to a more complex scenario originally introduced in [9]. We assume that in a grid-shaped area a contamination has been detected and a clean-up mission is to be started.

The most basic robot's activity depends on moving from one place to another. Action 2 provides a specification of this action: a ground robot `ID` can move to a new place `X` provided that:

- place `X` is safe or its safety is inconsistent or unknown, and
- robot `ID` is ready and it is a ground robot,

or

- air support is not needed or information as to its need is inconsistent.

The action results in setting the robot's status to occupied and moving it to place `X`.

```
action goTo(ID,X):
    preconditions:
        (safe(X)∈{t,i,u} ∧ status(ID, ready) ∧ type(ID, ground))
        ∨ airSupportNeeded(X) ∈ {i,f},
    postconditions:
        add:
            status(ID, occupied).
            position(ID, X).
        remove:
            status(ID, ready).
            position(ID, P) :- position(ID, P) ∧ P≠X.
end.
```

Action 2: Robot `ID` moves to place `X`.

Note that the action can be executed when its preconditions are true. This may happen even if its input belief base contains inconsistent information as to safety of place `X` or `X`'s safety is unknown, in which case safe$(X) \in \{t, i, u\}$ is true.

Consider a belief base Δ consisting of the following two sets of ground literals:

- C1={place(1), place(2), place(3), status(robot1, ready), position(robot1, 2), type(robot1, ground), safe(1), ¬safe(1) };
- C2={place(1), place(2), place(3), status(robot1, ready), position(robot1, 2), type(robot1, ground), ¬safe(1)}.

In C1, the value of safe(1) is $\mathfrak{i}$, status(robot1,ready) and type(robot1,ground) are **t**, airSupportNeeded(1) is unknown. Thus the precondition of action `goTo(robot1,1)` is **t**. Therefore the action is executable on C1.

In C2, the value of safe(1) is $\mathfrak{f}$, status(robot1,ready) and type(robot1,ground) are **t**, airSupportNeeded(1) is unknown. Thus the precondition of action `goTo(robot1,1)` is $\mathfrak{u}$. Therefore the action is not executable on C2.
The effects of executing `goTo(robot1,1)` on Δ is $\Delta' =$ {C1',C2}, where:

C1'={place(1), place(2), place(3), status(robot1, occupied), position(robot1, 1), type(robot1, ground), safe(3), ¬safe(3) }.

Two other examples of action specifications are Action 3 and 4.

```
action flyTo(ID, X):
  preconditions:
    airSupportNeeded(X) ∧ status(ID, ready) ∧ type(ID, uav)
  postconditions:
    the same as in Action 2
end.
```

Action 3: Robot `ID` flies to place `X`.

Action 3 assumes that:

- air support is needed in place `X`, and
- the robot `ID` is ready and it is an UAV (unmanned aerial vehicle).

The result of executing `flyTo` depends on making the UAV `ID` fly to place `X` and changing its status to occupied.

```
action pourL1(ID, X):
  preconditions:
    status(ID, ready) ∧ type(ID, ground) ∧ position(ID, X) ∧
    Bel(humidity(X,rain)) ∈ {t, i} ∧
    Bel(pressure(X, v_P) ∧ temperature(X, v_T) ∧ concentration1(X, v_C1)) ∧
    acceptable(v_P,v_T) ∧ v_C1 ≥ e_1
  postconditions:
    add:
      checkNeeded(X).
end.
```

Action 4: Robot `ID` pours liquid `L1` on place `X`.

Action 4 is a neutralization action depending on pouring a liquid on a contaminated place. It can be executed when:

- a ready ground robot ID is in position X, and
- ID's belief as to the rain is t or i, and
- ID believes that in place X: the pressure is v_P, the temperature is v_T and concentration of compound C1 is v_{C1}, and
- pressure together with temperature are acceptable, and
- the concentration of compound C1 is greater than a threshold given by constant e_1.

Actions specified that way can further be used in reasoning about their effects and in planning. Inconsistency or incompleteness of the underlying belief bases doesn't have to prevent actions form being executed. The reaction on I3 is left to the system designer, who can flexibly specify actions' behavior when I3 occurs. Moreover, such behavior can be specified in a highly contextual manner, which allows one to develop truly robust action specifications.

6 Conclusions

The paper presents ACTLOG, a novel extension of $4QL^{Bel}$ language, developed for specifying actions in distributed information-rich environments. ACTLOG competes with other approaches not only by providing comfortable tools for handling inconsistency and ignorance but also by setting the computational complexity on a polynomial level. Moreover, each of the distributed agents can have its own belief base or share beliefs in a group. The language permits to evaluate belief operators and other formulas on arbitrary belief bases, not necessarily on the global one. This makes ACTLOG suitable for applications in distributed intelligent systems.

Actions in ACTLOG do not act on sets of literals, like in other existing solutions but on belief bases. Each belief base stores multiple world representations which can be though of as feasible alternatives or different points of view. The action may be then referred to as a transformation between belief bases acting on many possible worlds at once. Beliefs are expressed using the Bel() operator.

As demonstrated in [3], $4QL^{Bel}$ allows for reasoning over belief structures [5–7], richer than belief bases. Belief structures model the process of transforming "raw" beliefs to "mature" beliefs. We currently work on extending ACTLOG to the full framework of belief structures.

ACTLOG, as presented in this paper, deals with atomic deterministic actions only. In our project we plan to extend ACTLOG to cover composite actions, too.

Acknowledgements This research has been supported by the Polish National Science Centre grant 2015/19/B/ST6/02589.

References

1. Bertossi, L., Hunter, A., Schaub, T.: Introduction to inconsistency tolerance. Bertossi et al. [2], pp. 1–14
2. Bertossi, L., Hunter, A., Schaub, T. (eds.): Inconsistency Tolerance, LNCS, vol. 3300. Springer (2005)
3. Białek, Ł., Dunin-Kęplicz, B., Szałas, A.: Rule-based reasoning with belief structures. In: Kryszkiewicz, M., Appice, A., Ślęzak, D., Rybiński, H., Skowron, A., Raś, Z. (eds.) Foundations of Intelligent Systems, Proceedings of ISMIS Conference. LNAI, vol. 10352, pp. 229–239. Springer (2017)
4. Doherty, P., Szałas, A.: Stability, supportedness, minimality and Kleene answer set programs. In: Eiter, T., Strass, H., Truszczyński, M., Woltran, S. (eds.) Advances in Knowledge Representation, Logic Programming, and Abstract Argumentation, LNCS, vol. 9060, pp. 125–140. Springer International Publishing (2015)
5. Dunin-Kęplicz, B., Szałas, A.: Epistemic profiles and belief structures. In: Proceedings KES-AMSTA 2012: Agents and Multi-agent Systems: Technologies and Applications. LNCS, vol. 7327, pp. 360–369. Springer (2012)
6. Dunin-Kęplicz, B., Szałas, A.: Taming complex beliefs. Trans. Comput. Collect. Intel. XI LNCS **8065**, 1–21 (2013)
7. Dunin-Kęplicz, B., Szałas, A.: Indeterministic belief structures. In: Jezic, G., Kusek, M., Lovrek, I., J. Howlett, Lakhmi, J. (eds.) Agent and Multi-Agent Systems: Technologies and Applications: Proceedings 8th International Conference KES-AMSTA, pp. 57–66. Springer (2014)
8. Dunin-Kęplicz, B., Verbrugge, R.: Teamwork in Multi-Agent Systems. A Formal Approach. Wiley (2010)
9. Dunin-Kęplicz, B., Verbrugge, R., Ślizak, M.: Teamlog in action: a case study in teamwork. Comput. Sci. Inf. Syst. **7**(3), 569–595 (2010)
10. Eiter, T., Faber, W., Leone, N., Pfeifer, G., Polleres, A.: Planning under incomplete knowledge. In: Lloyd, J., Dahl, V., Furbach, U., Kerber, M., Lau, K.K., Palamidessi, C., Pereira, L., Sagiv, Y., Stuckey, P. (eds.) Proceedings Computational Logic: 1st International Conference, pp. 807–821. Springer (2000)
11. Fagin, R., Halpern, J., Moses, Y., Vardi, M.: Reasoning About Knowledge. The MIT Press (2003)
12. Ferraris, P., Lifschitz, V.: On the minimality of stable models. In: Balduccini, M., Son, T. (eds.) Logic Programming, Knowledge Representation, and Nonmonotonic Reasoning. LNCS, vol. 6565, pp. 64–73. Springer (2011)
13. Fikes, R.E., Nilsson, N.J.: Strips: a new approach to the application of theorem proving to problem solving. In: Proceedings of the 2nd International Joint Conference on Artificial Intelligence, pp. 608–620. IJCAI'71, Morgan Kaufmann Publishers Inc. (1971)
14. Hewitt, C.: Formalizing common sense for scalable inconsistency-robust information integration using direct logic reasoning and the actor model. arXiv:0812.4852 (2008)
15. Hewitt, C., Woods, J. (eds.): Inconsistency Robustness. College Publications (2015)
16. Małuszyński, J., Szałas, A.: Living with inconsistency and taming nonmonotonicity. In: de Moor, O., Gottlob, G., Furche, T., Sellers, A. (eds.) Datalog 2.0. LNCS, vol. 6702, pp. 384–398. Springer (2011)
17. Małuszyński, J., Szałas, A.: Logical foundations and complexity of 4QL, a query language with unrestricted negation. J. Appl. Non-Classical Logics **21**(2), 211–232 (2011)
18. Małuszyński, J., Szałas, A.: Partiality and inconsistency in agents' belief bases. In: Barbucha, D., Le, M., Howlett, R., Jain, L. (eds.) KES-AMSTA. Frontiers in Artificial Intelligence and Applications, vol. 252, pp. 3–17. IOS Press (2013)
19. Mueller, E.: Commonsense Reasoning. Morgan Kaufmann (2006)
20. Sakama, C., Inoue, K.: An alternative approach to the semantics of disjunctive logic programs and deductive databases. J. Autom. Reason. **13**(1), 145–172 (1994)

21. Shieber, S.M.: Solving problems in an uncertain world. Bachelor's Thesis, Harvard College (1981)
22. Soininen, T., Niemelä, I.: Developing a declarative rule language for applications in product configuration. In: Gupta, G. (ed.) Proceedings PADL'99. LNCS, vol. 1551, pp. 305–319. Springer (1999)
23. Szałas, A.: How an agent might think. Logic J. IGPL **21**(3), 515–535 (2013)
24. Thrun, S., Burgard, W., Fox, D.: Probabilistic Robotics (Intelligent Robotics and Autonomous Agents). The MIT Press (2005)
25. Zadeh, L.: Fuzzy sets. Inf. Control **8**, 333–353 (1965)

A Modified Vickrey Auction with Regret Minimization for Uniform Alliance Decisions

Marin Lujak and Marija Slavkovik

Abstract We consider a supply chain management problem where a business alliance of small capacity retailers needs to collectively select a unique supplier considering the assignment's efficiency at both the alliance and retailers' level. We model the alliance as a multi-agent system. For this model, we present a modified Vickrey auction algorithm with regret minimization and compare it experimentally with aggregation of preferences by voting and standard Vickrey auction. Through simulation, we show that the proposed method on average reaches globally efficient and individually acceptable solutions. The solutions are evaluated in terms of different social welfare values.

Keywords Business alliance · Decision making · Task assignment · Vickrey auction · Voting · Regret minimization · Fairness · Social welfare

1 Introduction

As businesses continue to globalize and industries become increasingly complex, the creation of business alliances is ever more important to remain competitive at the global level. To assure the sustainability of the alliance, it is crucial to maintain a fine balance between the goals of the alliance as a group and the goals of the alliance members considering their individual interests and alternatives. The alignment of individual and overall alliance goals is a non-trivial problem. The alliance is formed when the objectives of self-interested alliance members are complementary and resulting in synergies. The main issue in alliances is how resources are leveraged and accumulated in the context where the utility function of an individual alliance member is not necessarily aligned with the utility function of the alliance as a whole.

M. Lujak (✉)
IMT Lille Douai, Douai, France
e-mail: marin.lujak@imt-lille-douai.fr

M. Slavkovik
University of Bergen, Bergen, Norway
e-mail: Marija.Slavkovik@uib.no

M. Ivanović et al. (eds.), *Intelligent Distributed Computing XI*,
Studies in Computational Intelligence 737, https://doi.org/10.1007/978-3-319-66379-1_6

Therefore, reaching an optimal resource assignment for such a heterogeneous group is a complex and challenging task.

Assuming that an alliance performs a uniform task, the globally optimal task is the one that minimizes the cost and/or maximizes the profit of the alliance as a whole while considering the *fairness* issues in respect to individual alliance members. An individual (rational) alliance member can perceive *fairness* as a measure of the quality of such a uniform task. Since a globally optimal task assignment is not necessarily optimal individually, there might be alliance members individually unsatisfied with the same. A high number of unsatisfied members can jeopardize the stability of the alliance and, thus, its existence.

Given that alliance members want to hide sensitive information from each other, we consider the following question in selecting a unique large supplier for an alliance of small retailers: Should the alliance members reveal cardinal information or revealing the ordinal preferences over alternatives is sufficient to achieve a fair and efficient assignment of a uniform task to the alliance?

We assume that retailers join the alliance attracted by the similarity of their utility functions and a common need for a large supplier whose requirements they cannot reach individually. Furthermore, we assume the existence of multiple large-sized suppliers, each endowed with complementary or substitutable products, and for delivery of products, they deliver only large-sized shipments whose orders can be formed only by a larger retailer alliance. In this context, an alliance-optimal, but unfair supplier assignment can serve as an incentive for unsatisfied retailers to break-up from the alliance and join a competitor one.

Market-based choice methods that use cardinal information for task assignment are auctions, while a social choice method that considers preference orders is voting. The key difficulty in social choice is that agents generally have conflicting preferences over the outcomes. To resolve this issue, classical coalition formation and matching algorithms might be inappropriate due to their focus on socially optimal solutions, while ignoring the issue of fairness, see, e.g., [21].

This paper is organized as follows. In Sect. 2, we present related work. In Sect. 3, we formulate the retailer-alliance supplier assignment problem. To achieve an efficient and fair assignment solution, in Sect. 4, we propose a modification of the Vickrey auction method by integrating it with the regret minimization mechanism. We also consider the method of preference aggregation in Sect. 5, where the preferences of retailers over the suppliers are aggregated using the Borda method. The experiments in Sect. 6 show both methods favor the least happy alliance members, thus resulting in favorable conditions for alliance stability. Section 7 concludes the paper.

2 Related Work

An alliance may be defined as a voluntary association that furthers the common interests of the constituent members, see, e.g., [16, 24]. It is established by an agreement with equitable risk and opportunity share among independent businesses established for a specific purpose and usually motivated by cost reduction and improved service for the customer, see, e.g., [16]. In [16, 24], some main collaboration strategies in business alliances are presented. The concept of an alliance is similar to the concept of a coalition in organization theory, which can be defined as a way for a group of agents with complementary goals to jointly determine their actions, see, e.g., [25]. It coordinates agreements among its members, while it interacts non-cooperatively with its non-members.

Coalition formation is a subject of different fields, e.g., game theory, where aspects like stability and payoff distribution are important elements [1], and economics, which is concerned with defining and analyzing criteria for distinguishing among strategic and social factors which propel many firms into coalition formation [6]. One of the most prominent criteria for coalition formation are the *efficiency* and the *equity* of assignments [23]. Assuming a preexistence of a lasting coalition, in this paper we are not concerned with the formation of coalitions *per se* but with the way the coalition makes collective decisions, which may put at stake their stability.

Methods for reaching globally efficient solutions have been in the research focus of economists [18, 27, 28], while notions of equality have been somewhat neglected [7]. However, with self-interested agents, each group member is crucial for the success of the group and a feasible locally acceptable solution is preferable to a globally optimal one. Coalition stability has been considered extensively in the literature. Assuming that there is substantial variability of preferences across states of nature, Pycia in [17] shows that there exists a core stable coalition structure in every state if and only if agents? preferences are pairwise-aligned in every state. Furthermore, core-stability was considered within various models of cooperative games with structure in [5].

In competitive contexts, it is difficult to incentivize agents to collaborate and unconditionally share information, especially if the collaboration might give a more costly result for an agent than its non-collaborative behavior. In economics, a common resource assignment approach for self-interested agents is auction. Common auction forms are English, Dutch, first-price sealed-bid, and Vickrey auction (VA). In the case of multiple items that have to be assigned to a team of collaborative agents, a combinatorial optimization auction-like Bertsekas algorithm can be used, see, e.g., [2]. Its performance depends on the quantity and quality of information exchanged, see, e.g., [14, 15].

Different incentive models can be integrated in auctions' rules so that a desirable fair outcome is chosen [19]. Assuming that collusion is not allowed and that agents have quasilinear utility functions, Vickrey auction (second-price sealed-bid auction) is strategy-proof or truthful, i.e., it gives bidders an incentive to bid their true values [26]. Moreover, in independent private-values context with symmetric risk-neutral bidders, Vickrey auctions are perfectly efficient economically while producing the same expected revenue for bid takers as equilibrium strategies in other common auctions. According to theorems by Green and Laffont in [8] and by Holmstrom in [10], it is essentially the only design to provide dominant strategy incentives and yield efficient auction outcomes.

In this auction, bidders place their bid in a sealed envelope and simultaneously submit them to the auctioneer without knowledge of other bids. The highest bidder wins, paying a price equal to the second-highest bid. If everyone bids their true value, the bidder with the highest valuation will receive the good. Since bidders have an incentive to bid truthfully in this type of auction, it is the quickest and most likely to achieve Pareto efficiency and profit maximisation as a result. Furthermore, in the jargon of game theory, bidding truthfully is a dominant strategy. If all players bid truthfully, then the VA produces an outcome that maximizes the social surplus. Finally, the VA is computationally tractable and it can be implemented in polynomial time.

As stated in [7], the quality of a group action can be assessed using tools from economics. Pareto optimality and utilitarianism are two of the most frequently used efficiency and equity criteria from economical theories that have found use in multiagent systems. Furthermore, in [7], Endriss and Maudet argue that in addition to *efficiency*, methods that produce egalitarian solutions are needed.

Regret is a concept in decision theory that calculates the difference between the utilities of two choices or outcomes. Regret theory [12] models how choices can be made under uncertainty by minimising the maximal possible regret that can be incurred by a choice. A *minimax regret* is a decision criterion first suggested by [20] in the realm of statistics. However, ever since, economists have been interested in it as well [22]. Minimax regret concept was used in multi-agent task-assignment in [11] while in [13] the minimization of voter's regret for a possible voting outcome was applied to reduce the elicitation requirements.

3 Problem Formulation

The problem of a unique supplier selection for a retailer alliance consists of selecting a supplier for an alliance of retailers out of a set of available suppliers. We assume that every supplier is willing to service those retailer alliances that satisfy minimal requirements, i.e., the payment of the reserve price (the price stipulated as the lowest acceptable by the supplier) and request at least a minimum number of products to be supplied. Furthermore, for simplicity, we assume that individual suppliers differ only in the costs of their services for the alliance, while the revenues for the alliance members of every supplier are equal. In this way, we can concentrate only on the minimization of the alliance's costs.

Let $A = \{a_1, \ldots, a_r\}$ be a set of r retailer agents in an alliance. Moreover, let $\Theta = \{\theta_1, \ldots, \theta_s\}$ be a set of all available suppliers where s is the cardinality of set Θ. Each retailer $a_i \in A$ associates a *cost* $c_{a_i,\theta_j} > 0$ for each of suppliers $\theta_j \in \Theta$. The cost c_{a_i,θ_j} is the price retailer a_i pays for services when supplier θ_j is selected by alliance A. The profile of the costs that agents $a_i \in A$ attribute to suppliers $\theta_j \in \Theta$ can be represented through the alliance cost matrix $\mathbf{C}_A = [c_{a_i,\theta_j}]^{r \times s}$ whose elements are strictly positive rational numbers.

Let supplier θ_j be a *qualifying supplier* if for every agent $a_i \in A$, the cost c_{a_i,θ_j} of having θ_j as a supplier of alliance A is such that $c_{a_i,\theta_j} \leq c^q_{a_i}$, where $c^q_{a_i} = \beta_i \cdot c_{a_i,\theta_j^{min}}$ is a qualifying supplier cost for agent a_i and $c_{a_i,\theta_j^{min}} = \min_j c_{a_i,\theta_j}$, is the cost of individually optimal supplier θ_j^{min} for agent $a_i \in A$, and $\beta_i \geq 1$ is a tolerance factor of agent a_i representing the maximum acceptable cost ratio between alliance's assigned and individual agent's locally optimal supplier. Therefore, θ_j is a qualifying supplier for alliance A if for each alliance member $a_i \in A$, $c_{a_i,\theta_j} \leq c^q_{a_i}$ holds. For every agent for which the latter does not hold, we will call him β-unsatisfied. The β_i factor should be agreed within the alliance based on different individual agent's i characteristics.

Regarding the information available to each agent, we assume that each retailer agent $a_i \in A$ has at its disposal the information vector regarding its cost c_{a_i,θ_j}, for the procurement of every of suppliers $\theta_j \in \Theta$, and that agents have no insight into each other's information regarding supplier cost values.

In this setting, the objective is to assign alliance A to a unique supplier $\theta_j \in \Theta$, such that the alliance assignment cost $\sum_{a_i \in A} c_{a_i,\theta_j}$ is minimal and that $c_{a_i,\theta_j} \leq c^q_{a_i}$ holds for all $a_i \in A$. Furthermore, regarding the solution equity, we consider the following four social welfare functions:

Utilitarian social welfare: $u(\theta, \mathbf{C_A}) = \sum_{a_i \in A, \theta_j} c_{a_i,\theta_j}$,
Egalitarian social welfare: $e(\theta, \mathbf{C_A}) = \max_{a_i \in A, \theta_j} c_{a_i,\theta_j}$,
Elitist social welfare: $el(\theta, \mathbf{C_A}) = \min_{a_i \in A, \theta_j} c_{a_i,\theta_j}$, *and*
Nash social welfare: $n(\theta, \mathbf{C_A}) = \prod_{a_i \in A, \theta_j} c_{a_i,\theta_j}$.

For alliance A, supplier $\theta \in \Theta$ is better than supplier $\theta' \in \Theta$ if and only if a social welfare function $\mathrm{SWF}(\theta, \mathbf{C}_A) < \mathrm{SWF}(\theta', \mathbf{C}_A)$, since the lower the cost a retailer alliance has to pay, the better off the alliance is.

While the utilitarian, egalitarian and elitist welfare functions are well known, the Nash social welfare is perhaps less familiar. A low Nash value, when it is defined in terms of costs, is an indication of both good utility value and a good egalitarian value, i.e., assignment solutions with a low Nash value are both locally and globally good solutions. It can be seen as a kind of a compromise between the collective and individual optimality of an assignment.

4 Vickrey Auction with Regret Minimization

Since the retailer-alliance choice of an alliance supplier depends on the individual alliance members' costs, the Vickrey auction is a straightforward method of resolving this problem in a distributed way for self-interested agents.

Vickrey auction is performed in two stages: bidding and assignment. Bidders are individual retailers within the alliance whose objective is to minimize their total individual costs. In a **bidding phase**, bidders $a_i \in A$ submit to the auctioneer in a sealed bid their full list of costs c_{a_i,θ_j} for a set of suppliers $\theta_j \in \Theta$. In the **assignment phase**, auctioneer calculates alliance total cost $C_{(A,\theta_j)}$ for every supplier $\theta_j \in \Theta$, and assigns the alliance to the supplier with the least total cost. Group total cost $C_{(A,\theta_j)}$ is measured as the sum of the individual bidder costs for each $\theta \in \Theta$.

Vickrey rules. For the services of the assigned supplier to the alliance, each member of the alliance $a_i \in A$ pays an individual price p_i which is calculated based on the Vickrey rules. The auctioneer computes price p_i as follows:

1. Let C denote the total cost from the efficient allocation ($C = \sum_{a_i \in A} c_{a_i,\theta_j}$)

2. Let C^{-i} denote the total cost that could be generated if a_i did not participate, and the auctioneer allocated (not necessarily the same) supplier $\theta_j \in \Theta$ to the rest of the bidders to minimize total alliance assignment cost.

3. Then, a_i's payment in the regular Vickrey auction is $p_i = c_{a_i,\theta_j} + (C - C^{-i})$.

Here, c_{a_i,θ_j} represents an individual cost of agent a_i for supplier θ_j, while $(C - C^{-i})$ represents the cost of agent a_i's participation in the alliance. However, the regular Vickrey auction does not consider the fairness issues and there might be alliance members who have to accept the alliance assignment even though they are β-unsatisfied. If all alliance members are considered necessary for contracting with a supplier, it is in the interest of an alliance to keep every alliance member. The latter is stable as long as all the members are β-satisfied with the assignment and the alliance profit is strictly positive. To consider the fairness in the alliance, we propose modified Vickrey rules in the following.

Modified Vickrey rules. We propose the modification of Vickrey rules by the integration of *regret* which is seen here as an opportunistic cost of β-unsatisfied agents for the alliance assignment. In this context, the regret of agent $a_i \in A$ with respect to supplier θ_A assigned to an alliance, is:
$r(a_i, \theta_A) = c_{a_i,\theta_A} + (C - C^{-i}) - c^q_{a_i}$, $\forall a_i \in A$.

Let $\Psi \subset A$ be a set of all β-unsatisfied agents for which $r(a_i, \theta_A) > 0$ and $\Phi = A \backslash \Psi$ be a set of β- satisfied agents. Payment $p(a_i)$ for each β-unsatisfied agent $a_i \in \Psi$ is lowered by the difference of value sufficient to reach its minimally acceptable alliance assignment solution:
$p(a_i) = c_{a_i,\theta_A} + (C - C^{-i}) - r(a_i, \theta_A)$, $\forall a_i \in \Psi$.

After all the individual β-unsatisfied agents' payments are transferred, their total regret $r(\Psi, \theta_A) = \sum_{a_i \in \Psi} r(a_i, \theta_A)$ is then distributively paid by satisfied agents $a_i \in \Phi$ as an additional cost $\delta(a_i) \geq 0$ to their Vickrey payment $p(a_i)$. The latter is calculated as $\delta(a_i) = r(\Psi, \theta_A)/|\Phi|$.

For the distribution of additional cost $\delta(a_i)$ over satisfied agents $a_i \in \Phi$, we apply a heuristic approach of ordering satisfied agents $a_i \in \Phi$ in a non-increasing order of their individual payments $p(a_i)$, $p_i = c_{a_i,\theta_j} + (C - C^{-i})$, and iteratively charge each one of them in rounds additional unitary payment $\eta_{a_i}(t)$ until total regret $r(\Psi, \theta_A)$ isn't distributed over all β-satisfied agents.

To avoid that satisfied agents $a_i \in \Phi$ become β-unsatisfied, we limit additional unitary payment $\eta_{a_i}(t)$ of each satisfied agent $a_i \in \Phi$ in round t as follows: $\eta_{a_i}(t) = \max\left(p(a_i)(t) + \delta(a_i) - c^q_{a_i}, 0\right)$, $\forall a_i \in \Phi$ since $p(a_i)(t) + \delta(a_i) - c^q_{a_i}$ may be lower than zero. We assume that the assignment is known only within the same alliance and any time the alliance structure changes, the alliance members recalculate the assignment. Next we present the voting-based approach.

5 Voting with Borda Count

Voting is a general group option-choosing method for societies of self-interested agents [4] and is also used for the purpose of fair distribution of desired items. Compared to auctions, as means to distribute items of interest among interested parties, voting can be seen as less "competitve" collective decision making approach. In an auction, the bidders compete directly with each other and each agent is concerned with only maximizing her own utility. Voting, in contrast, is concerned with finding a collective choice, in our case an allocation, that is least as bad for every participating agent, or voter. In this sense, we can see the voting method as "cooperative"-inducing.

Formally, a voting problem is specified by a non-empty set of social options O and a set $A = \{a_1, \dots, a_n\}$ of at least two agents. Each agent $a_i \in A$ reports his/her preferences over elements in O, which are represented by a complete, transitive preference relation $\succcurlyeq_i$. A profile $P = \{\succcurlyeq_i \mid a_i \in A\}$ is the set of the preference orders of the agents A. A *voting rule* is function F, that assigns to each tuple of n preference orders a non-empty sub-set of options from O. The choice of voting rule is determined by the nature of the problem.

The problem we are considering, finding a common supplier for an alliance (coalition) of retailers, can be naturally represented as a voting problem by setting $O = \Theta$. Each retailer in the coalition $a_i \in A$ constructs a preference order over the available supplers in the following manner: $\theta_j \succ_i \theta_k$ if and only if $c_{i,j} < c_{i,k}$ and $\theta_j \sim_i \theta_k$ if and only if $c_{i,j} = c_{i,k}$. Thus a retailer a_i prefers a supplier θ_j over supplier θ_k if and only if the cost of the supplier θ_k is lower than that of θ_j. We use the Borda count scoring rule that considers not only who the top ranked candidate is, like the plurality rule and the fallback bargaining rule [3], but also how strongly a candidate is preferred in respect to other candidates. An additional advantage of the Borda count rule is the low computational complexity of the winner determination [9].

According to the Borda count rule, each supplier $\theta_j \in \Theta$ is given a score based on its position in the individual preference orders in P. The scores for the $\theta \in \Theta$ are defined as $sb(\theta) = \sum_{a_i \in A} \#\{(\theta') | \theta' \in \Theta \text{ and } \theta \succcurlyeq_i \theta'\}$. The number $\#\{(\theta') | \theta' \in \Theta \text{ and } \theta \succcurlyeq_i \theta'\}$ is effectively the position of the option θ in the retailer i's preference order. For example, in the order $\theta_1 \succ_i \theta_2 \succ_i \theta_3$, the top ranked option θ_1 is assigned a value 3 because it at least as good as 3 other options including itself. The Borda count rule $F_B(P)$ returns the option with the highest score as a winner of the election $F_B(P) = arg\max_{\theta \in \Theta} sb(\theta)$.

As other voting rules, Borda can sometimes produce tied alternatives. We use the lexicographic ordering over suppliers to break ties. Here is our supplier-selection algorithm based on the Borda count voting rule.

Let the set of retailer agents be A, and a set of suppliers Θ, with $\succcurlyeq$ being a tie-breaking order over Θ. Each retailer $a \in A$ does the following steps:

1. it keeps in its memory a set of suppliers $\theta_u \in \Theta$ and associates for each of the suppliers, the cost $c(\theta)$ if that supplier is selected as the supplier for the group.
2. it constructs the preference order over Θ as described.
3. it receives preference orders from other retailer agents and constructs a preference profile P to which he applies the F_B rule. If $F_B(P)$ produces a tie, the supplier who is ranked the highest according to $\succcurlyeq$ is chosen.

6 Simulation Setup and Results

We simulate a multi-agent system with retailer alliance and supplier agents applying the modified Vickrey auction method with regret minimization and the voting method in MatLab. In the following, the results of 10 different instances are presented for the problem with 100 retailer agents and 100 suppliers. We tested different scenarios with up to 100 agents in a discrete simulation setting where the initial retailer agent costs are based on the Euclidian distances from their to the suppliers positions, and the positions are generated uniformly randomly in the range $[0, 100]^2 \in \mathbb{R}^2$. The experiments with less agents have similar result dynamics but are not presented here due to the lack of space.

We concentrate on the minimization of the group assignment cost, and, therefore, measure the sum of differences between agents $a_i \in A$ and $\theta \in \Theta$ and calculate utilitarian, egalitarian, elitist, and Nash social welfare values for the best group supplier as negative values of the total, maximum, minimum, and product of distances in the multi-agent system. Moreover, the cost of assignment is proportional to the measured distance. Therefore, the elitist welfare is measured as the utility of the agent that is currently best off as negative distance cost $-\min_i \sum_{t=1}^{T} dist_{ai}(t)$. The utilitarian social welfare is the sum of individual utilities $-\sum_{i=1}^{n} \sum_{t=1}^{T} dist_{ai}(t)$, while the egalitarian social welfare is given by the utility of the agent that is currently worst off $-\max_i \sum_{t=1}^{T} dist_{ai}(t)$.

From Fig. 1 it can be seen that the utilitarian welfare is the highest for the Vickrey auction algorithm without considering fairness issues even though the proposed two solutions follow quite well the best Vickrey solution. However, the egalitarian (Fig. 2) and the utilitarian welfare in both modified auction and modified voting cases are very close in all the experimented instances to the regular Vickrey algorithm, except that for Egalitarian welfare, the modified auction algorithm performes in average better than the other two. However, due to the inclusion of regret in the bid calculation, the egalitarian welfare of the auction algorithm with regret is in average (7 out of 10 instances) better than the one of the regular Vickrey auction algorithm.

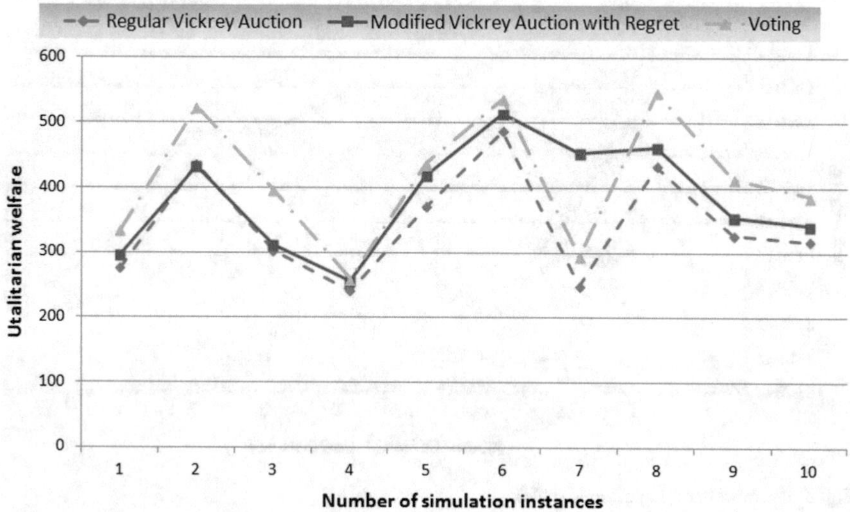

Fig. 1 Utilitarian welfare

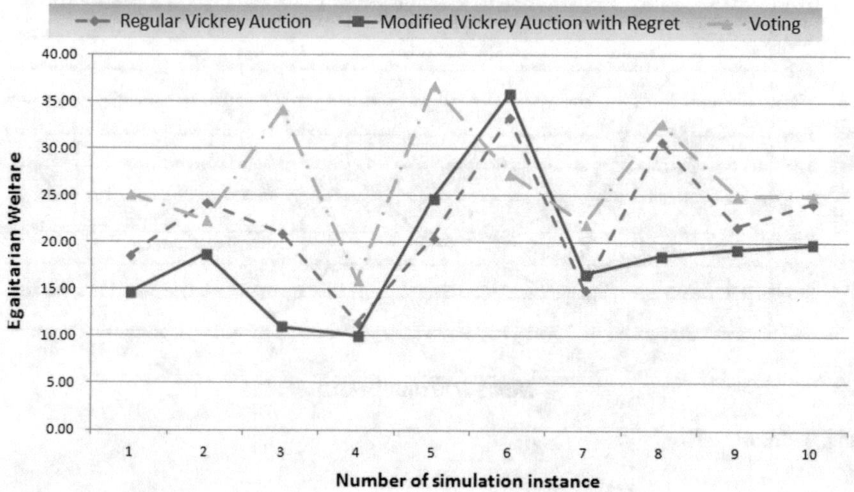

Fig. 2 Egalitarian welfare

The elitist welfare of the regular Vickrey auction algorithm (Fig. 4) is the best in 4 and second best in 2 out of 10 instances, followed by the auction algorithm with regret which only in 2 instances has a lower result than the voting method.

We also measured Nash welfare (Fig. 3) which shows the superiority of the regular Vickrey auction algorithm, followed by the modified Vickrey auction algorithm and the Borda count method. Please note, that, since the objective is to lower the total cost, lower values show better group performance.

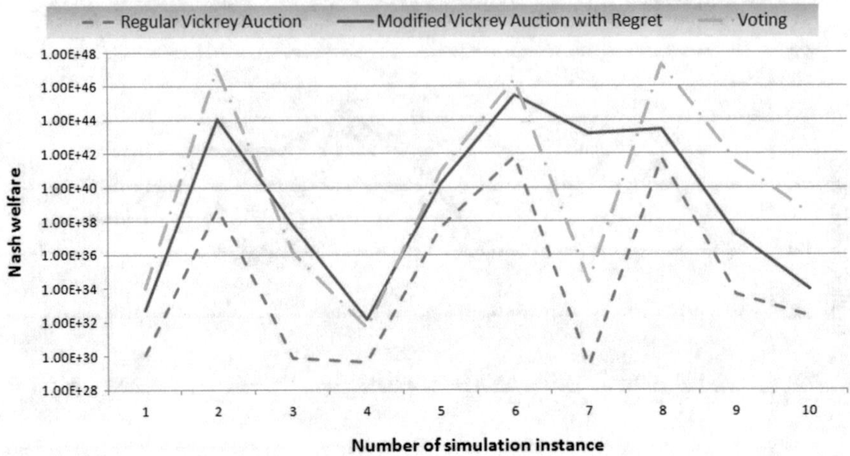

Fig. 3 Nash welfare (logarithmic scale)

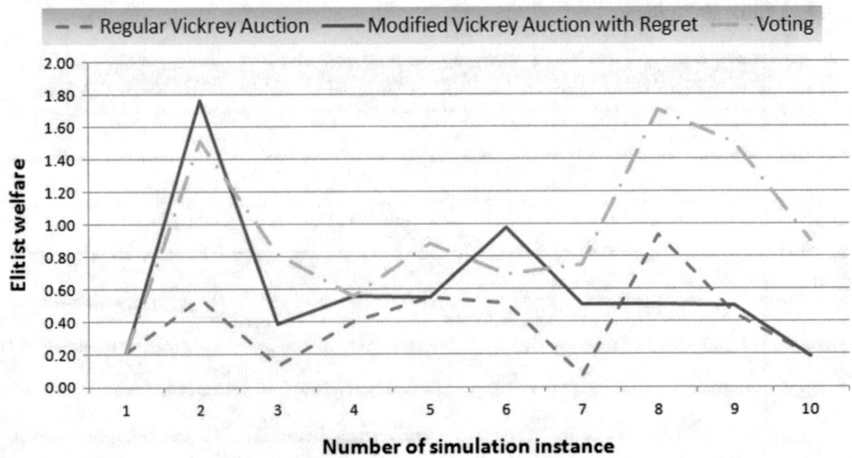

Fig. 4 Elitist welfare

The inferior behavior of the voting method can be explained by the ordinal and not cardinal order of the options which doesn't give a mathematical basis based on which fine-tuning and the mathematical optimization can be made. Furthermore, the greedy policy in the ordering of the preferences sent from one agent to others influences strongly other agents and directs them to less preferred targets, thus minimizing individual cost and, on average, worsening the global and other individual solutions. The individual assignment process of the voting method does not take into consideration individual agent bias and is performed taking into consideration all the information at disposal with the same weight. Furthermore, since a random factor is introduced through lexicographic ordering of equivalently valued suppliers, the distribution of the quality of the solution on the best-case and worst-case retailer agent cannot be achieved.

7 Conclusions

In this paper, we proposed a modified Vickrey auction with regret minimization for the retailer group supplier assignment problem. We showed in simulations that our proposed method reaches solutions that are globally cost efficient and individually acceptable. We compared the proposed method with the voting method with Borda count and showed that they both favor the least happy retailer alliance members. A disadvantage of the voting approach is its sequentiality, on which the solution depends. In our experiments, we used a regret minimization sequence.

In our future work, we intend to analyse in detail the proposed method and compare it with an improved voting procedure exploring different agent orders in voting.

References

1. Banerjee, S., Konishi, H., Sönmez, T.: Core in a simple coalition formation game. Soc. Choice Welfare **18**(1), 135–153 (2001)
2. Bertsekas, D.: The auction algorithm for assignment and other network flow problems: a tutorial. Interfaces 133–149 (1990)
3. Brams, S.J., Kilgour, D.M.: Fallback bargaining. Group Decis. Negotiat. **10**(4), 287–316 (2001)
4. Brandt, F., Conitzer, V., Endriss, U.: Computational social choice. In: Weiss, G. (ed.) Multiagent Systems. MIT Press (2012)
5. Chalkiadakis, G., Markakis, E., Jennings, N.R.: Coalitional stability in structured environments. In: Proceedings of AAMAS 2012, vol. 2, pp. 779–786 (2012)
6. Eisenhardt, K.M., Schoonhoven, C.B.: Resource-based view of strategic alliance formation: strategic and social effects in entrepreneurial firms. Organ. Sci. **7**(2), 136–150 (1996)
7. Endriss, U., Maudet, N.: Welfare engineering in multiagent systems. In: Omicini, A., Petta, P., Pitt, J. (eds.) Engineering Societies in the Agents World IV, LNCS, vol. 3071, pp. 93–106. Springer (2004)
8. Green, J.R., Laffont, J.J.: Incentives in Public Decision Making (1979)
9. Hemaspaandra, E., Hemaspaandra, L.A., Rothe, J.: Exact analysis of Dodgson elections: Lewis carroll's 1876 voting system is complete for parallel access to NP. J. Assoc. Comput. Mach. **44**(6), 806–825 (1997)
10. Holmström, B.: Groves' scheme on restricted domains. Econom.: J. Econom. Soc. 1137–1144 (1979)
11. Koenig, S., Zheng, X., Tovey, C., et al.: Agent coordination with regret clearing. In: Proceedings of AAAI'08, vol. 1, pp. 101–107 (2008)
12. Loomes, G., Sugden, R.: Regret theory: an alternative theory of rational choice under uncertainty. Econ. J. **92**(368), 805–824 (1982)
13. Lu, T., Boutilier, C.: Robust approximation and incremental elicitation in voting protocols. In: Proceedings of the 22nd IJCAI, vol. 1, pp. 287–293 (2011)

14. Lujak, M., Giordani, S.: On the communication range in auction-based multi-agent target assignment. In: IWSOS'11: Proceedings of the 5th International Conference on Self-organizing Systems. LNCS, vol. 6557, pp. 32–43 (2011)
15. Lujak, M., Giordani, S.: Value of incomplete information in mobile target allocation. In: MATES'11: Proceedings of the 9th German Conference on Multiagent System Technologies. LNCS, vol. 5774, pp. 89–100 (2011)
16. Morden, T.: Principles of Strategic Management. Routledge (2016)
17. Pycia, M.: Stability and preference alignment in matching and coalition formation. Econometrica **80**(1), 323–362 (2012)
18. Sambasivan, M., Siew-Phaik, L., Mohamed, Z.A., Leong, Y.C.: Impact of interdependence between supply chain partners on strategic alliance outcomes: role of relational capital as a mediating construct. Manag. Decis. **49**(4), 548–569 (2011)
19. Sandholm, T.: Making markets and democracy work: a story of incentives and computing. In: Proceedings of the 18th IJCAI, pp. 1649–1671. Morgan Kaufmann Publishers Inc. (2003)
20. Savage, L.J.: The theory of statistical decision. J. Am. Stat. Assoc. **46**(253), 55–67 (1951)
21. Sethuraman, J., Teo, C.P., Qian, L.: Many-to-one stable matching: geometry and fairness. Math. Oper. Res. **31**(3), 581–596 (2006)
22. Stoye, J.: Axioms for minimax regret choice correspondences. J. Econ. Theory **146**(6), 2226–2251 (2011)
23. Tadenuma, K.: Efficiency first or equity first? Two principles and rationality of social choice. J. Econ. Theory **104**(2), 462–472 (2002)
24. Tjemkes, B., Vos, P., Burgers, K.: Strategic Alliance Management. Routledge (2013)
25. Van Deemen, A.M.: Coalition Formation and Social Choice, vol. 19. Springer Science & Business Media (2013)
26. Vickrey, W.: Counterspeculation, auctions, and competitive sealed tenders. J. Finance **16**(1), 8–37 (1961)
27. Wu, W.Y., Shih, H.A., Chan, H.C.: The analytic network process for partner selection criteria in strategic alliances. Expert. Syst. Appl. **36**(3), 4646–4653 (2009)
28. Yang, J., Wang, J., Wong, C.W.Y., Lai, K.H.: Relational stability and alliance performance in supply chain. Omega **36**(4), 600–608 (2008)

Lightweight Cooperative Self-Localization as Support to Traffic Regulation for Autonomous Car Driving

Assia Belbachir, Marcia Pasin and Amal El Fallah Seghrouchni

Abstract Self-localization is a basic service for Intelligent Transportation Systems (ITS) such as traffic regulation services. Most of the used techniques are based on integration of Inertial Navigation System (INS) and Global Positioning System (GPS). However, navigation through areas such as tunnels, where GPS coverage is vulnerable, obliges the use of a different approach. Based on this observation, we designed and implemented a lightweight cooperative positioning algorithm based on Adaptive Localization Protocol (ALP). In this paper, we apply our method as support to an intersection service for traffic regulation, in which a group of concurrent cars shares an intersection/critical section. We found that our algorithm improves car position and regulates the traffic.

Keywords Vehicle localization · Adaptive protocol

1 Introduction

Due to the increase of cars in the road and the stabilization of the infrastructure, traffic jam appears. Given two flows of cars situated in two concurrent incoming flows for intersection cross and since the intersection is a critical section, it is not possible to have both flows crossing at the same time. Only one flow may cross the critical section at a time. Actually, this decision is made by traffic lights or by the drivers

A. Belbachir (✉)
Institut Polytechnique des Sciences Avancées (IPSA), Ivry-sur-Seine, France
e-mail: assia.belbachir@ipsa.fr

M. Pasin
Universidade Federal de Santa Maria, Santa Maria, Brazil
e-mail: marcia@inf.ufsm.br

A. Belbachir · A.E.F. Seghrouchni
University Pierre and Marie Curie (Paris 6), Paris, France
e-mail: amal.elfallah@lip6.fr

M. Ivanović et al. (eds.), *Intelligent Distributed Computing XI*,
Studies in Computational Intelligence 737, https://doi.org/10.1007/978-3-319-66379-1_7

themselves. Thus, Vehicular Ad-hoc Networks (VANETs) might be a solution and a new challenge to traffic management and transportation networks. For instance, cars would be able to share information and decide which car will cross an intersection first. If a car receives information from an approaching car about a traffic jam, the car will be able to avoid this traffic jam. With the full implementation of VANETs, new alternatives to regulate intersection control [7], lane merging [2] and route/speed guidance can be established.

To accomplish all these missions, localization methods are required. In the intersection and in the lane merging problem, it is required to know at least which is the car in the head of the incoming flow. In the route guidance service, the car localization must be computed using a localization system. Usually, the used method is based on integration of Inertial Navigation System (INS) and Global Positioning System (GPS). However, GPS coverage is vulnerable and can interfere with other signals which obliges an alternative approach. Thus, there is a strong motivation to develop a back-up localization solution to GPS failure. In this paper, we designed and implemented a lightweight cooperative positioning algorithm based on trilateration and an intelligent noise learning technique called Adaptive Value Tracking (AVT). We used our localization service as a building block for traffic regulation. Based on self-localization, we developed a traffic regulation strategy. The remainder of this paper is organized as follows. In Sect. 2 we review related work. Then, in Sect. 3 we explain the implemented algorithm for self-localization using Adaptive Localization Protocol (ALP). The target scenario and the obtained results from the experimental evaluation using the AVT approach are reported in Sect. 4. Finally, Sect. 5 concludes the paper.

2 Related Work

Several techniques of localization are developed [6]. Proximity is mostly used in sensor networks. This technique uses fixed nodes (i.e. antenna) that know their position and have a fixed range wireless communication. If a moving node (i.e. a car) is under coverage of the fixed node, then the position is known; otherwise the position is unknown. This technique is useful when several nodes have a fixed and known position. Another employed localization technique is *Trilateration and Angulation*. This technique needs at least three nodes. The connectivity between two nodes can allow them to exchange information about their geometric information (angle of distance). Our work aims to improve localization techniques for cars using positioning from the infrastructure in a cooperative way. Qu et al. [8] developed a localization method based on information synchronization for a ring communication topological structure. However, it imposes a rigid topology to be able to make the localization service. Shame et al. [9] considered the problem of cooperative self-localization of mobile agents and proposed a solution based on graph theory. Both works did not consider the computational time and energy consumption. We are interested to integrate self-localization as an input to intersection control. There is a lot of works in which intersection control and self-localization can be combined, such as in Dresner and Stone [4]. Self-localization can be also useful to leader election [5]. Finally, self-localization can also be applied as a key service to other transportation network problems such as lane merging [2].

3 Adaptive Localization Protocol (ALP)

In this scenario, when a car gets out of GPS coverage, it first broadcasts a "GPS failure"message to all its neighbors and starts its adaptive localization protocol (ALP). A neighbor can be a traffic light, a car, an antenna, etc. When a neighbor receives a message, it first checks/stabilizes itself in its current position. Then it propagates the message through the network. The propagation continues until all the cars stabilize themselves. After, the failing car localizes itself, it broadcasts a "localization success" message to all its neighbors. In this phase, the failing car still goes on localizing itself using ALP. Upon having started receiving GPS signal, the failing car stops executing ALP and goes on its mission using GPS. In ALP, we take a different approach to localize cars and treat it as a *search process* that finds the actual position of a mobile car in a swarm of cars traffic where the collected information and measurements are noisy. We propose a robust and efficient adaptive position tracking technique which the goal is to localize cars without GPS coverage by communicating with the infrastructure (stable position). The proposed approach handles single-hop localization as a two steps *search* procedure using Adaptive Value Tracking (AVT) [1] in two steps.

Step 1: Ranging As we stated previously, the first step of the localization procedure is ranging. In this step, the car communicates with the infrastructure and estimates its relative distances by using a particular ranging method. However, due to network dynamics, the quality of its estimations are affected by the measurement noise. In order to get more robust estimates, we propose to handle the ranging as search process during which the relative distance d^* is searched inside the search space $[d_{min}, d_{max}]$ using an AVT, where d_{min} and d_{max} are the minimum and maximum distances that can be measured physically by the ranging method.

Algorithm 1 : Estimates the range between a mobile and stable node.

1: Obtain a new range estimate $\hat{d}$ using any ranging method
2: *error* = *avt.getValue()*$-\hat{d}$
3: **if** *error* < 0 **then** *avt.adjust(f* $\uparrow$)
4: **else if** *error* > 0 **then** *avt.adjust(f* $\downarrow$)
5: **else** *avt.adjust(f* $\approx$)

Algorithm 2 : Estimates the (x, y) position of the car.

1: Calculate $(\hat{x}, \hat{y})$ by trilateration
2: $x_{err} = avt_x.getValue - \hat{x}$
3: $y_{err} = avt_y.getValue - \hat{y}$
4: **if** $x_{err} > 0$ **then** $avt_x.adjust(f \uparrow)$
5: **else if** $x_{err} < 0$ **then** $avt_x.adjust(f \downarrow)$
6: **else** $avt_x.adjust(f \approx)$
7: **if** $y_{err} > 0$ **then** $avt_y.adjust(f \uparrow)$
8: **else if** $y_{err} < 0$ **then** $avt_y.adjust(f \downarrow)$
9: **else** $avt_y.adjust(f \approx)$

The ranging steps between a car and a particular infrastructure are summarized in Algorithm 1. If the distance value $\hat{d}$, which is estimated by using the available hardware e.g., ultrasonic transmitters and receivers, is higher than the range value proposed by the *avt* used for tracking the relative distance of the mobile node, an increase feedback $f \uparrow$ for increasing the distance, if it is smaller then a decrease feedback $f \downarrow$ for decreasing the distance, otherwise a good feedback $f \approx$ for indicating that the current range value is good is sent. Employing this algorithm, the value proposed by the *avt* of the mobile node will converge to the actual distance value in finite amount of time.

Step 2: Adaptive Localization The second step of the localization procedure is the localization algorithm that estimates the relative position of the car. This estimation is based on trilateration, hence position and ranging data from three infrastructures are required. Thus, we assume that the car is within the range of three infrastructures and it applied Algorithm 1 to obtain its relative distance estimates. Therefore, the car requires three AVTs in order to track these relative distances. Moreover, we also assume that the position information of all three infrastructure are obtained via communication. The localization steps of the car is summarized in Algorithm 2. Having estimated its relative distances and obtained the positions of the infrastructure, a trilateration is sufficient to estimate the (x, y) coordinate of the car. However, the estimation error of the ranging step affects the quality of this estimation. In order to have a robust and stable estimation, we propose to estimate these coordinates as a search process during which the actual coordinate (x^*, y^*) is searched for within the search spaces $[x_{min}, x_{max}]$ and $[y_{min}, y_{max}]$ respectively by using two AVTs. At any time, the avt_x and avt_y can propose the (x, y) coordinates. The error between the estimated $\hat{x}$ and $\hat{y}$ values calculated by the trilateration and the values proposed by avt_x and avt_y are calculated to inform the AVTs about the current feedback. With these steps, the value proposed by avt_x and avt_y of the mobile node will converge to the actual position in finite amount of time.

As we mentioned before, we assumed that a car that applied Algorithm 1 can track its relative distance to the infrastructure. However, it is worth to underline that Algorithm 2 can also be applied to the raw distance measurements, i.e. the distance measurements that are not tracked with AVTs. This could also be very relevant since the requirement of three distance tracking AVTs is eliminated, which reduces CPU and memory overhead but also it is decreasing robustness.

4 Implementation and Evaluation

We implemented the AVT algorithm for car self-localization to deal with intersection control using the Arduino and Raspberry Pi platforms. The self-localization service requires at least three anchors (i.e., fixed nodes). Each anchor represents a node that knows its position and does not change along the time, for instance an antenna. Each antenna a_i is able to estimate the distance (e.g. d_i) between itself and a specific car. When a car is in the coverage area of an antenna, it also estimates its distance to the antenna. Using both information (car distance estimation and antenna distance estimation), ALP computes car position. The method is based on noise learning. Figure 1 defines three antennas, and several cars. The first car is able to get different estimated distances d_1, d_2 and d_3 from respectively a_1, a_2 and a_3. The second information is the position of each antenna that is important for ALP to compute the trilateration. We implemented two experiments to show the efficiency of our approach. The first one with the ALP and the second one with a simple trilateration algorithm. As first experiment, we used a car that contains an Arduino Due board running the ALP algorithm and a communication sensor ESP8622.

Each antenna contains a Raspberry Pi and an ultrasonic sensor to estimate the distance between the car and the antenna. Figure 2a represents the obtained result of an ALP implementation using a car v located at the position (50, 50) and the antennas a_1 at (0, 0), a_2 at (100, 0) and a_3 at (0, 100). After 60 iteration, the self-localization value using ALP converges to the required position (50, 50). However, the obtained self-localization without ALP does not converge (see Fig. 2a). Several experiments were given by changing the car position, however the same results where observed. We can conclude that ALP estimates the car position in a better way than trilateration algorithm without ALP.

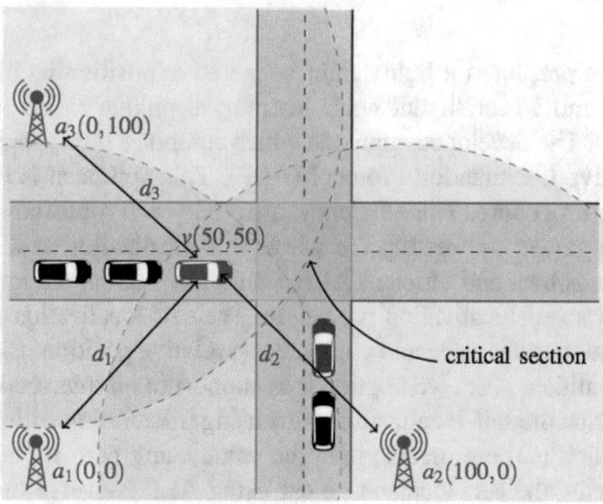

Fig. 1 The target scenario with one intersection, two roads and three antennas a_1-a_3

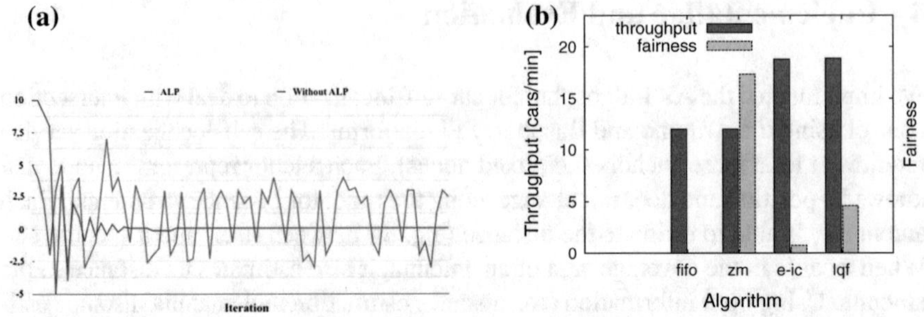

Fig. 2 **a** Illustration of the obtained results for car self-localization using ALP at position (50, 50) and **b** throughput and fairness by car (average)

As a second part of our experiments and to validate traffic regulation strategies based on ALP, we also implemented an intersection scenario using SUMO (Simulation of Urban MObility) [3] and different traffic regulation strategies [7]. The simulation lasts 2 h using a congested urban scenario (max. speed allowed in each road 60 km/h). Results can be found in Fig. 2b. Metrics used for evaluation include throughput and fairness. Throughput means the number of cars that completed the journey by the duration of a journey. As expected, platoon-based algorithms (e-ic and lqf) achieved the best throughput. We defined fairness following [2]. Again, as expected, FIFO got the best value to fairness but not zipper merge (zm) since FIFO definitions are based on the same metric used to calculate fairness. The e-ic got the best trade-off regarding throughput and fairness.

5 Conclusion

In this work, we presented a lightweight cooperative positioning algorithm based on trilateration and an intelligent noise learning technique called Adaptive Value Tracking (AVT). The developed approach which combines both previous techniques is called Adaptive Localization Protocol (ALP). This approach is considered as a dynamic search procedure. The efficiency, adaptivity and robustness of this procedure have been gained by treating the whole self-localization as a search process and exploiting a robust and efficient ALP. Using this notion, we implemented this approach for cars self-localization procedure. The self-localization procedure uses information related to distance and computes its relative position using trilateration with ALP. To validate ALP, we applied it as support to an intersection control service. We found that the self-localization value using trilateration with ALP converges into a value which is more precise than the value using pure trilateration without ALP. Additionally, the self-localization car using ALP converge faster than trilateration. Finally, we regulate traffic using the Efficient Intersection Control approach (e-ic). This technique is an adaptation of FIFO which got the best trade-off regarding throughput and fairness.

References

1. Agliamzanov, R., Gürcan, Ö., Belbachir, A., Yildirim, K.S.: Robust and efficient self-adaptive position tracking in wireless embedded systems. In: Proceedings of the 8th IFIP Wireless and Mobile Networking Conference, pp. 152–159. Munich (2015)
2. Baselt, D., Knorr, F., Scheuermann, B., Schreckenberg, M., Mauve, M.: Merging lanes—fairness through communication. Veh. Commun. **1**(2), 97–104 (2014)
3. Behrisch, M., Bieker, L., Erdmann, J., Krajzewicz, D.: SUMO—Simulation of Urban MObility (An Overview). In: Proceedings of the 3rd International Conference on Advances in System Simulation, pp. 63–68, Oct 2011
4. Dresner K., Stone, P.: Multiagent traffic management: a reservation-based intersection control mechanism. In: Proceedings of the 3rd International Joint Conference on Autonomous Agents and Multiagent Systems, pp. 530–537 (2004)
5. Ferreira, M., Fernandes, R., Conceição, H., Viriyasitavat, W., Tonguz, O.K.: Self-organized traffic control. In: Proceedings of the 7th ACM International Workshop on VehiculAr InterNETworking, pp. 85–90 (2010)
6. Karl, H., Willig, A.: Protocols and Architectures for Wireless Sensor Networks. Wiley (2005)
7. Pasin, M., Scheuermann, B., Fao de Moura R.: VANET-based intersection control with a throughput/fairness tradeoff. In: Proceedings of the 8th IFIP Wireless and Mobile Networking Conference, pp. 208–215. Munich (2015)
8. Qu, Y., Zhang, Y., Zhou, Q.: Cooperative localization of UAV based on information synchronization. In: International Conference on Mechatronics and Automation, pp. 225–230 (2010)
9. Shames, I., Fidan, B., Anderson, B., Hmam, H.: Cooperative self-localization of mobile agents. IEEE Trans. Aerosp. Electron. Syst. **47**(3), 1926–1947 (2011)

A New Approach for Vertical Handover Between LTE and WLAN Based on Fuzzy Logic and Graph Theory

Zlatko Dejanović

Abstract Today, two major wireless technologies exist, LTE and WLAN. Greatest challenge is to decide which one to use when both are available. Process of moving a device from LTE to WLAN network is called vertical handover. Many factors could impact the handover decision, and a lot of algorithms exist with their advantages and disadvantages. Here, a new algorithm is proposed which is based on fuzzy logic and a new combination of input factors. Also, a large number of vertical handover could produce ping-pong effect and reduce user performance. According to this, a new solution for reducing ping-pong effect is proposed. A solution is based on graph theory and finding bridges in a graph.

Keywords Vertical handover · WLAN · LTE · Fuzzy logic · Graph theory

1 Introduction

Today, the two most popular ways to access a wireless network are WLAN (Wireless Local Area Network), also known under the commercial name Wi-Fi, and WWAN (Wireless Wide Area Network), which is mostly done via LTE (Long Term Evolution) network, as the main representative of the fourth generation (4G) of mobile networks. Both of these methods have their advantages and disadvantages. Benefits of the WLAN are higher speed, higher bandwidth and lower price. Disadvantage of WLAN is that they provide services only on limited distances. In LTE and similar networks users have less speed, less bandwidth and higher prices, but they can communicate over longer distances. By development of both technologies, challenge was met about a decision on which network user needs to connect when both are available.

Z. Dejanović (✉)
Faculty of Electrical Engineering, University of Banja Luka, Patre 5, 78000 Banja Luka, Bosnia and Herzegovina
e-mail: zlatko.dejanovic@etf.unibl.org

M. Ivanović et al. (eds.), *Intelligent Distributed Computing XI*,
Studies in Computational Intelligence 737, https://doi.org/10.1007/978-3-319-66379-1_8

Managing transition of users from one network to another of the same type is quite convenient, as it depends on a small number of factors—often only on the strength of the signal and capacity. This kind of the transition is called horizontal handover. Unlike that, moving users between different types of networks is called a vertical handover and algorithms to manage moving users between different networks are a lot more complex because the decision depends on several factors. One approach could be to design such an algorithm using artificial intelligence, and the fuzzy logic. The most challenging part of the job when such an algorithm is being created presents a selection of input variables that will influence the decision. The available references have proposed various combinations of input values [1–6]. This paper proposes an algorithm for vertical handover with a new set of input parameters.

Today, a user wants to be seamlessly mobile between all available networks. During his movement, he will have to perform a large number of vertical handovers. Due to sensitive timing of handover execution, the ping pong effect may lead to unsuccessful handovers, destroying the purpose of seamless connectivity. In [7] a new history-based communication graph scheme is presented to perform vertical handover. The proposed scheme has shown a greater number of successful handovers thus reducing the ping pong effect in heterogeneous networks. This paper describes an improvement of proposed algorithm. An improvement is based on finding bridges in a graph.

2 Graph Theory

Despite the wide use of graph theory, there are not many software solutions that deal with visualization of its basic principles. Some of them are Mathematica, Graphviz and NetworkX. Their cost and lack of usability were the main motivation for the realization of a simple desktop application which parts are later used in this paper. With the implemented software solution it is possible to perform basic operations on graphs. The main motive for the implementation lies in the significant usage of graph theory principles in the field of computing. Practical implementation of trees, Euler and Hamilton graphs are programmed among others. All of the techniques in the application are implemented in a graphical way.

The most important operation for a model presented later in this paper is searching for bridges in the graph according to Tarjan algorithm [8, 9]. A bridge is an edge of a graph whose deletion increases its number of connected components.

3 Algorithm for Vertical Handover

Vertical handover allows the continuity of the session for existing user. User wants to be switched to the WLAN if the relevant requirements are met. Existing algorithms, that are relied on Mobile IP, SIP (Session Initiation Protocol) and SCTP

(Stream Control Transmission Protocol), have shown certain shortcomings, particularly the inefficiency and the introduction of too much load when handover is performed [10]. Because many factors affect the decision (bandwidth, latency, BER (Bit Error Rate), RSS (Received Signal Strength), battery status of mobile device…), the most appropriate approach for the development of an algorithm, which would decide when to carry out the handover, are MADM (Multi Attribute Decision Making) methods, which are based on fuzzy logic. Generally, the vertical handover process can be divided into three stages:

1. System detection—measurements of RSS, QoS, security aspects, battery status.
2. Handover decision—selecting the most appropriate available network.
3. Handover execution.

If the mobile device connected to a LTE network detects existing WLAN network within range, handover factor is calculated. That factor determines whether handover needs to be carried out. The collected input values are delivered in fuzzificator of Mamdani FIS editor, which transforms them into fuzzy sets on the basis of defined membership functions. After that, a set of IF-THEN rules is applied to get sets of decisions. Fuzzy sets of outputs are aggregated into a single set. Obtained set is sent to the defuzzification process. As a result of defuzzification, quantitative handover factor is obtained. Obtained factor determines whether a handover is required.

Since WWAN is in most cases always present, and WLAN is optional, target of vertical handover from WWAN to WLAN is QoS (Quality of Service) improving. User connected to the WWAN network mainly wants to move to the WLAN, primarily in order to achieve higher speed transmission for a smaller price. The decision to make handover may be left to the network or mobile device. To simplify implementation and, therefore, to make algorithm faster, it is decided that this decision is completely up to mobile device.

For input parameters of the algorithm there is no predefined combination. There are different combination of input parameters in available references. However, in the references available to the author combination of the following parameters has not been proposed:

- WLAN network signal strength,
- bandwidth,
- jitter,
- mobile device battery status.

Each input size is assigned to one of three fuzzy sets—low, medium or high.

Assuming that the signal of LTE network is constant, an important factor during the handover may be a signal strength of the WLAN network. Almost all mobile devices have indicator of signal strength and, therefore, the entrance to the algorithm would be relatively easy to bring. Also, the advantage of this information is that it is modelled as a fuzzy in the real system. In this paper range of -100 to -60 dBm is modelled because standard signal strength in the 802.11 networks is between -90 and -70 dBm.

1. If (RSS is low) and (Bandwidth is low) and (Jitter is low) and (Battery is low) then (Handover is NO) (1)
2. If (RSS is low) and (Bandwidth is high) and (Jitter is medium) and (Battery is low) then (Handover is Probably) (1)
3. If (RSS is medium) and (Bandwidth is low) and (Jitter is low) and (Battery is low) then (Handover is Probably) (1)
4. If (RSS is medium) and (Bandwidth is high) and (Jitter is low) and (Battery is low) then (Handover is YES) (1)
5. If (RSS is medium) and (Bandwidth is medium) and (Jitter is medium) and (Battery is medium) then (Handover is YES) (1)
6. If (RSS is high) and (Bandwidth is low) and (Jitter is high) and (Battery is high) then (Handover is Probably) (1)
7. If (RSS is high) and (Bandwidth is medium) and (Jitter is medium) and (Battery is medium) then (Handover is YES) (1)
8. If (RSS is high) and (Bandwidth is high) and (Jitter is low) and (Battery is low) then (Handover is YES) (1)

Fig. 1 IF-THEN rules

The bandwidth must be one of the most important factors when deciding on handover. In the paper scope of 0–100 Mbit/s is modelled.

Delay is one of the most important parameters in measuring the performance of any network. However, since the delay is closely connected to the bandwidth, in this paper it is decided not to measure the delay, but the jitter. Jitter has an important role in determining the QoS and here the fuzzy set is modelled in scale of 0–12 ms.

Battery status of mobile device is a factor that is also taken into consideration, because its rapid discharge is one of the biggest problems smartphones have. it is shown that the 4G network has more influence on battery power than the WLAN network. Each mobile device has an indicator that shows how much battery is left available, so that value is easily transformed into another fuzzy set.

The next step is to define the IF-THEN rules. This step has most influence on how well the algorithm will work. Number of rules should not be too big. On the other hand, it needs to cover all major input combinations. In the paper, number of rules is reduced from possible 81 (three phase values for the four possible inputs) to 8. Rules in MATLAB are shown in Fig. 1.

4 Improved Solution for Reducing Ping-Pong Effect

One solution to minimize the ping-pong effect is presented in [7]. In mentioned solution it is recommended to remember the paths that the mobile user often traverses in the form of a graph. Then, decision whether to perform handover would be based on already known performed decisions through history. In case that a decision should be made, and that the path is not known, handover is done in the classical manner, which is based on RSS. Authors presumed that mobile users often traverse the same paths during regular day.

This paper proposes a new model for improving the history-based algorithm previously described in a short manner. The graph we use is used for crucial decisions.

The nodes of the graph are the coordinates at which the vertical handover was successfully executed. In the case that a new location in which the algorithm was initiated appears, a vertex in the graph is created only if the handover was successfully executed. Every time handover is successfully initiated, three new vertices are added to the graph, one for both heterogeneous networks and one inter-vertex which will be used for reducing ping-pong effect.

Fig. 2 Example of adding edges to the graph

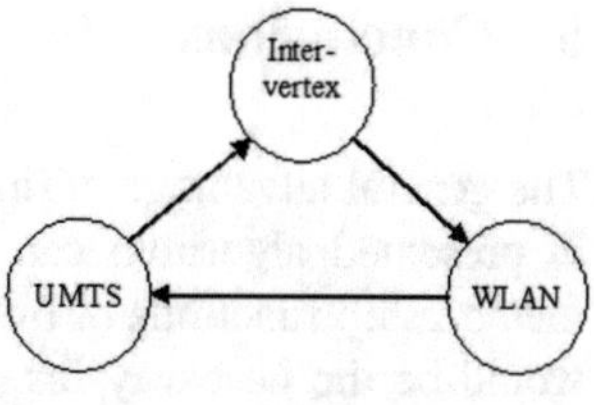

Edges in the graph are formed when handover is successfully executed. The percentage of successful handovers at certain point defines the weight of an edge from present node to the other heterogeneous node. If the handover is performed from the LTE network to the WLAN, two new edges will be added to the graph as shown in Fig. 2. In the opposite direction two nodes will be directly connected.

It's obvious that a graph will have cycles (or loops) between vertices where a ping-pong effect is often expressed.

If handover is to be performed between two vertices in a graph which are connected with a bridge edge and the weight of a bridge is over 0.8 (percentage of successful handovers through history was over 80%), then the handover occurs immediately. Otherwise, it is necessary to perform fuzzy algorithm described in the previous chapter. It is easily to conclude that presented model will reduce the ping-pong effect in the graph as fuzzy algorithm will not be initiated again following the handover if the edge of a graph is the bridge.

It is necessary that only subgraph of a main graph be taken into consideration during the process because it is likely that a huge main graph has a cycle that is not visible at first. So, subgraph needs to include vertices between whose handover is initiated, their neighbours and neighbours of their neighbours. It is trivial to prove the fact that model on subgraph is mathematically correct if we look at the definition of a bridge in a graph.

An example of finding bridges in application mentioned in Chap. 2 is given in Fig. 3. Bridges are marked in purple. Weights given on edges of a left graph in Figure are not relevant when decision whether the edge is a bridge is made.

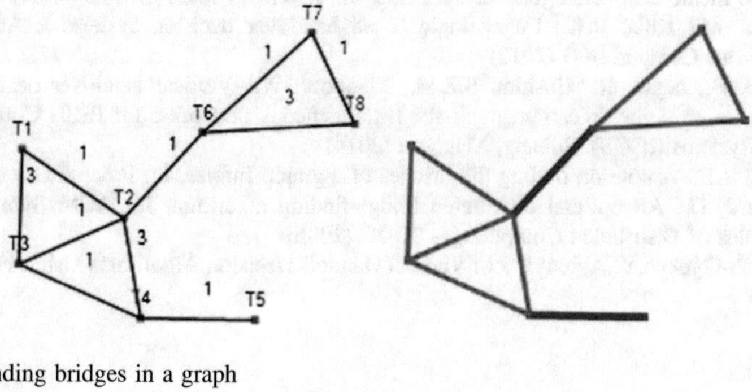

Fig. 3 Finding bridges in a graph

5 Conclusions

The general advantage of fuzzy logic is its flexibility so that the situation described in presented algorithm can easily be written otherwise, simply by changing the membership functions or by adding new IF-THEN rules. Practical use of algorithm would be the best way for developing ideas for its modifications that would lead towards its improvement.

Proposed algorithm in Chap. 3 is tested for situation when handover is made from LTE to WLAN. Since WLAN networks have significantly less coverage, there must be accurate decisions about vertical handover when the user is exiting the coverage of the WLAN network. Because of the already mentioned characteristics of fuzzy system, the process of its construction could be modelled in a very similar manner as in the described procedure.

A proposed model for reducing ping-pong effect presents a new direction that could be used for better performance when vertical handover is done. Graph theory has, once again, proved very useful for solving engineering problems.

In this paper it is not considered the possibility of vertical handover when one of the network WiMAX network. However, its inclusion in the algorithm would only represent upgrade of a model, i.e. it is not needed to write it from scratch.

References

1. Mohamed, L., Leghris, C., Abdellah, A.: An intelligent network selection strategy based on MADM methods in heterogeneous networks. IJWNM **4**(1) (2012)
2. Vasu, K., Maheshwari, S., Mahapatra, S., Kumar, C.S.: QoS-aware fuzzy rule-based vertical handover decision algorithm incorporating a new evaluation model for wireless heterogenous networks. EURASIP J. Wirel. Commun. Netw. (2012)
3. Kausar, M.S., Cheelu, D.: Context aware fuzzy rule based vertical handoff decision strategies for heterogeneous wireless networks. Int. J. Eng. Sci. **3**(7) (2013)
4. Singhrova, A., Prakash, N.: Adaptive vertical handoff decision algorithm for wireless heterogeneous networks. In: 11th IEEE International Conference on High Performance Computing and Communications (2009)
5. Lahby, M., Cherkaoui, L., Adib, A.: An intelligent network selection strategy based on MADM methods in heterogeneous networks. Int. J. Wirel. Mob. Netw. (IJWMN) **4**(1) (2012)
6. Sharma, M., Khla, R.K.: Fuzzy logic based handover decision system. J. Adhoc Sens. Ubiquitous Comput. **3**(4) (2012)
7. Naeem, B., Ngah, R., Hashim, S.Z.M., Maqbool, W.: Vertical handover decision using history-based communication graph for heterogeneous networks. In: IEEE Conference on Open Systems (ICOS), Subang, Malaysia (2014)
8. Tarjan, R.E.: A note on finding the bridges of a graph. Inform. In: Process Lett (1974)
9. Pritchard, D.: An optimal distributed bridge-finding algorithm. In: ACM Symposium on Principles of Distributed Computing—PODC (2006)
10. Nkansah-Gyekye,Y.,Agbinya, J.I.: Vertical Handoff Decision Algorithms Using Fuzzy Logic (2007)

Part III
Multi-agent Systems

Programming the Interaction Space Effectively with ReSpecTX

Giovanni Ciatto, Stefano Mariani and Andrea Omicini

Abstract The lack of a suitable toolchain for programming the interaction space with coordination languages hinders their adoption in the industry, and limits their application as core calculus, proof-of-concept frameworks, or rapid prototyping/simulation environments. In this paper we present the ReSpecTX language and toolchain as a first step toward closing the gap, by equipping a core coordination language (ReSpecT) with tools and features commonly found in mainstream programming languages, improving likelihood of adoption in real-world scenarios.

Keywords Coordination · Multi-agent systems · Tools · TuCSoN · ReSpecTX

1 Introduction

Many efforts are being devoted, in both the industry and the academia, to deal with the issue of enabling and governing the *interaction space* [17], that is, the dimension of computation—orthogonal to the purely algorithmic one—defining the admissible interactions among the components of a (concurrent, distributed) system. While in the industry they often take the form of *communication protocols* tailored to the particular business domain—e.g. MQTT versus CoAP for the IoT landscape, FIPA[1] protocols for multi-agent systems (MAS), REST versus SOAP for micro-services—in the academia they usually fall under the research umbrella branded as *coordination*

[1]http://www.fipa.org/.

G. Ciatto · A. Omicini
DISI—Università di Bologna, Bologna, Italy
e-mail: giovanni.ciatto@unibo.it

A. Omicini
e-mail: andrea.omicini@unibo.it

S. Mariani (✉)
DISMI—Università degli Studi di Modena e Reggio Emilia, Reggio Emilia, Italy
e-mail: stefano.mariani@unimore.it

M. Ivanović et al. (eds.), *Intelligent Distributed Computing XI*,
Studies in Computational Intelligence 737, https://doi.org/10.1007/978-3-319-66379-1_9

models and languages [22], that is, the set of abstractions and mechanisms enabling the governance of dependencies amongst computational activities.

In spite of the number of coordination languages available to date, they are mostly either core calculus, proof-of-concept frameworks, or domain-specific languages for rapid prototyping/simulation, rather than full-fledged programming languages. Also, no suitable *toolchain* for supporting the increasingly complex task of programming the interaction space is usually provided, resulting in the lack of features typical of state-of-art programming languages—debugging, static-checking, code-completion, etc.

Instead, agent-oriented programming (AOP) frameworks are nowadays mostly integrated with mainstream programming languages and come equipped with all sorts of development tools. JADE [3] for instance, is a Java-based AOP and infrastructure equipped with a GUI for remote *monitoring* of the agents' lifecycle, an Introspector agent (with a GUI) to *debug* agents' inner working cycle, and a Sniffer agent (again, with a GUI) to *observe* agents' messaging protocols.

The aim of this paper is to close the gap between the forthcoming maturity of AOP languages and the weaknesses of coordination frameworks w.r.t. supporting the engineering of distributed systems, by presenting the ReSpecT𝕏 language and toolchain for programming coordination of distributed multi-agent systems (MAS henceforth). ReSpecT𝕏 builds upon the ReSpecT language [20] while pushing it beyond the limits of other coordination languages through features such as modularity, reactions ordering, imperative-style syntactic sugar, and an Eclipse IDE plugin[2] for static-checking, auto-completion, and code generation.

Accordingly, the remainder of the paper is organised as follows: Sect. 2 provides the background context required to motivate our work, Sect. 3 presents ReSpecT𝕏 and describes its main features, Sect. 4 showcases a few of them through practical examples, and Sect. 5 provides conclusive remarks and an outlook on further developments.

2 Developing MAS: Computation Versus Interaction

Two prominent examples of agent development frameworks born in the academic world and proficiently transferred to the industry are JADE [3] and *Jason* [6]: the former is an object-based framework (JADE agents are Java objects) for developing agent-oriented distributed applications in compliance with FIPA standard specifications; the latter is a Java-based implementation of an extension of the AgentSpeak(L) language [26] as well as a BDI agent runtime. Other notable mentions among the many are JADEX [25] and JACK [30], which are both industry-ready platforms for developing and running MAS featuring BDI agents. Among the application context where the aforementioned platforms have been actually deployed there

[2] http://www.eclipse.org/ide.

are autonomous guidance of unmanned vehicles [29], smart homes security [7], surveillance [8], and healthcare [27].

Conversely, examples of coordination languages and infrastructures proficiently transferred to and steadily exploited in the industrial world are more difficult to find, despite the abundance of well-known and expressive models. Among the many we mention the few we found to have some degree of maturity w.r.t. either supporting developers or enabling deployment in real-world systems:

Reo [1] is a *channel-based* coordination model implemented as a Java library, which comes with a set of Eclipse-based development tools [2]; however, we found no evidence of industrial applications in the literature

KLAIM [10] is a kernel coordination language for mobile computing, distributed through its Java implementation KLAVA [4], and recently extended by X KLAIM [5]; however, no real-world deployments exist as far as we know

LIME [24] is an extension of LINDA [16] aimed at mobile agent coordination, distributed as a Java library providing adaptation layers to different mobile code frameworks and tuple space implementations; its only actual deployment is a variation used in the monitoring of heritage buildings [9]

JavaSpaces/Jini [15] is Oracle's implementation of LINDA aimed at coordinating distributed Java programs, allowing them to write/read/consume objects to/from object-spaces, that is, tuple spaces storing Java objects. The Jini technology is still alive as part of the Apache River project[3]

TuCSoN [23] enriches the LINDA tuple space abstraction with *programmability*; it is distributed as a Java middleware and is actively exploited, for instance, in the healthcare field [13]

In the following, we focus on TuCSoN, and in particular on the ReSpecT language therein exploited to program tuple centres, since ReSpecTX is built on top of it.

2.1 Structuring the Interaction Space with TuCSoN

TuCSoN [23] is a model and infrastructure providing *coordination as a service* [28] to a MAS in the spirit of the archetypal LINDA model. TuCSoN provides to interacting agents a set of *coordination primitives*—the original LINDA ones extended with bulk, predicative, and probabilistic versions—they may use to coordinate by reading/producing/consuming first-order logic tuples within *tuple centres* [21], leveraging LINDA's suspensive semantics [16]. Since TuCSoN makes no assumption on agents' inner architecture nor capabilities—besides being able to invoke its primitives—any Java program can exploit its coordination services. In this sense, TuCSoN is a general purpose coordination medium for distributed systems in general. The aforementioned tuple centres are essentially LINDA's tuple spaces enhanced with a *behaviour specification*, that is, a program specifying how the tuple space must

[3] http://river.apache.org.

react to coordination-related events happening therein—and possibly the pro-active activities the tuple centre should carry on autonomously. Tuple centres' behaviour specifications are expressed in the ReSpecT language [20] shortly described in next subsection.

TuCSoN is fully integrated with JADE and *Jason* by properly harmonising LINDA suspensive semantics and TuCSoN invocation modes with JADE and *Jason* concurrency models [19], and comes equipped with a few tools for monitoring, debugging, manual testing, and inspection of the interaction space. Thus, TuCSoN represents a seldom case of mature-enough coordination infrastructure actually viable as a solid option for coordinating real-world industrial applications—for instance, to replace message-based with stigmergic coordination in those scenarios where loose coupling of interacting entities is required (e.g. in smart homes [11] and eHealth scenarios [13])—, with an added benefit for those already exploiting JADE or *Jason*.

2.2 *Programming the Interaction Space with* ReSpecT

ReSpecT [20] is a Prolog-based declarative language for the expression of tuple centres' behaviour specification. Each specification is composed by one or multiple *specification tuples*, which are a special kind of first-order logic tuples of the form `reaction` ($\langle E\rangle$, $\langle G\rangle$, $\langle R\rangle$), where: $\langle E\rangle$ is the *triggering event* of the reaction, that is, the coordination-related event—represented by the coordination primitive invoked—whose occurrence triggers evaluation of the reaction; $\langle G\rangle$ is the (set of) *guard predicate*(s) which must evaluate to `true` for the reaction to actually execute—enabling fine-grained control over reactions execution; $\langle R\rangle$ is the *reaction body*, that is, the set of Prolog computations and ReSpecT primitives to execute to bring about the reaction's effects.

For instance, the following reaction allows agents to consume an unbounded amount of tuples matching template `inf(T)`, which may be useful, for instance, in a master/worker scenario to model the low-priority activity any worker should execute when idle:

```
reaction( in(inf(T)), invocation , ( no(inf(T)), out(inf(T)) ) ).
```

Such a reaction is *triggered* by an `in(inf(T))` primitive (event), then, because the `invocation` guard evaluates to `true` *before* the operation is served—thus before any tuple is actually removed from the tuple centre—, the reaction body produces a matching tuple if one does not exist yet (primitive `no`).

Each reaction is executed *sequentially* (according to a non-deterministic order), *atomically*, and with a *transactional* semantics. In short, this implies that reactions execute one at a time—in a given tuple centre—with no overlapping whatsoever (sequentially), that they either succeed or fail as a whole (atomically), and that a failed reaction causes no effects at all (they are transactions). The ReSpecT Virtual Machine (VM henceforth) is the Prolog-based engine responsible for on-the-fly interpretation (triggering, evaluation, and execution) of specification tuples, which may be either statically programmed by human developers at design-time, or injected in a running TuCSoN system through dedicated TuCSoN operations, either by coordinating agents or tuple centres themselves.

Despite ReSpecT being a Turing-powerful language [12] capable of capturing most of other coordination models and languages and actively exploited in academic and industrial projects [11, 13], the lack of features typical of mainstream programming languages—e.g. modularity, syntactic sugar, imperative-style syntax, etc.—as well as of a suitable toolchain assisting developers through coding and debugging—with services such as static-checking and code generation—hinders its diffusion and adoption in industrial environments. Accordingly, ReSpecTX extends ReSpecT dealing with both the aforementioned issues by re-designing the language and providing suitable IDE tools.

3 ReSpecTX: eXtended ReSpecT

ReSpecTX empowers ReSpecT—remaining as the underlying language actually exploited for coordination by the TuCSoN middleware—with a few crucial features, enhancing the language itself and adding the necessary tooling, thoroughly described in the upcoming subsections:

modularity ReSpecTX program definitions can be split in different *modules* to be imported in a root *specification* file, enabling and promoting code reuse as well as development of code libraries

development tools ReSpecTX programs are written through an editor distributed as an *Eclipse IDE plugin* and featuring syntax highlighting, static error checking, code completion, and code generation (ReSpecT specification files and Prolog theories)

syntactic sugar ReSpecTX adds special guard predicates testing presence/absence of tuples *without* side effects (e.g. actual consumption of tuples), and adopts a more *imperative style* syntax for the benefit of developers not familiar with declarative languages such as Prolog

Though still in beta stage and not yet available as a ready-to-use Eclipse plugin package, ReSpecTX technology is already publicly available as open source code[4]—installation instructions are also provided.

3.1 Syntax Overview

A ReSpecTX script consists of a single file containing a ⟨`Module`⟩ definition. Modules are of two sorts: *library modules*, conceived to be reused by other modules, and *specifications*. Both contain the definition of the ⟨`Reaction`⟩s implementing

[4]http://bitbucket.org/gciatto/respectx.

Table 1 ReSpecT𝕏 language grammar

<table>
<tr><td>⟨Module⟩</td><td>::=</td><td>module ⟨QualifiedName⟩ { ⟨ModuleBody⟩ } |
specification ⟨QualifiedName⟩ { ⟨ModuleBody⟩ }</td></tr>
<tr><td>⟨ModuleBody⟩</td><td>::=</td><td>include ⟨QualifiedName⟩ | ⟨PrologExpr⟩ . |
⟨Reaction⟩ | ⟨ModuleBody⟩ ⟨ModuleBody⟩</td></tr>
<tr><td>⟨Reaction⟩</td><td>::=</td><td>⟨OptName⟩ reaction ⟨ReactionBody⟩</td></tr>
<tr><td>⟨OptName⟩</td><td>::=</td><td>ε | @⟨ReactionName⟩(⟨PrologVarList⟩)</td></tr>
<tr><td>⟨OptVirtual⟩</td><td>::=</td><td>ε | virtual</td></tr>
<tr><td>⟨InlineReaction⟩</td><td>::=</td><td>react ⟨ReactionBody⟩ | @⟨ReactionName⟩(⟨PrologExprList⟩)</td></tr>
<tr><td>⟨ReactionBody⟩</td><td>::=</td><td>⟨Event⟩ { ⟨PrologExpr⟩ } | ⟨Event⟩ : ⟨GuardsList⟩ { ⟨PrologExpr⟩ }</td></tr>
<tr><td>⟨Event⟩</td><td>::=</td><td>⟨PrimitiveExecution⟩ | time ⟨Instant⟩</td></tr>
<tr><td>⟨PrimitiveExecution⟩</td><td>::=</td><td>⟨Primitive⟩ ⟨TupleTemplate⟩ | ⟨ListSetterPrimitive⟩ ⟨ListTemplate⟩ |
⟨ListGetterPrimitive⟩ ⟨TupleTemplate⟩ returns ⟨ListTemplate⟩ |
⟨SpecificationPrimitive⟩ ⟨InlineReaction⟩</td></tr>
<tr><td>⟨Primitive⟩</td><td>::=</td><td>out | rd | rdp | in | inp | no | nop |
urd | urdp | uin | uinp | uno | unop</td></tr>
<tr><td>⟨ListSetterPrimitive⟩</td><td>::=</td><td>out_all</td></tr>
<tr><td>⟨ListGetterPrimitive⟩</td><td>::=</td><td>in_all | rd_all | no_all</td></tr>
<tr><td>⟨SpecificationPrimitive⟩</td><td>::=</td><td>out_s | rd_s | rdp_s in_s | inp_s | no_s | nop_s</td></tr>
<tr><td>⟨GuardsList⟩</td><td>::=</td><td>⟨Guard⟩ | ⟨Guard⟩, ⟨GuardsList⟩</td></tr>
<tr><td>⟨Guard⟩</td><td>::=</td><td>⟨Existence⟩ | ⟨Absence⟩ | invocation | completion | success |
failure | endo | exo | intra | inter | from_agent | from_tc</td></tr>
<tr><td>⟨Existence⟩</td><td>::=</td><td>?⟨TupleTemplate⟩ | ??⟨TupleTemplate⟩</td></tr>
<tr><td>⟨Absence⟩</td><td>::=</td><td>!⟨TupleTemplate⟩</td></tr>
</table>

the coordination mechanisms and policies for a particular application domain. Each module may declare an arbitrary amount of reactions and Prolog clauses (⟨`PrologExpr`⟩). Finally, as in ReSpecT, reactions have a triggering event (⟨`Event`⟩), a set of guards (⟨`GuardsList`⟩), and a ⟨`ReactionBody`⟩—composed by ReSpecT primitives and Prolog predicates/functors.

Table 1 shows a detailed description of ReSpecT𝕏 grammar. The most interesting features are thoroughly discussed in the following subsections.

3.2 Modularity, Re-usability, Composability

The ReSpecT VM expects reactions to be loaded on a tuple centre as a single monolithic script, lacking modularity. Further reactions can be dynamically added to (or removed from) a tuple centre by means of the `out_s` (or `in_s`) meta-coordination primitive, but reusability is nonetheless hindered. ReSpecT in fact provides no *linguistic abstractions* to partition specifications.

ReSpecT𝕏 overcomes such a limitation by providing two explicit scoping mechanisms at the language level, namely, *modules* and *specifications*:

- ReSpecT𝕏 programs can be split into different modules (each corresponding to a single file) each one grouping logically-related reactions. A module definition contains an arbitrary number of: (i) statements of the form `include` ⟨`QualifiedName`⟩ which import all the reactions defined in the referenced module; (ii) Prolog facts and rules; (iii) ReSpecT𝕏 reactions
- ReSpecT𝕏 programs must have a single specification file that the ReSpecT𝕏 compiler parses and translates into the aforementioned monolithic file expected

by the ReSpecT VM, by composing all the ReSpecTX reactions defined therein and in each included module

The above mechanisms straightforwardly support *modularity*, by enabling the creation of libraries of modules implementing general coordination mechanisms and policies suitable to be used, and composed together, in different contexts.

Furthermore, reactions in ReSpecTX can be decorated by a `@`⟨*ReactionName*⟩ tag, making them *referenceable* by ReSpecT *meta-coordination primitives*. Besides names, tagged reactions may specify unbound Prolog variables as arguments, which are unified to actual Prolog atoms or structures when those reactions are referenced in some meta-coordination primitives. Such a feature enables the creation of parametric reactions.

Finally, reactions labeled with keyword `virtual` are explicitly meant to be referenced by other reactions' bodies, so they are not active until "activated" by an `out_s`—an example is provided in Listing 1.1.

3.3 *Toolchain: Static-Checking, Code Completion, Code Generation*

ReSpecT lacks *development tools*, thus, for instance, ReSpecT programmers become aware of syntactic or semantic errors only at *run-time*. ReSpecTX overcomes the issue by empowering ReSpecT with Eclipse IDE integration (in the form of a plugin) featuring static-checking, code completion, and generation.

The Eclipse IDE plugin is implemented by exploiting the `Xtext` framework[5], which provides a few handy features common in mainstream programming languages, such as syntax coloring, code completion, static-checking while writing code, and automatic generation of ReSpecT code—there included Prolog predicates and functors. Syntax coloring and code completion straightforwardly move ReSpecTX closer to mainstream programming languages. The static-checker detects: (i) *repeated* reactions within the same specification, e.g. reactions triggered by the same triggering event and enabled by the same guards; (ii) inconsistent temporal constraints; (iii) bad-written URLs or TCP port numbers (e.g. reserved ones); (iv) singleton variables within a reaction, that is variables appearing only once, which may hide a typo; (v) contradictory ReSpecT guards preventing reaction execution regardless of the context, as defined in table below.

`invocation, completion`	`endo, exo`
`intra, inter`	`success, failure`
`from_agent, from_tc`	`to_agent, to_agent`
`?X, !Y` if `X = Y, ground(X)`	`before(T1), after(T2)` if `T1 >= T2`

[5] http://eclipse.org/Xtext/.

3.4 Syntax Enhancements

A potential barrier for ReSpecT adoption is represented by its declarative syntax: the vast majority of mainstream programming languages follow the imperative programming style. Accordingly, ReSpecT𝕏 provides an imperative-style syntax:

- primitive invocations are unary prefix operators: `out T` equals `out(T)`
- the `if C then T else F` construct is introduced as a more familiar alternative to Prolog `(C -> T ; F)` expressions
- ReSpecT𝕏 reactions now resemble Java methods. For instance, the following snippet shows the ReSpecT𝕏 version of the reaction allowing agents to infinitely consume a tuple presented in Sect. 2.2:

```
reaction in inf(T) : invocation , !inf(T) { out inf(T) }
```

Furthermore, ReSpecT𝕏 also provides some syntactic sugar reducing the boilerplate code w.r.t. ReSpecT specifications. For instance, *special* guards checking the presence (`?`⟨*`TupleTemplate`*⟩) or absence (`!`⟨*`TupleTemplate`*⟩) of a tuple are provided, also when tuple consumption is required (`??`⟨*`TupleTemplate`*⟩).

4 Guided Tour on Reusability

We now focus on ReSpecT𝕏 modularity feature to showcase how *reusability* of reactions is straightforwardly enabled by *encapsulation* and *composition*.

Accordingly, in Sect. 4.1 we describe two ReSpecT𝕏 modules encapsulating the logic for scheduling of periodic activities and handling of tuples multiplicity, provided as ready-to-use coordination mechanisms from ReSpecT𝕏 standard library. Then, in Sect. 4.2, we describe the "decay" module—implementing a mechanisms consuming tuples periodically to decrease their multiplicity over time—which needs to be composed with the two aforementioned ones to properly work: on the one hand, scheduling of periodic activities is required since it triggers decay, on the other hand handling multiplicity of tuples comes in hand to seamless operate on decorated tuples and not decorated ones.

Other ready-to-use ReSpecT𝕏 modules are available in its standard library, such as for application-specific overlay networks configuration ("neighbourhood" module) and information dissemination ("spreading" module), while others are currently under development and testing.

4.1 Building Reusable Libraries

Scheduling Periodic Activities. Listing 1.1 shows the ReSpecT𝕏 code implementing module `rsp.timing.Periodic`, making it possible to schedule a periodic activity, which is a building block for several distributed design patterns, i.e. decay or resilient spreading [14]. Activities are represented by an `Activity` tuple: the module takes care of emitting the `Activity` tuple once every `Period` milliseconds; then, if a reaction has `out(Activity)` as triggering event, its body would be executed periodically.

```
module rsp.timing.Periodic {
  reaction out start_periodic(Period, Activity) : completion {
    inp start_periodic(Period, Activity),
    if nop periodic_context(_, _, _, Activity) then (
      current_time(Now), out periodic_context(Period, 0, Now, Activity),
      out tick(Activity)
    )
  }
  reaction out tick(Activity) : endo, ?periodic_context(Period, _, _,
     Activity) {
    current_time(Now), NextTickInstant is Now + Period,
    out_s @next_tick(NextTickInstant, Activity)
  }
  @next_tick(T, A)
  virtual reaction time(T) : ??periodic_context(_, TickNumber, _, A), ??
     tick(A) {
    NextTickNumber is TickNumber + 1, current_time(Now),
    out periodic_context(Period, NextTickNumber, Now, A),
    out A, out tick(A)
  }
  reaction out stop_periodic(Activity) : completion {
    in_all stop_periodic(Activity) returns _,
    in_all periodic_context(_, _, _, Activity) returns _,
    in_all tick(Activity) returns _
  }
}
```

Listing 1.1 The `Periodic` module

By emitting tuple `start_periodic(Period, Activity)` / `stop_periodic(Period, Activity)`, the periodic activity is started / stopped, respectively, causing (i) reification of a `periodic_context` (if none already exists) tracking the period, number of executions carried out, last execution instant, and the `Activity` tuple—to allow for several periodic activities to be executed concurrently; (ii) emission of the `tick(Activity)` tuple to trigger scheduling of the next insertion, thus creating the desired periodic loop—through insertion of a new instance of the *virtual* reaction `schedule_next_tick`. Whenever `schedule_next_tick` is executed: (i) the `periodic_context` is updated; (ii) the `Activity` tuple is emitted; (iii) tuple `tick(Activity)` is emitted to (re)trigger the loop.

Decorating Tuples with Multiplicity. There are application contexts for which it is convenient to decorate tuples with their multiplicity, increasing performance of getter operations; for instance, in the case tuple spaces are used as biochemical solutions simulators [18]. In that context tuples are considered as molecules floating in a chemical solution (the tuple centre), tuple templates as chemical species, and multiplicity of tuples their chemical concentration.

Listing 1.2 shows the `rsp.lang.Concentration` module, providing library support to such a form of decorated tuples: (i) the tuple centre is forced to behave like a set instead of a multi-set for tuples matching the `conc(Tuple)` template, which are stored as `conc(Tuple, Concentration)`; (ii) whenever a tuple `conc(Tuple)` is emitted (consumed) the corresponding `Concentration` is increased (decreased).

```
module rsp.lang.Concentration {
    put_one(Tuple) :-
        if nop conc(Tuple, _) then out conc(Tuple, 1)
        else if inp conc(Tuple, CurrentConcentration) then (
            NextConcentration is CurrentConcentration + 1,
            out conc(Tuple, NextConcentration),
            if (NextConcentration > 1) then inp conc(Tuple)
        ) else fail.
    @on_conc_tuple_insertion
    reaction out conc(Tuple) : completion, exo {
        put_one(Tuple)
    }
    @on_conc_tuple_blocking_removal
    reaction in conc(Tuple) : invocation, exo {
        if nop conc(Tuple, _) then ( out conc(Tuple, -1)
        ) else if inp conc(Tuple, CurrentConcentration) then (
            NextConcentration is CurrentConcentration - 1,
            out conc(Tuple, NextConcentration),
            if (NextConcentration > 0) then out conc(Tuple)
        ) else fail
    }
    remove_one_if_any(Tuple) :-
        if inp conc(Tuple, CurrentConcentration) then (
            if (CurrentConcentration > 0) then (
                NextConcentration is CurrentConcentration - 1,
                out conc(Tuple, NextConcentration),
                in_all conc(Tuple) returns _,
                if (NextConcentration > 0) then out conc(Tuple)
            )
        ).
    @on_conc_tuple_removal
    reaction inp conc(Tuple) : completion, exo {
        remove_one_if_any(Tuple)
    }
}
```

Listing 1.2 The `Concentration` module

Essentially, the module makes ordinary primitives (e.g. `out`, `in`, `rd`, `no`, etc.) conform to their usual contract despite tuples' decoration:

- if a species `conc(Tuple)` already exists, a tuple `conc(Tuple, Concentration)` and *exactly one* copy of `conc(Tuple)` are stored until `Concentration` =< 0—so as to make `rd`, `rdp`, `no`, and `nop` function as usual. An invocation of either `inp` or `in` `conc(Tuple)` would just decrease the `Concentration` value
- if `Concentration` =< 0 for a given species no `conc(Tuple)` tuple is stored, to preserve usual functioning of `rd`, `rdp`, `no`, and `nop`. An invocation of `inp` `conc(Tuple)` would fail, whereas `in` `conc(Tuple)` would decrease the `Concentration` of that species while tracking down waiting agents

4.2 Composition: Decay = Periodicity + Multiplicity

As a simple yet paradigmatic example of reusability through encapsulation and composition, the `rsp.land.Decay` module shown in Listing 1.3 implements the "decay" mechanism often found in nature-inspired and/or adaptive coordination models [22] whenever the relevance of some information must decrease as time progresses. It relies on the other modules just described: periodically, tuples are consumed regardless of whether they are individual or decorated ones.

```
module rsp.lang.Decay {
    include rsp.lang.Concentration
    include rsp.timing.Periodic
    decay_one(Something) :-
        if (Something = conc(Tuple)) then (
            remove_one_if_any(Tuple)
        ) else (
            inp Something
        ).
    @decay_species_once
    reaction out decay(Something) {
        inp decay(Something),
        decay_one(Something)
    }
}
```

Listing 1.3 The `Decay` module

The module works as follows: (i) an agent or a tuple centre emits the `start_periodic(Period, decay(TT))` tuple to trigger periodic emission of therein defined tuple `decay(TT)`; (ii) as a consequence, reaction `@decay_species_once` start re-triggering in loop, creating the decay effect. In a similar way, ReSpecTX standard library favours composition of the modules mentioned in Sect. 4 (neighbourhood and spreading) to build increasingly complex coordination patterns, such as gossiping in a dynamic and mobile network of devices, pheromone-based stigmergic coordination, and others [14].

5 Conclusions and Further Work

In this paper we present the ReSpecTX language and toolchain for programming the interaction space of distributed systems, aimed at closing the gap between the conceptual advancement of coordination languages and their technological maturity, so as to promote their adoption in the industry. To this end, ReSpecTX is equipped with a few crucial features paving the way toward full integration with mainstream programming languages and toolchain: modularity, static error checking, and automatic code generation being the most notable mentions. Next steps to further improve ReSpecTX should include the development of a rich standard library of ready-to-use composable coordination mechanisms, and the distribution of ReSpecTX as a ready-to-install Eclipse IDE plugin.

References

1. Arbab, F.: Reo: a channel-based coordination model for component composition. Math. Struct. Comput. Sci. **14**(3), 329–366 (2004). doi:10.1017/S0960129504004153
2. Arbab, F., Koehler, C., Maraikar, Z., Moon, Y.J., Proença, J.: Modeling, testing and executing Reo connectors with the Eclipse coordination tools. In: International Workshop on Formal Aspects of Component Software (FACS 2008) (Sep, 2008)
3. Bellifemine, F.L., Caire, G., Greenwood, D.: Developing Multi-agent Systems with JADE. Wiley, Feb (2007)
4. Bettini, L., De Nicola, R., Pugliese, R.: Klava: a Java package for distributed and mobile applications. Softw. Pract. Exp. **32**(14), 1365–1394 (2002). doi:10.1002/spe.486
5. Bettini, L., Nicola, R.D., Pugliese, R.: X-Klaim and Klava. Electron. Notes Theor. Comput. Sci. **62**, 24–37 (2002). doi:10.1016/S1571-0661(04)00317-2
6. Bordini, R.H., Hübner, J.F., Wooldridge, M.J.: Programming Multi-Agent Systems in AgentSpeak using Jason. Wiley, Oct (2007)
7. Bos, B., Chmielewski, L., Hoepman, J.H., Nguyen, T.S.: Remote management and secure application development for pervasive home systems using *Jason*. In: 3rd International Workshop on Security, Privacy and Trust in Pervasive and Ubiquitous Computing (SecPerU 2007), pp. 7–12. Jul (2007). doi:10.1109/SECPERU.2007.9
8. Castanedo, F., Patricio, M.A., García, J., Molina, J.M.: Extending surveillance systems capabilities using BDI cooperative sensor agents. In: 4th ACM International Workshop on Video Surveillance and Sensor Networks (VSSN '06), pp. 131–138. ACM (2006). doi:10.1145/1178782.1178802
9. Ceriotti, M., Mottola, L., Picco, G.P., Murphy, A.L., Guna, S., Corra, M., Pozzi, M., Zonta, D., Zanon, P.: Monitoring heritage buildings with wireless sensor networks: the Torre Aquila deployment. In: 2009 International Conference on Information Processing in Sensor Networks (IPSN 2009), pp. 277–288. IEEE Computer Society (2009). http://ieeexplore.ieee.org/document/5211924/
10. De Nicola, R., Ferrari, G., Pugliese, R.: KLAIM: a kernel language for agent interaction and mobility. IEEE Trans. Softw. Eng. **24**(5), 315–330 (1998). doi:10.1109/32.685256
11. Denti, E., Calegari, R.: Butler-ising homemanager: a pervasive multi-agent system for home intelligence. In: 7th International Conference on Agents and Artificial Intelligence (ICAART 2015), pp. 249–256. (10–12 Jan 2015). doi:10.5220/0005284002490256
12. Denti, E., Natali, A., Omicini, A.: On the expressive power of a language for programming coordination media. In: 1998 ACM Symposium on Applied Computing (SAC'98), pp. 169–177. ACM (27 Feb–1 Mar 1998). doi:10.1145/330560.330665
13. Dubovitskaya, A., Urovi, V., Barba, I., Aberer, K., Schumacher, M.I.: A multiagent system for dynamic data aggregation in medical research. BioMed. Res. Int. (2016). doi:10.1155/2016/9027457
14. Fernandez-Marquez, J., Marzo Serugendo, G., Montagna, S., Viroli, M., Arcos, J.: Description and composition of bio-inspired design patterns: a complete overview. Nat. Comput. **12**(1), 43–67 (2013). doi:10.1007/s11047-012-9324-y
15. Freeman, E., Arnold, K., Hupfer, S.: JavaSpaces Principles, Patterns, and Practice, 1st edn. Addison-Wesley Longman Ltd., Essex, UK (1999)
16. Gelernter, D.: Generative communication in Linda. ACM Trans. Program. Lang. Syst. **7**(1), 80–112 (1985). doi:10.1145/2363.2433
17. Gelernter, D., Carriero, N.: Coordination languages and their significance. Commun. ACM **35**(2), 96–107 (1992). doi:10.1145/129630.376083
18. González Pérez, P.P., Omicini, A., Sbaraglia, M.: A biochemically-inspired coordination-based model for simulating intracellular signalling pathways. J. Simul. **7**(3), 216–226 (2013). doi:10.1057/jos.2012.28
19. Mariani, S., Omicini, A.: multi-paradigm coordination for MAS: integrating heterogeneous coordination approaches in MAS technologies. In: WOA 2016—17th Workshop "From

Objects to Agents". CEUR Workshop Proceedings, vol. 1664, pp. 91–99. (29–30 Jul 2016). http://ceur-ws.org/Vol-1664/w16.pdf
20. Omicini, A.: Formal ReSpecT in the A&A perspective. Electron. Notes Theor. Comput. Sci. **175**(2), 97–117 (2007). doi:10.1016/j.entcs.2007.03.006
21. Omicini, A., Denti, E.: From tuple spaces to tuple centres. Sci. Comput. Program. **41**(3), 277–294 (2001). doi:10.1016/S0167-6423(01)00011-9
22. Omicini, A., Viroli, M.: Coordination models and languages: from parallel computing to self-organisation. Knowl. Eng. Rev. **26**(1), 53–59 (2011). doi:10.1017/S026988891000041X
23. Omicini, A., Zambonelli, F.: Coordination for Internet application development. Auton. Ag. Multi-Agent Syst. **2**(3), 251–269 (1999). doi:10.1023/A:1010060322135
24. Picco, G.P., Murphy, A.L., Roman, G.C.: LIME: Linda meets mobility. In: 21st International Conference on Software Engineering (ICSE '99), pp. 368–377. ACM Press (16–22 May 1999). doi:10.1145/302405.302659
25. Pokahr, A., Braubach, L., Lamersdorf, W.: Jadex: a BDI reasoning engine. In: Multi-Agent Programming: Languages, Platforms and Applications, pp. 149–174. Springer US (2005). doi:10.1007/0-387-26350-0_6
26. Rao, A.S.: AgentSpeak(L): BDI agents speak out in a logical computable language. In: Agents Breaking Away, Springer (1996). doi:10.1007/BFb0031845
27. Su, C.J., Wu, C.Y.: Jade implemented mobile multi-agent based, distributed information platform for pervasive health care monitoring. Appl. Soft Comput. **11**(1), 315–325 (2011). doi:10.1016/j.asoc.2009.11.022
28. Viroli, M., Omicini, A.: Coordination as a service. Fundam. Inform. **73**(4), 507–534 (2006). http://content.iospress.com/articles/fundamenta-informaticae/fi73-4-04
29. Wallis, P., Ronnquist, R., Jarvis, D., Lucas, A.: The automated wingman—using JACK intelligent agents for unmanned autonomous vehicles. In: 2002 IEEE Aerospace Conference, vol. 5, pp. 2615–2622 (2002). doi:10.1109/AERO.2002.1035444
30. Winikoff, M.: Jack™ intelligent agents: an industrial strength platform. In: Bordini, R.H., Dastani, M., Dix, J., El Fallah Seghrouchni, A. (eds.) Multi-Agent Programming: Languages, Platforms and Applications, pp. 175–193. Springer, US (2005). doi:10.1007/0-387-26350-0_7

Multi-agent System to Design Next Generation of Airborne Platform

Ludovic Grivault, Amal El Fallah-Seghrouchni and Raphaël Girard-Claudon

Abstract Remote Piloted Aircraft Systems (RPAS) are operating in highly critical contexts and carry out a wide collection of complex mission tasks through the use of sensors. In this paper, we present a new agent-based architecture that handles sensors of these platforms. Today, the requirements of the platform in terms of autonomy, modularity, robustness and reactivity as well as the industrial constraints call for the design of a new multifunction system architecture. Such a design may rely on multi-agent paradigm since it is modular by design and the agents naturally bring autonomy and pro-activity to the system. This paper presents new and original contributions: (1) an original agentification of the system in the form of a multi-agent architecture that helps to capture the dynamic of the environment; (2) firsts results of the architecture's simulation for autonomy and scheduling evaluation.

Keywords Sensor suite · Autonomous system · RPAS · Multi-agent systems · Multi-function · Agent

1 Introduction

Nowadays, airborne platforms are used worldwide as a strategic asset during different kinds of operations including conflicts, surveillance and rescue. These operations occur in highly dynamic environments with a low predictability under scenarios combining up to a thousand entities. The involved entities all have their own behaviors, speeds and trajectories. In this context, onboard instruments (i.e. sensors) allow the platform, hence the mission manager, to collect knowledge from the field. Throughout the years, sensors have become complex systems, able to share

L. Grivault (✉) · A. El Fallah-Seghrouchni
Laboratoire d'Informatique de Paris 6 - LIP6, Sorbonne Universités, Paris, France
e-mail: ludovic.grivault@lip6.fr

A. El Fallah-Seghrouchni
e-mail: amal.elfallah@lip6.fr

L. Grivault · R. Girard-Claudon
Thales Airborne Systems, Elancourt, France

M. Ivanović et al. (eds.), *Intelligent Distributed Computing XI*,
Studies in Computational Intelligence 737, https://doi.org/10.1007/978-3-319-66379-1_10

data, communicate and, since recently, collaborate. Sensors are all specific to various physical dimensions and different range. Because of this variety, collaboration between sensors allows to deduce new data concerning the environment by overlapping outputs coming from many sensors. This operation is called track merging (a track being a set of data received from an object on the field). In this chapter, we will study the management of resources onboard Remote Piloted Aircraft Systems (RPAS). Our approach aims to design a suitable architecture to deal with resources, i.e. various sensors in our target application. We adopt the multi-agent paradigm by using an agent-based architecture for the multi-sensor system (MSS) [5, 6]. We will show in this article how the sensors' coordination can be ensured by temporal scheduling within this architecture.

From a MSS point of view, the orthogonal constraints brought by low cost versus high autonomy objectives lead to look for a new architecture able to enhance the MSS' autonomy and resilience while optimizing the sensors' use [5].

This paper is organized as follows: Sect. 2 presents the related work and emphasizes the originality of our contributions. Section 3 presents our framework including the multi-agent architecture we propose for the design of the next generation of airborne platforms (NGAP); Sect. 4 details the scheduling mechanism; Sect. 5 provides our experimental results based on the scenario given by our industrial partner. Finally Sect. 6 concludes this paper and presents our perspectives.

2 Related Work

Many studies evaluated the use of multi-agent architectures to automatize Airports' Air Traffic Controllers (ATC) and discharge operators [3, 12]. These agent-based ATC researches demonstrated the advantages brought by agents in term of autonomy. The task of following aircraft and gather flight data is usually done by one or many human operators who can be potentially overburdened depending on area attendance [8]. In this context, agents can be used to follow the location of aircraft in a geographical area and assist/alert the operator along different situations.

In agent-based ATC, agents are mainly used as secondary area operators assisting the main system's user with automatic treatments, discharging the operator from a certain workload [15]. ATCs have many constraints in common with a MSS, especially complex visualization of the field, data overloads, high criticality and low delays requirements. Also, the objectives of MSSs and ATCs are similar: to detect and follow objects in a real situation. Researches for the project *OASIS air traffic management system* [11] represented each real aircraft by an *aircraft-agent* in their agent-based ATC, global agents where responsible for inter-aircraft interactions (e.g. Coordinator and Trajectory Manager agents). Mirroring the real aircraft by an agent in the system allows to predict, monitor and plan the physical aircraft's behavior.

The main differences between ATC and MSS lies in: (1) the presence of more complex sensors, working in various physical domains; (2) in the non-static and highly-constrained natures of RPAS. Sensors onboard RPAS are highly complex instruments, multifunction and continuously expecting precise requests to work in real-time (frequency, orientation, duration, power, tracking trajectories, etc.) [2]. Furthermore, all requests, treatments and products should be treated in a real-time manner, leading to a highly responsive and predictive handling of sensors.

Architectures to assist combat aircrafts pilots were studied in the *"Copilote Electronique" project* [9] and were trying to solve work overload issues by managing planing through a knowledge based system.

Driving sensors through a multi-agent system was studied in the context of sensor-mission assignment [14]. In this previous architectures, sensors were agentified (agentification is the process of allocating systems functions to agents) and were sharing missions which were given by a mission manager in order to improve the sensors loads and consumptions. In work [10], an agent-based architecture for network of simple sensors is addressed and relies on agents argumentations to converge on a high-level objective. The agents' argumentations being too slow for handling the MSS in a real-time way, argumentation was not retained as a solution for sensors coordination. In our system, mission goals are not only coming from the Mission Manager but are also generated by agents after analyzing the data coming from the field and making sensors plans in consequence. This feature leads the MSS to support low-level sensors' requirements as well as high-level autonomy goals simultaneously.

From a scheduling point of view, our scheduler manages plans of tasks feasible in a particular time window. Each task is specified by precedence and duration constraints. The plans are weighted by a priority coefficient operationally determined and the industrial need requires to take principally this coefficient in input. In our architecture, the objective is not to balance the use of resources since each task is dedicated to one precise resource but to have all priority plans scheduled in the end of the scheduling process. This approach is close to RCPSP problems [1] and Job-Shop scheduling problems.

3 The MSS Framework

At first sight, the MSS acts as an interface between the Mission Manager (or MM, the operator's station to control the RPAS) and the sensors heads,the hardware parts of the sensors interacting with environment. The MSS, as an abstraction layer between the mission manager and the sensors, helps to provide high-autonomy features as well as an accurate control of sensors and efficient use of limited available resources (sensors, power, frequencies, dedicated processing units, etc.) [4].

To realize the MSS architecture, we will use a multi-agent architecture since the agents are suitable to bring the flexibility and the autonomy required by the MSS.

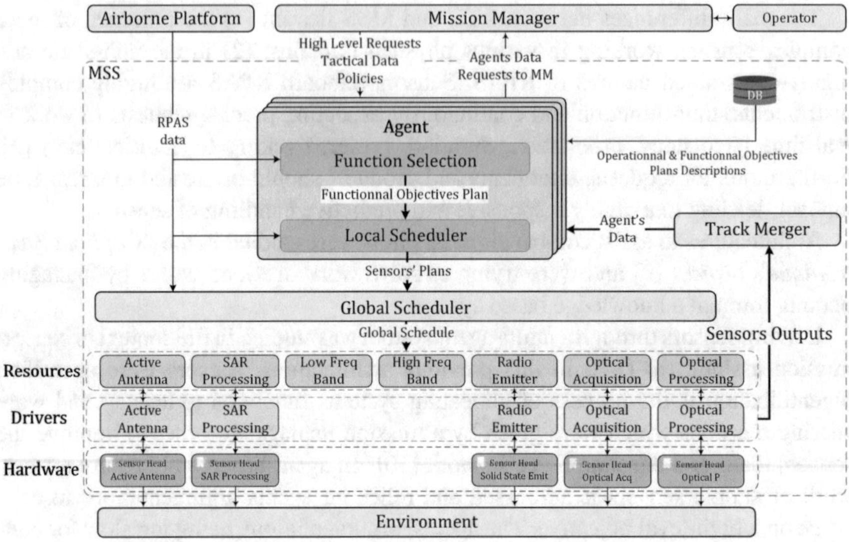

Fig. 1 The agent-based MSS architecture

The following section will describe our proposed architecture given in Fig. 1 as well as the inputs and outputs of our MSS architecture. Between high-level decisions and sensor management, agents play an important role in the MSS' architecture.

3.1 Agent Design

One of the most important contributions of our architecture is the conceptual meaning of the agents. Indeed, in the MSS, agents are not additional software mission managers but virtual instances of a field objects. Each agent is the mirror of a real field object as in the OASIS project [11]. As an example, when a sensor detects a dangerous vehicle, a new generic agent, called a *tactical agent*, is created in the system reflecting the real vehicle and producing sensors plans to watch and follow the detected object. Agents have a unique objective: collect as much data as possible about a unique field object through the use of sensors. To achieve his goal, an agent will try to select and execute one of the available sensors plans in his plan database. In practice, a function will rely on a pre-compiled plan of tasks while the local scheduler is in charge of time instantiation (tasks durations, deadlines, etc.). In our framework each task is associated with a resource and sensors are assimilated to material resources.

To enlarge the architecture's potential we consider three classes of resource: (1) sensors (e.g. an antenna); (2) any type of equipment that can be reserved for the functioning of the sensors (e.g. an image processing unit); and (3) any physical magnitude necessary for the proper functioning of the sensors (e.g. frequency). This allows the agents to make plans of tasks involving sensors as well as the resources needed by sensors. This classification of resources has proved to be useful for the exclusive use of sensors when they need access to the same non-material or material resource. With this classification, the tasks dependencies are reduced, and the allocation process is faster due to fewer exchanges between the scheduler and the agents. In our architecture, all the resources are considered as artifacts [13]. In this context an agent has a double role: creating high-level sensors' objectives and generating, for a given function, a feasible plan of tasks with an accurate allocation of resources. Each agent is equipped with communication modules, memory and a core. Communications are needed for exchanging data with agent's environment while the memory provides necessary variables for agent's operations and rational behavior. The core is hosting all running algorithms supported by the two previous features in order to exchange and store computed data. As an agent reflects a real object from the field, it has in his memory a map of all variables standing for the real objects such as speed, altitude, position, attitude and vital signs. This map is empty at the creation of the agent and get filled over the time with collected data to be aggregated in order to provide knowledge about the object.

Please refer to the article [5] for more details about agent's design and MSS architecture.

3.2 Generic Tactical Agents

Agents are generic when created and become specialized along the platform flight after receiving data about the corresponding field object. It should be specified that the MSS can detect an object without knowing either which kinds of object it is or the object's position. Therefore, it should also be specified that all sensors cannot be used with all kind of object. In case the agent is not specified because of a weak data supplying, available functions for this agent would refer to a very large set of sensors. When knowledge about this object expands, functions become more specific and more precise to this kind of agent. Each agent is the mirror of an actual object from the field. This approach creates a complete matching between the tactical situation and the agents' group. This connection between the agents and field objects brings many advantages in phase with operational and industrial requirements:

- A natural virtual embedded vision of the field with a network of active objects.
- An easy access to behavior analysis and learning functions versus in-field unexpected event.
- A strong modularity of development: an advantage for systems designers.
- A high autonomy of the MSS provided by the agents' proactiveness.

- An easy modeling of an open system with objects that appear or disappear dynamically.
- A first step for a full decentralized tactical situation architecture, bringing new opportunities for representing operations in virtual systems, a higher granularity of data from the field, a possibility for sharing objectives between platforms and a better cooperation on missions.

Representing the tactical situation by agents brought us to call them the *tactical agents*.

4 Scheduling

4.1 MSS Efficiency

The efficiency of the MSS relies on the consistency of achieved tasks according to environment's parameters:

- Events from the field (e.g. weather changes).
- Platform condition (e.g. platform's speed and attitude).
- MSS state (e.g. sensor failure).
- Field objects' behaviors (e.g. object's appearance or attitude changes).
- Operators' instructions (e.g. specific operating policies given by different operators).

The highest efficiency is reached if the MSS collected the maximum volume of significant information about the field regarding all the previous parameters.

To answer these constraints, one solution is to attribute to each agent a priority level. The agent's priority reflects the potential interest of the field object from an operational point of view and hence allows a proportional access to sensors.

The great number of objects present on the field implies a big amount of sensors' plans created by the agents. Many of these plans can be insignificant from an operational point of view. As an example, we can imagine a scenario in which the platform is tracking an important object on the field through the radar sensor, the importance of the object implies a high level of priority. After sorting by priority order, all the requests will not be achievable by the same sensor. A part would be achieved by another one (e.g. camera sensor) while the other part will be simply unachieved. In spite of a partial realization of agents' requests the resulting efficiency is optimal in the given situation.

The determination of the agents' priority level is an important point of the scheduling coherence.

4.2 Plan of Tasks

Year after year, the number of functions (e.g. *take a picture* or *listen to signals on M-band*) achievable by a MSS multiplied. Today, sensors allow to realize many different functions. Each function is achieved through a specific plan of tasks.

A task is an indivisible action achieved by a resource. A task can be identified as T_k, of duration D_k and scheduled on the timeline of an indivisible resource r_j. The task is starting at t_s and finishing at $t_s + D_k$.

A plan of tasks is an ordered set of tasks to achieve a sensor function (e.g. *Take a picture* requires the use of two resources: a *Optical camera* and an *Optical image processing unit* in a specific order and with precise delays). The order of tasks' set is described by the constraints and are specifying the start-start, start-end, min/max start-end delays and precise start-start delays for each task of the plan.

The plan P_k is defined such that $P_k = (\gamma, T_r, T_d, C, T)$ where γ is the plan's priority with $\gamma \in \mathbb{N}$, T_r the release time of the plan, T_d the plan's execution deadline and C the set of constraints which specifies the order of the set of tasks $T_k = \{T_1, T_2, T_3\}$.

4.3 Scheduler

The scheduler receives input the plans issued by the agents and the plans already scheduled on the timelines, their priorities and define a global schedule. After sorting all the plans by priority, the scheduler's algorithm is calculating the start time for each task contained within the plans. The result is a global schedule constituted of interleaved tasks. This scheduling is achieved for a *temporal horizon* T_H, meaning that all plans should be scheduled in the time window $[t; t + T_H]$, where t is the scheduling start time. The plans which were not accepted within the temporal horizon are not scheduled and will be processed later when the average priority of all the plans will be lower. If a plan is not scheduled, the agent is advised about the failure and is able to submit a new plan on less busy resources. The scheduling algorithm is detailed in [7].

5 Experimental Results

Since test in real situations is complex and very expensive to be achieved with this kind of platform and MSS, we implemented this architecture in simulation. Hence, we developed a special test scenario based on real common situations and many missions feedbacks, able to show the main decisions an operator takes during a mission. This scenario gathers up to 10 sub-scenarios where the platform is deployed in different critical contexts. Thanks to this scenario we can now compare decisions

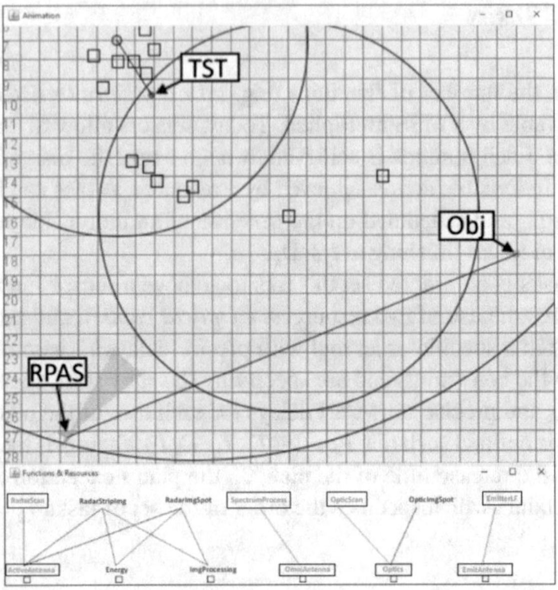

Fig. 2 Visualization of simulator's main frames

taken by our MSS architecture face to the behavior of traditional ones. Figure 2 is the visualization of the main window of the simulation engine.

The bottom frame represents functions and resources available in the MSS. Framed resources and functions are currently working and unframed ones are not. The links between functions depict the functions' dependencies of sensors. At this step of the scenario, the vigil mode of the RPAS, which was turned on at the power-on of the MSS, planned and executed the use of an electromagnetic detector. It detected the presence of a radar ("*Obj*" on the figure) and is heading to the emitting object to get more data about it. Like in reality, the MSS is not managing the platform attitude (neither deciding the platform maneuvers nor controlling RPAS surfaces) but the Mission Manager is deciding to go toward the object after the MSS shared data received and proposed an identification procedure (proposition emitted by the corresponding agent) on that point. The detection of this object led to activation of different other sensors. The cone around the RPAS is the visualization of an optical sensor (i.e. camera) turning around the platform. This sensor was also activated by the vigil function. After many tasks, an objective is given to the platform: "search the object *TST*" (i.e. Time Sensitive Target) in a particular area. After 2 min and many achieved tasks the TST was found as expected without human control on the MSS' sensors. Some functions were implemented to enhance the robustness of the MSS including agent death and replication for avoiding blocking agents issues by detecting and killing blocked agents and creating a new agent with the tactical data backup from the previous one. The MSS' global behavior matched our expectations during simulations and sensors' tasks were scheduled in time with coherency regarding the

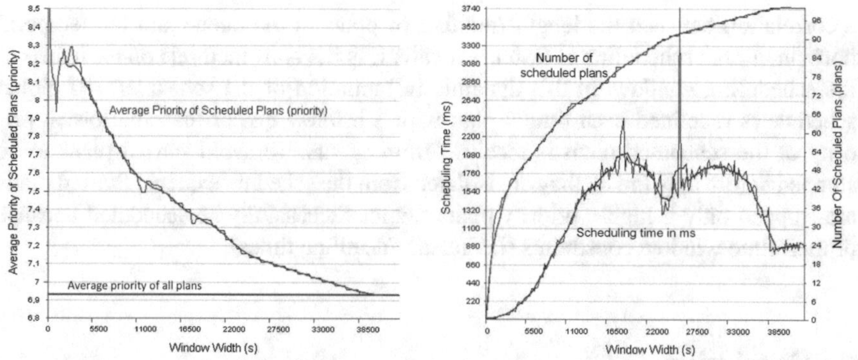

(a) Average priority of scheduled plans depending on temporal horizon (also called window width)

(b) Number of scheduled plans and scheduling time depending on the temporal horizon.

Fig. 3 Scheduling output characteristics depending on the window width

simulated field. The modularity of the MSS is improved by this architecture and the agent nature allows to specify architecture's characteristics block by block. Concerning the system autonomy, the simulation showed the ability the MSS has for managing high-level objectives depending on its own observations, without intervention from the operator.

Agents have submitted 100 plans to the global scheduler. As done by our algorithm, the scheduling results are given in Fig. 3a, b.

Figure 3a shows that the plans with the highest priority are scheduled as soon as possible even for a close temporal horizon. In addition, the average priority of scheduled plans converges to the average priority of all the plans. The operational requirements are met by the scheduling we propose since most of the time the MSS faces situations that need short time scheduling with few plans of high priority. Sorting plans before scheduling explains why the highest global priority is reached for a tiny time window.

Figure 3b shows that the number of scheduled plans increases with the size of schedule window. In our simulation, whatever the priority of the plan, its deadline coincides with the temporal horizon. In case the plans are quite temporally constrained then giving more time to the scheduler is not useful to increase the number of scheduled plans. Figure 3b also shows the scheduling time depends on the temporal horizon. The larger is the horizon, the less reactive is the scheduler. From an operational point of view, a scheduling time above 120 ms is not acceptable since many quick actions should be applied faster than this delay. For this reason, a temporal horizon around 1 or 2 min will be preferred. The Fig. 3b shows also that with a temporal horizon high enough all the plans are scheduled. The scheduling was implemented in Java and executed on a 3 GHz processor computer and may reach higher results on a different hardware. The trough in 25,000 s can be explained by a correlation between the lengths/number of plans in the queue and the temporal horizon, a plan scheduling is faster if it calls less the shift methods on each step of the scheduling window. In this dynamic instantiation of the scheduler, the global schedule is redefined each time a plan with a priority higher than the lowest priority of the scheduled plans is received from agents. To avoid started plans to be stopped before they finish, they are isolated from the schedule's queue. Started plans are stopped only if higher priority plans cannot successfully be scheduled because of their time window constraints (i.e. release/deadline times).

6 Conclusion

Our study aims to deal with sensors coordinations in the context of NGAP. We presented in this article an agent-based architecture able to handle high-levels expectations as well as real-time sensors management. We are interested in the scheduling of plans of tasks instead of the classical scheduling of tasks. This implies several differences with existing algorithms. For instance, removing unfeasible plan of tasks releases a set of resources which strongly impacts the ongoing scheduling. We have also to deal with a flow of requests from the agents. This can be roughly viewed as an online scheduling, but at this stage we have no informations about the probabilities of agents requests. Our application domain is hardly constrained and we have to deal with real-time scheduling. From the architecture point of view, our design of multi-agent system allows to consider dynamic and open theaters. All the new objects from the field are taken in charge by tactical agents at the runtime. The dynamic of the architecture, its flexibility and the first results of our scheduling mechanism provide promising solution for the next generation of airborne platform. Indeed, the multifunction and multi-sensor features of this platform are fully exploited by the multi-agent system.

The generic characteristic of this architecture allows it to be potentially adapted to less constraining platforms like underwater vehicles, piloted aircrafts, or land vehicles.

References

1. Artigues, C., Demassey, S., Neron, E.: Resource-Constrained Project Scheduling: Models, Algorithms, Extensions and Applications. Wiley (2013)
2. Bezouwen, H.V., Feldle, H.P., Holpp, W.: Status and trends in AESA-based radar. In: 2010 IEEE MTT-S International Microwave Symposium (May 2010)
3. Callantine, T.J.: CATS-Based Air Traffic Controller Agents. San Jose State University (2002)
4. Chabod, L., Galaup, P.: Shared Resources for Airborne Multifunction Sensor Systems. IET International Conference on Radar Systems (2014)
5. Grivault, L., El Fallah-Seghrouchni, A., Girard-Claudon, R.: Agent-Based Architecture for Multi-sensors system deployed on airborne platform. In: 2016 IEEE International Conference on Agents (ICA), pp. 86–89 (Sep 2016)

6. Grivault, L., El Fallah Seghrouchni, A., Girard-Claudon, R.: Coordination of sensors deployed on airborne platform: a scheduling approach. In: EUMAS-AT2016. Valencia, Spain (Dec 2016)
7. Grivault, L., El Fallah Seghrouchni, A., Girard-Claudon, R.: Next generation of airborne platforms: from architecture design to sensors scheduling. In: IEEE ICA-2017, Beijing, China (July 2017)
8. Ibrahim, Y., Higgins, P., Bruce, P.: Evaluation of a collision avoidance display to support pilots' mental workload in a free flight environment. In: IEEE International Conference on Industrial Engineering and Engineering Management (2013)
9. Joubert, T., Salle, S.E., Champigneux, G., Grau, J.Y., Sassus, P., Le Doeuff, H.: The Copilote Electronique Project: First Lessons as Exploratory Development Starts, p. 188 (1995)
10. Lesser, V., Ortiz, C.L., Jr., Tambe, M.: Distributed Sensor Networks: A Multiagent Perspective, vol. 9. Springer Science & Business Media (2012)
11. Ljungberg, M., Lucas, A.: The OASIS Air Traffic Management System, vol. 92. Seoul, Korea (1992)
12. Nguyen-Duc, M., Guessoum, Z., Marin, O., Perrot, J.F., Briot, J.P.: A multi-agent approach to reliable air traffic control. In: 2nd International Symposium on Agent Based Modeling and Simulation, Vienna, Austria (March 2008)
13. Omicini, A., Ricci, A., Viroli, M.: Artifacts in the A&A meta-model for multi-agent systems. Auton. Agents Multi-Agent Syst. **17**(3), 432–456 (2008)
14. Thao Le, Timothy J. Norman, Wamberto Vasconcelos: Agent-based Sensor-Mission Assignment for Tasks Sharing Assets. IFAAMA (2009)
15. Tumer, K., Agogino, A.: Distributed agent-based air traffic flow management. In: The Sixth International Joint Conference on Autonomous Agents and Multi-Agent Systems—AAMAS (2007)

A Drone-Based Building Inspection System Using Software-Agents

Jun Jo, Zahra Jadidi and Bela Stantic

Abstract Regular building inspections are a key means of identifying defects before getting worse or causing a building failure. As a tool for building condition inspections, Unmanned Aerial Vehicles (UAVs) or drones offer considerable potential allowing especially high-rise buildings to be visually assessed with economic and risk-related benefits. One of the critical problems encountered in automating the system is that the whole process involves a very complicated and significant amount of computational tasks, such as UAV control, localisation, image acquisition and abnormality analysis using machine learning techniques. Distributed software agents interact and collaborate each other in complicated systems and improve the reliability, availability and scalability. This research introduces a ubiquitous concept of software-agents to a drone-based building inspection system that is applied to crack-detection on concrete surfaces. The architecture and new features of the proposed system will be discussed.

Keywords Unmanned aerial vehicle · Software agents · Distributed system · Building inspection · Deep learning

1 Introduction

Concrete structures often experience fatigue stress leading to cracks on the surface. The cracks might cause serious damages, and hence, early detection prevents a possible failure. Traditional inspection methods based on humans are slow, costly and associated with hazard [1]. The safety and accuracy can be improved by

J. Jo (✉) · Z. Jadidi · B. Stantic
Institute for Integrated and Intelligent Systems, Griffith University, Nathan, Australia
e-mail: j.jo@griffith.edu.au

Z. Jadidi
e-mail: zahra.jadidi@griffith.edu.au

B. Stantic
e-mail: b.stantic@griffith.edu.au

M. Ivanović et al. (eds.), *Intelligent Distributed Computing XI*,
Studies in Computational Intelligence 737, https://doi.org/10.1007/978-3-319-66379-1_11

automating the methods. However, performing a large number of simultaneous tasks on a centralised system requires a great deal of time and resources. Agent-based methods can reduce the amount of resources required for data collection, analysis and action coordination [2]. They are very efficient for repetitive tasks and change their behaviour to adapt to the changed conditions. These methods have been employed in various studies [3]. Some [2] developed a multi agent system to prevent bridge disasters. The hazard rescue system [4] is another example, which uses multi-agent processing to make rational selection and improve the accuracy. This paper proposes a distributed intelligent monitoring system (DIMS) for autonomous building inspection. Multiple agents communicate with each other and precisely monitor a structure and detect abnormalities [5]. Controlling data, captured images and analysis data are exchanged among the multiple agents. The UAV used in this paper interacts with the ground station to receive control messages and transmit images. These images will be processed by deep learning to detect cracks, and identify their locations and severity. Deep learning algorithms have shown high performance in image processing, and they significantly reduce the error rate of image detection. This paper is organised as follows. Section 2 discusses the architecture of DIMS. Section 3 explains the system operations, and Sect. 4 is conclusion.

2 Architecture of the Distributed Intelligent Monitoring System

The system is composed of five major agents: the Event Manager (EM); the UAV Control (UC) agent; the Camera Module (CM) agent; the Vision Analysis (VA) agent; and Machine Learning (ML) agent, Fig. 1.

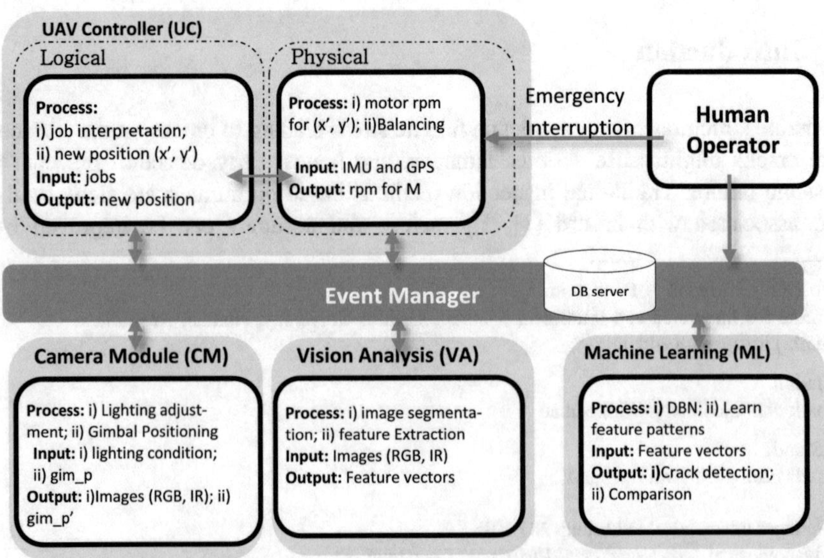

Fig. 1 The system architecture of the distributed intelligent monitoring system

2.1 The Event Manager (EM)

The EM allows DIMS to identify any changes that demand a reaction. Each agent is programmed to send messages at an interval or when an event is triggered. The EM has a rulebook about how to react to each event. For instance, if the UAV is malfunctioning and an 'emergency' event is triggered, the system may call the UAV to return to home. The user can also interrupt the operation and stop the operation at any time.

Physical agent (UC^p): is responsible for low-level control of the UAV movement interacting with the environment through sensors. This module is composed of the main controller, a set of sensors such as a GPS and an Inertial Measurement Unit (IMU), and actuators. UC^p identifies the current position and orientation of the UAV, changes the speed and the direction of the movement, and monitors the amount of battery left. In addition, this agent embeds a transmitter and interacts with other agents to receive instructions and transmit data.

Logical agent (UC^l): is used for high-level management. It is responsible for interpreting the jobs given by the central manager into the behaviour of the UAV. UC^l communicates with UC^p to send the instructions and receive information about the status of the current environment and the UAV. According to the task assigned by EM, UC^l provides information about the flying path, flight time and energy status. UC^l is also responsible to specify the sensors required to perform each mission tasks.

2.2 Camera Module (CM) Agent

This agent is equipped with a DSLR camera and an infrared (IR) thermal camera to capture crack images from concrete surfaces. The gimbal is mounted on the drone and allows a balanced movement of a camera making the rotation about two axes. CM stores the visible and invisible images captured by the two cameras in the local SD card. It also transfers the images to the DB server that is located in EM in real-time. The cameras can receive some management data from ML. The input data are the current illumination level and settings required to obtain the best images, Fig. 1. The outputs of the camera agent are Iris, gimbal movement.

2.3 Vision Analysis (VA) Agent

VA receives RGB and IR images taken by the CM agent as inputs. Initially, a shadow removal method is used to remove noises from the captured images. Afterward, a segmentation method extracts the features from the optimised images. Some examples of Segmentations are:

$Seg^c = \{Seg_1^c, Seg_2^c, \ldots, Seg_l^c\} // Seg_i^c$ shows segmentation of ith image in a camera image.
$Seg^i = \{Seg_1^i, Seg_2^i, \ldots, Seg_m^i\} // Seg_j^i$ shows segmentation of jth image in an infrared image.

2.4 Machine Learning (ML) Agent

This agent contains a Deep Belief Network (DBN) method, which learns the features extracted by VA. The trained DBN can classify the images into pre-defined categories. The DBN also learns the balancing features and sends a feedback about balancing requirements to the UC^l of the UAV.

3 The System Operation

The DIMS is operated by the distributed software agents. When the inspection job is given by EM, the UAV moves to the target position, captures images and transfers the images to the DB. VA analyses the images and the features are sent to ML. If ML detects any abnormality, it may ask the UAV to capture more detailed information and send the report to the human expert.

3.1 Balancing

The movement of UAV is affected by various forces, such as wind, from many directions. These forces can change the position and rotation rate of the UAV. Various sensors are embedded into the UC^p for stabilization, for example, accelerometers, gyroscopes and IMUs. Information about UAV status (such as acceleration, vibration, angular velocity and rotation) can be obtained from these sensors. Four agents control UAV's stability: UC^p; UC^l; CA; and ML (see Fig. 1). The messages exchanged among the agents are: (a) Enquiry about the location (current situation) (UC^l and CA); (b) Send the controlling commands in the emergency situations (from the control interface to UC^l); These emergency commands are overwritten on the existing operating commands; (c) Analyse and interpret the control command from EM to the operable control, for example the desired angle and speed (UC^l); (d) Send the instructions from UC^l to the UC^p; (e) Request images for balancing (UC^l and CA); (f) Send Gimbal control and switch control message from UC^l to UC^p; (g) Send the angle value (θ) from ML to UC^l, $IMU = \{\theta^1, \theta^2, \theta^3, \theta^4\}$. Controlling and balancing information of the UAV can be used to train ML. Then, the trained ML can guide this balancing task of the UAV.

3.2 Capturing Images

The target area can be photographed in a specific amount of time with high-resolution images. Real-time images can help to identify cracks in the structure. In addition, the camera location for each image can be controlled by the autonomous module. Three agents are responsible for capturing images: CM, VA and ML. The exchanged messages are: (a) Requests for the captured images from VA to CM; (b) Request for re-capturing blurred images from VA to CM; (c) Store the captured images in the camera SD card; (d) Request for images about the location of the UAV from UCl to CM for stabilization; (e) Transfer the captured images with their geographical information to the DB server in EM. In addition to the DSLR camera, the thermal infrared camera has the capability of detecting temperature changes caused by cracks occluded by surface object such as grass or moss. Therefore, the use of IR images helps the detection rate of the classifier increase, as invisible cracks in RGB images can be detectable by processing IR images. Figure 2a shows samples RGB and IR images.

3.3 Image Analysis

The collected images should be analysed by image analysis tools to detect the desired objects, and compare them with their previous status. The critical issues should be quickly shared with the control centre using a mobile application. VA and ML are responsible for image analysis. ML analyses images received from CA, and extracts features (see Fig. 1). Afterward, an intelligent classifier learns these features and classifies images to crack and non-crack groups. The DB server stores all classified images. Three agents communicate in this stage: CA, VA and ML. The requests are: (a) request images from the specific location from ML to CA; (b) search for images from the same location in the DB server to compare and identify changes in the situation (VA to EM); (c) send balancing requirements from ML to CA; (d) request for re-capturing blurred images from VA to CA. Image analysis stage extracts features and trains the learning agent. The analysis task of the trained ML, t^s, consists of two groups: balancing analysis, t^{sb}, and classification analysis, t^{sc}[6]. The balancing analysis is for controlling the speed and the rotation

(a) (b)

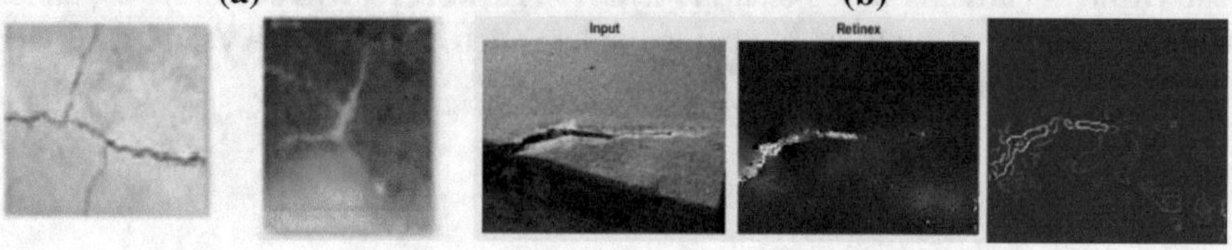

Fig. 2 **a** Captured RGB and IR images, **b** shadow removal and colour segmentation

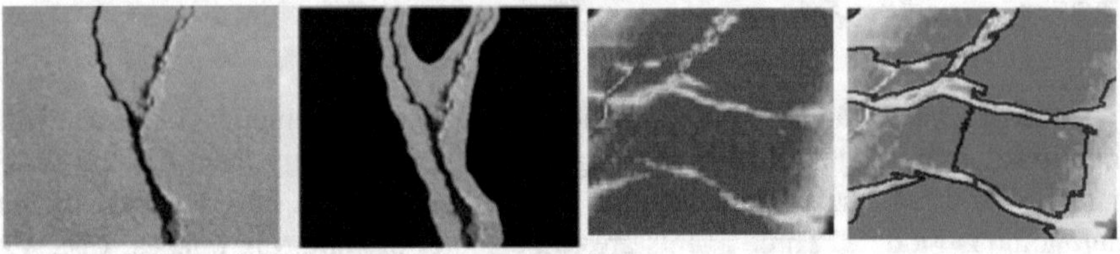

Fig. 3 Detected crack in the learning agent

of the UAV. The classification analysis is used to classify images. $t^s = \{t^{sb}, t^{sc}\}$. The captured images normally have noises like shadow. A non_local retinex method [7] is deployed in this paper to remove noise from concrete images. Afterward, the images are sent to the segmentation module to extracts the image features. Figure 2b shows the original and the optimised images, and the colour segmentation output.

3.4 Machine Learning (ML) Agent

The ML agent in the DIMS analyses the RGB and IR images taken by CM and detect cracks on concrete surfaces. The images are also used to train a deep learning method (DBN) embedded in ML. DBN is a powerful hierarchical classification method. The input layer receives the image, and then, the image features will be transferred through multiple hidden layers. The output layer is responsible to detect the class of features using a linear classifier. DBN provides an accurate classifier which can detect cracks based on features extracted by a segmentation method. Normalized cut (Ncut) segmentation is used in this paper. Figure 3 shows the crack detected in RGB and IR images.

4 Conclusion

This paper introduced the DIMS using the concept of software agent-based distributed systems. The individual agent modules and their cooperative operations within the system were described. A case study with sample cracks was introduced. The DIMS is currently based on a single UAV. However it will be further expanded with the use of various additional modules, such as extra UAVs, sensors and cameras fixed on the ground.

References

1. Giri, P., Kharkovsky, S.: Detection of surface crack in concrete using measurement technique with laser displacement sensor. IEEE Trans. Instrum. Meas. **65**, 1951–1953 (2016)
2. Cheng, M.Y., Wu, Y.W.: Multi-agent-based data exchange platform for bridge disaster prevention: A case study in Taiwan. Nat. Hazards **69**, 311–326 (2013)
3. Paolucci, M., Sacile, R.: Agent-based Manufacturing and Control Systems: New Agile Manufacturing Solutions for Achieving Peak Performance, CRC Press (2016)
4. Mocanu, A., Bădică, C.: Scrutable multi-agent hazard rescue system. In: Intelligent Distributed Computing, IX PP. 37–47, Springer (2016)
5. Chanda, S., Bu, G., Guan, H., Jo, J., Pal, U., Loo, Y.C., Blumenstein, M.: Automatic bridge crack detection–a texture analysis-based approach. In: IAPR Workshop on Artificial Neural Networks in Pattern Recognition, pp. 193–203, Springer (2014)
6. Jo, J., Tsunoda, Y., Sullivan, T., Lennon, M., Jo, T., Chun, Y.S.: BINS: blackboard-based intelligent navigation system for multiple sensory data integration. In Proceedings of the International Conference on Image Processing, Computer Vision, and Pattern Recognition (IPCV) 1 (The Steering Committee of The World Congress in Computer Science, Computer Engineering and Applied Computing (WorldComp)) (2013)
7. Zosso, D., Tran, G., Osher, S.: A unifying retinex model based on non-local differential operators. In: IS&T/SPIE Electronic Imaging 865702-865702-865716. International Society for Optics and Photonics (2013)

Part IV
Data Analysis, Mining and Integration

Connecting Social Media Data with Observed Hybrid Data for Environment Monitoring

Jinyan Chen, Sen Wang and Bela Stantic

Abstract Environmental monitoring has been regarded as one of effective solutions to protect our living places from potential risks. Traditional methods rely on periodically recording assessments of observed objects, which results in large amount of hybrid data sets. Additionally public opinions regarding certain topics can be extracted from social media and used as another source of descriptive data. In this work, we investigate how to connect and process the public opinions from social media with hybrid observation records. Particularly, we study Twitter posts from designated region with respect to specific topics, such as marine environmental activities. Sentiment analysis on tweets is performed to reflect public opinions on the environmental topics. Additionally two hybrid data sets have been considered. To process these data we use Hadoop cluster and utilize NoSql and relational databases to store data distributed across nodes in share nothing architecture. We compare the public sentiments in social media with scientific observations in real time and show that the "citizen science" enhanced with real time analytics can provide avenue to nominatively monitor natural environments. The approach presented in this paper provides an innovative method to monitor environment with the power of social media analysis and distributed computing.

Keywords Social media · Sentiment analysis · Hybrid data

J. Chen
Griffith Institute for Tourism, Brisbane, Australia
e-mail: jinyan.chen@griffith.edu.au

S. Wang · B. Stantic (✉)
Institute for Integrated and Intelligent Systems, Griffith University, Brisbane, Australia
e-mail: b.stantic@griffith.edu.au

S. Wang
e-mail: sen.wang@griffith.edu.au

M. Ivanović et al. (eds.), *Intelligent Distributed Computing XI*,
Studies in Computational Intelligence 737, https://doi.org/10.1007/978-3-319-66379-1_12

1 Introduction

More and more people are now connected to social media networks such as Twitter, Facebook, Instagram, and Sina Weibo. Posts on these social networks can be used for understanding opinions about an event, organization, product or service. Significant amount of data can be collected, for example Twitter on average has about 6,000 tweets every second which total to over 500 million tweets per day. These posts can provide valuable information of diverse topics. A stunning story of a successful prediction of the 2017 USA Presidential Election by Griffith University Big Data lab has shown that social media is playing a more and more important role in obtaining social opinions[1,2]. Power and promptness of social media was also harnessed by the U.S. Geological Service as they enhanced network of seismological sensors with real time analytic of twitter posts.

As one of the world most iconic World Heritage Areas, the Great Barrier Reef (GBR) attracts several millions tourists from all over the world each year. Unfortunately, the GBR is suffering serious environmental issues, like coral bleaching. There is need to better integrate existing monitoring programs and also address gaps by implementing new approaches, including citizen science. Particularly, huge volumes of social opinions on attracted tour sites towards specific topics can offer another effective and efficient way to explore the trends of environmental changes. Sentiment mined from the social media data using Natural Language Processing (NLP) techniques immediately reflect public response to the environmental changes, which can complement the traditional monitoring method. To process such a volume and such veracity of data new approaches are required.

In this paper, we propose a real time social network-based system to monitor the GBR by harnessing Big data analytics in distributed environment. The approach presented in this paper provides an innovative method to monitor environment with the power of social media analysis. In Big Data environment, there are two types of data processing engines dedicated for different purposes, namely batch and streaming engines. Batch processing is concerned with the handling of massive volume of data while streaming is concerned with processing data of high velocity. Several platforms and tools have been developed to support big data environments, of which the most widely known is Hadoop[3] which utilizes MapReduce batch processing. It is obvious that some applications require near real-time analysis and this type of analysis can be supported by streaming engines, such as Storm and Spark Streaming which are fault tolerant and guarantee message delivery [5] and in this work we rely on both. Proposed concept is able to provide not just indication of environmental changes but also number of visitors in specific areas, which can be very valuable in

---

[1] https://theconversation.com/can-big-data-studies-know-your-thoughts-and-predict-who-will-win-an-election-63110.

[2] https://phys.org/news/2016-11-big-analyticsnostradamus-21st-century.html.

[3] Apache Hadoop, http://hadoop.apache.org/.

normal situation for planning but more importantly in crisis situations such as disaster management. Proposed concept relies on distributed architectures and algorithms which are able in real time to manage, process, and evaluate high volume and high velocity of hybrid data.

2 Related Work

In relation to sentiment analysis literature elaborates methods which can be allocated into three main groups: Machine learning, Dictionary based, and Hybrid approaches. Typical representative of Supervised machine learning approach requires annotation of data as well as pre-processing, extraction of feature followed by learning process and finally classification [1, 2]. Different statistical based machine learning, such as Support Vector Machine (SVM) and Naive Bayes, are some methods used for sentiment analysis [11]. A dictionary based sentiment analysis methods have been used for the detection of subjectivity versus objectivity but it can be also used for sentiment polarity detection. Dictionary based methods utilize dictionaries with words along their polarities as well as set of rules. A default sentiment dictionary is generally created by humans, however, it can be also appended automatically over the time in the process of using the system [10]. For instance, a dictionary may contain words, such as excellent, better, goodbad, worse, terrible, with their respective polarities. Clear advantage of dictionary based methods for sentiment analysis is that there is no need for annotation of text for training. But there is still need to create dictionary and annotate polarity of words, however, it needs to be done only once. This is obviously less costly considering that the annotation for supervised machine learning method needs to be undertaken for each new context [2, 9].

In hybrid approaches both dictionary and machine learning are used to independently compute sentiment polarities and then individual results are combined to provide sentiment polarity [4, 7]. Specific hybrid model which is using keywords and Naive Bayes algorithm has been recommended to calculate sentiment polarities of social media tweets [3].

Recent literature also addresses the issues with regard to trust and reputation measures in social network systems, especially in presence of thematic social groups [8].

3 Methodology

3.1 Dataset and Environment

Twitter has 1.3 billion accounts and 3 million monthly active users. There are 500 million Tweets sent each day.[4] Twitter provides streaming APIs, which can be

[4]https://www.brandwatch.com/blog/44-twitter-stats-2016/.

connected with a streaming of tweets.[5] Streaming APIs give access to world-wide twitter users posts, including tweet itself, tweet language, location where account was opened, and place from which the tweet was sent, images/videos, etc. To obtain twitter data in our experiment we utilized public API provided by Twitter, specifically we looked into posts sent from Great Barrier Reef (GBR) area. To store and process huge amount of tweets, we use MongoDB[6] database as it is able to efficiently manage diversity of unstructured data.

To simplify management, access and aggregation data are stored in individual collections organized by day. Furthermore, in order to ensure efficient access, data are processed, aggregated and stored in Redis in-memory key value NoSQL database, which can serve web applications at very fast speed. It is evident that the relational database is struggling in handling large amount of unstructured data [12]. High speed is particularly required for real-time visualization of data. In addition, we utilize Neo4j, which is a highly scalable native graph database, to leverage data relationships as first-class entities. In Neo4j everything is stored in form of an edge, a node or an attribute. Each node and edge can have any number of attributes. Both the nodes and edges can be labeled, which enables labels to narrow the searches. This ensures efficient management of attributes and their relationships (i.e. which words are typically connected in text), which is particularly valuable in applications for sentiment analysis. Our experiments were conducted on a cluster with 20 nodes, each node is equipped with quad core Intel(R) Core(TM) i5-2400 CPU @ 3.10 GHz with 4GB RAM. We used Hadoop (2.6.0), MongoDB (3.2.9) and Redis (3.0.1). Scientific structured data are stored in relational MySql database, while unstructured data are stored in MongoDB.

To compare twitter posts with scientific observations, specifically, we considered only geo-tagged tweets posted from the GBR region. Geo-tagged tweets are tweets associated with geographic coordinates measured by either an exact coordinate or an approximate coordinate (polygon). To determine an approximate region of the GBR a rectangular bounding box was considered: 141.459961, −25.582085 and 153.544922, −10.69867. This bounding box is good approximation of the Great Barrier Reef region because most data come from the coastal areas of above specified bounding box. To ensure that only relevant tweets are considered we filter all captured posts with keywords that are relevant to Great Barrier Reef. These keywords are related to Marine life (Whale and different species of fish) and activities (snorkelling, diving) as well as specific tourist destinations in GBR. The filtered data set resulted in 17,622 out of 378,898 tweets from 18/03/2016 to 20/04/2017.

3.2 System Concept

In Fig. 1 the methodological flow chart of our concept has been shown.

[5]https://dev.twitter.com/streaming/overview.

[6]http://www.mongodb.org/.

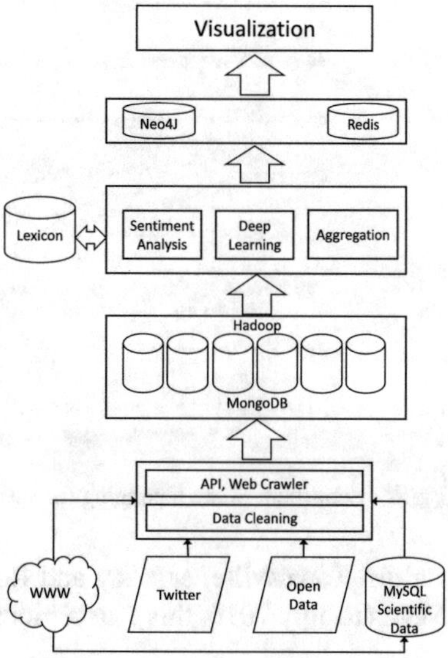

Fig. 1 Flow chart for collecting and analyzing data, and ultimately visualizing useful outputs for the end user

3.3 Sentiment Analysis

Sentiment analysis is method which can be used for analyzing social media content. This process basically converts subjective text into quantitative data. It also can extract information about events and patterns as well as determine the emotional tone of text. Sentiment analysis is a challenging and computationally demanding task because of vast volume of data, their speed of collection, common grammatical errors and misspelling, slang and abbreviations, additionally social media post can contain emoticons that cannot be simple discarded as they also carry valuable information for sentiment.

There are many sentiment analysis methods for English text presented in literature. Considering that in this work we analyze sentiment of social media post we utilized Valence Aware Dictionary for Sentiment Reasoning (VADER) approach for sentiment analysis because it is purposely developed for sentiment analysis calculation of short text found in social media posts [6]. VADER relies on dictionary but it also has set of rules, which takes into consideration punctuation, emoticons, and many other heuristics. Dictionary contains items with associated sentiment intensity which are annotated by humans. Intensity score is normalized and ranges from minus one, representing negative, to positive denoted by plus one. Because dictionary only contains English words, as a result VADER can only provide sentiment for English part of the post while content written in other language is not taken into consideration for sentiment calculation. We are in the process of developing machine learning sentiment analysis method which will also rely on domain specific lexicons. We run the Vader Sentiment analysis in the Hadoop System as it has been shown in Fig. 1. In Fig. 2 we present the result of sentiment scores. The reason why we have chosen

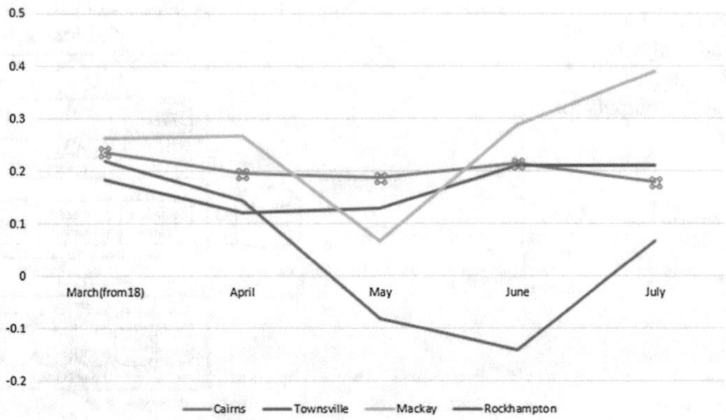

Fig. 2 Sentiment Score Changing for several Regions

Cairns, Townsville, Mackay and Rockhampton to calculate result is because from March to July 2016, this four regions had the most posts sufficient for correlation.

3.4 Scientific Data

We look into how the public opinions response in real time to environmental changes as well as disaster situations such as happened during the recent Cyclone Debbie in GBR area. Particularly, we analyze tweets sentimental results with respect to the environmental topics during a certain period. Meanwhile, we use quantitative values that measure scales of environmental changes from hybrid data that contain various types of data, such as geo-information (latitude and longitude), time-stamp, survey records, etc. We are working on utilizing Convolutional Neural Networks (CNN) and aim to extract aesthetic values as measurements of environmental conditions with intention to compare public opinions from social media with quantitative values from evidence-based data.

We mainly consider two data sets as follows:

- "*Eye on the Reef*"[7]: is a reef monitoring and assessment program run by the Great Barrier Reef Marine Park Authority. The program enables anyone who visits the Great Barrier Reef to contribute to its long-term protection by collecting valuable information about reef health, marine animals and incidents. This data also contain details where and when someone saw specific marine life or reported state of the coral bleaching, which can be indication of the environmental changes of the reef. The data is in CSV format. In this collection of sighting reports, we count the number of sighted object types as measurement of environmental aesthetics. Because of geo-information of each report, we assigned the nearest city label to each report according to its distances to different cities that were collected. We also

[7]http://www.gbrmpa.gov.au/managing-the-reef/how-the-reefs-managed/eye-on-the-reef.

obtained the sighting score matrix, denoted by $P \in \mathbb{R}^{n \times m}$, where n is the number of total months and m is the number of cities.

- "*CoralWatch*"[8]: provides a simple way for people to quantify coral health and contribute to the CoralWatch global database. It uses scale from 1–6 to demonstrate what is the condition of the coral. In CoralWatch data both geo-information and timestamps are recorded. Similarly as for 'Eye on the Reef' data, we assigned a city label to each sensor according to the shortest distance to the selected cities. The overall coral health points of all records with the same city label in the same month period were then averaged as the observed measurements. Similarly, we obtained CoralWatch health score matrix, denoted by $Q \in \mathbb{R}^{n \times m}$, which was generated to reflect coral health conditions in the GBR regions.

P and Q were used as observation-based data, sighting report scores and CoralWatch overall health scores respectively, in the remaining experiments for comparisons against the public sentiments towards the relevant environmental topics.

4 Results and Analysis

In this section, we show results obtained by analyzing hybrid data set. Specifically we show sentiment changes during the different months (March to July) and we compare Scientific data (From Coral Watch and Eyes on the reef) as well as HeapMap of correlation of Twitter data and Scientific data.

In addition to measurements or comments both Eye on the Reef and Coral Watch data has Longitude and Latitude coordinate information. We imported this data in MongoDB database in JSON format, which is the same as Twitter dataset. As can be seen in Fig. 3, Twitter data correlate with scientific observations data. In second experiment we compared the sentiment change over the months as well as geographical location by cities of interest for GBR area with regard to scientific observations report.

We have normalized sentiment scores, sighting scores and CoralWatch scores into a unique range (0, 1), and compared them with respect to four cities in Fig. 4. Note that the data in July for sighting reports and data in June for CoralWatch are missing.

From the observations in Fig. 4a, we have found city Mackay have the highest averaged sentiment scores across five months, followed by Cairns and Townsville. The worst public opinions on environmental topics were posted for Rockhampton in June and July. We checked the tweets and found the reason for negative sentiment scores were due to many negative posts. For instance, one person posted *'Sad truth': greatbarrier Reef may never rebound to previous health* at Rockhampton in May, which was very concerned to the GBR health conditions. We monitored also sentiment during recent cyclone Debbie, which significantly influenced not only sentiment level but also number of visitors evident from number of tweets posted from

[8] http://www.coralwatch.org/web/guest/home1.

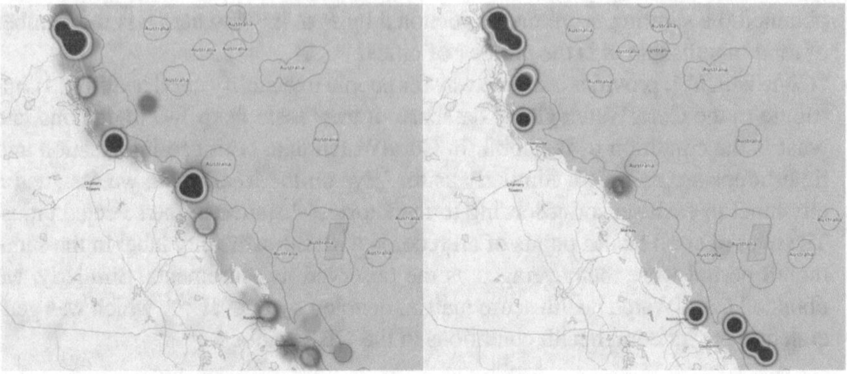

Fig. 3 Heat map of twitter data left and Scientific data (*right*)

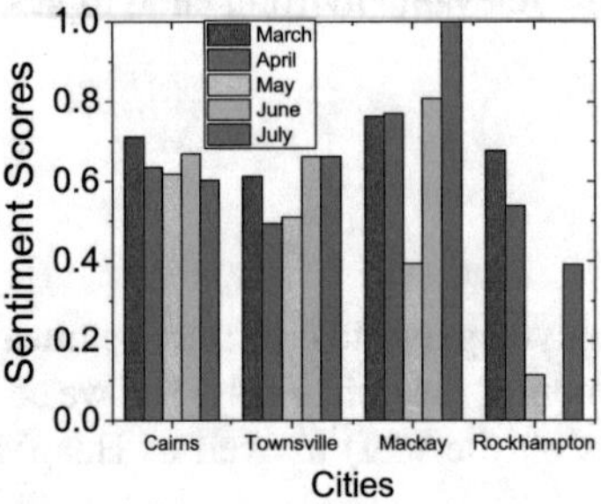

(a) Sentiment Scores

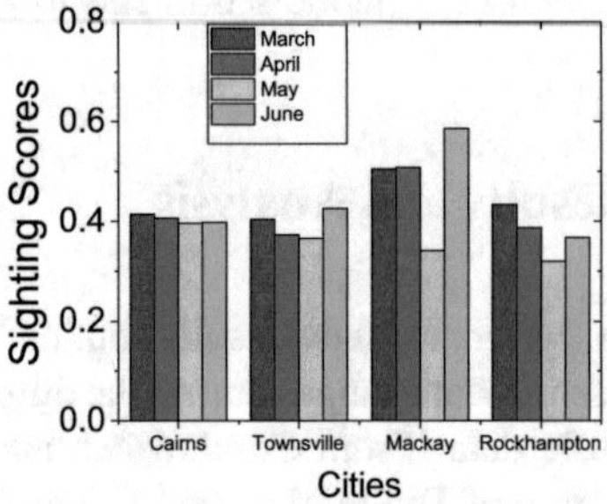

(b) Sighting Scores

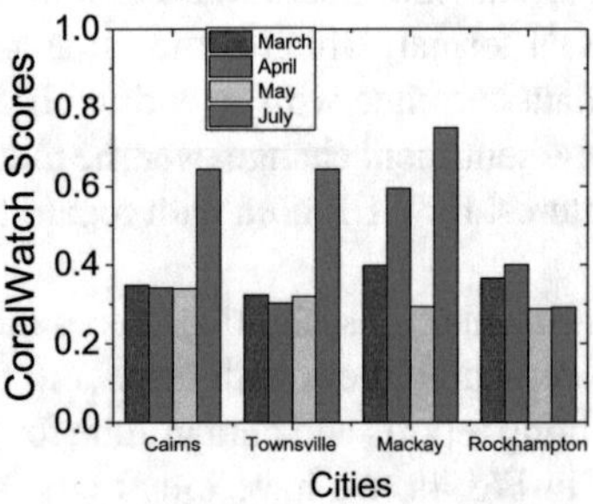

(c) CoralWatch Scores

Fig. 4 City comparisons in terms of different scores. **a** Sentiment Scores **b** Sighting Scores **c** CoralWatch Scores

the area. In Fig. 4b, Mackay also achieved the highest averaged scores across four months which reflects better biological diversity. The last one, Rockhampton, does not have too much difference comparing to other peer cities. For CoralWatch scores in Fig. 4c, the ranking order in July is quite similar to the ones in Fig. 4a in the same month.

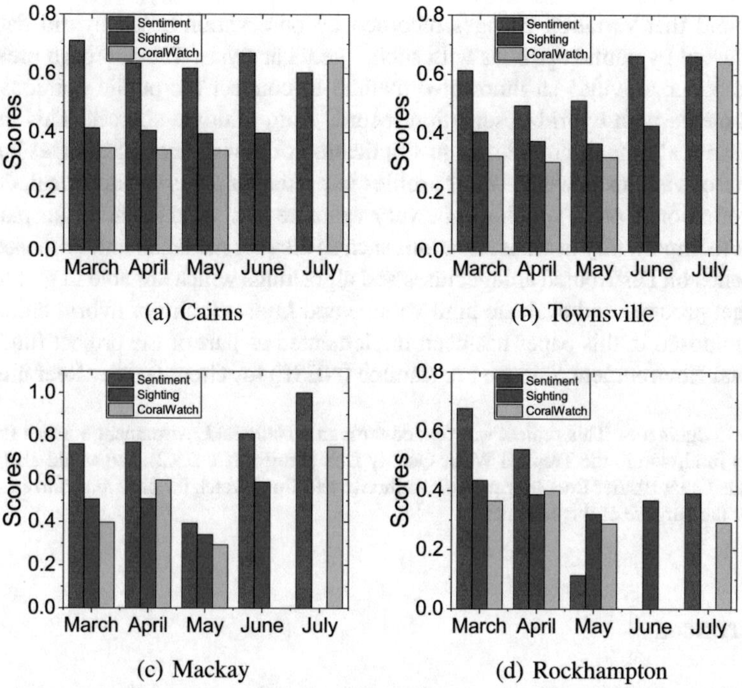

Fig. 5 Score comparisons across different months in four cities

In Fig. 5, we illustrated the comparisons of different three scores for each city between March and July. Note that there are missing data for sighting reports and CoralWatch in July and June, respectively. We can see variations of public sentiment with correlations with corresponding hybrid data changes. For example, Townsville has relatively low sentiment scores in April and May comparing to other months. Meantime, we can also witness the similar trends on sighting scores and CoralWatch scores. Interestingly, the changing scales are much smaller on two hybrid data sets. Another example can be found in Rockhampton, where sentiment scores significantly drop from March to May while both sighting and CoralWatch scores mildly decreased at the same period.

5 Conclusion

In this paper, we used the latest database technologies with a distributed computing architecture to investigate how the public opinions through social media response to environmental changes. Particularly, we analyzed tweets sentimental results with respect to the environmental topics during the certain period. We compared public opinions from social media with quantitative values from evidence-based data.

We found that variation changes recorded by observation-based hybrid data can be reflected by public opinions with social media analysis. The approach presented in this paper provides an innovative method to connect the public opinions from social media with hybrid observation records from multiple sources. This method is able in real time to provide not just indication of environmental changes but also number of visitors as well as their profiles in particular areas (visitos/local, domestic/international, etc.), which can be very valuable in normal situation for planning but more importantly in crisis situations such as disaster management. Proposed concept relies on Distributed architectures and algorithms which are able in real time to manage, process, and evaluate high volume and high velocity of hybrid data. Concept proposed in this paper has been implemented as part of the project funded by National Environment Science Programme (NESP) for Great Barrier Reef area.

Acknowledgements This project was funded through a National Environment Science Program (NESP) fund, within the Tropical Water Quality Hub (Project No: 2.3.2). We would also like to thank the Great Barrier Reef Marine Park Authority and CoralWatch for providing citizen science data for the purpose of this research.

References

1. Bjorkelund, E., Burnett, T., Norvag, K.: A study of opinion mining and visualization of hotel reviews. In: Proceedings of the 14th International Conference on Information Integration and Web-based Applications and Services, pp. 229–238
2. Brob, J.: Aspect-oriented sentiment analysis of customer reviews using distant supervision techniques. Ph.D. Thesis, Department of Mathematics and Computer Science, University of Berlin
3. Claster, W., Dinh, Q., Cooper, M.: Naive bayes and unsupervised artificial neural nets for caneun tourism social media data analysis. In: Nature and Biologically Inspired Computing, in Proceedings of the Second World Congress on Nature and Biologically Inspired Computing, pp. 158–163
4. Claster, W., Dinh, Q., Cooper, M.: Thailand-tourism and conflict modelling sentiment from twitter tweets using nave bayes and unsupervised artificial neural nets. In: Proceedings of the second International Conference on Computational Intelligence, Modelling and Simulation, pp. 89–94
5. Franciscus, N., Milosevic, Z., Stantic, B.: Influence of parallelism property of streaming engines on their performance. In: ADBIS (Short Papers and Workshops) Communications in Computer and Information Science, vol. 637, pp. 104–111. Springer (2016)
6. Hutto, C., Gilbert, E.: Vader: A parsimonious rule-based model for sentiment analysis of social media text. In: Proceedings of the 8th International AAAI Conference on Weblogs and Social Media
7. Kasper, W., Vela, M.: Sentiment analysis for hotel reviews. In: Proceedings of the Computational Linguistics-Applications Conference, pp. 45–52
8. Meo, P., Messina, F., Rosaci, D., Sarne, M.: Combining trust and skills evaluation to form e-learning classes in online social networks. Inf. Sci. **405**, 107–122 (2017)
9. Ribeiro, F., Araujo, M., Goncalves, P., Goncalves, M.F.B.: A benchmark comparison of state-of-the-practice sentiment analysis methods (2015). arXiv151201818N
10. Sharma, D., Kulshreshtha, A., P. PaygudeShahrivari, S.: Tourview: sentiment based analysis on tourist domain. Int. J. Comput. Sci. Inf. Technol. **6**(3), 2318–2320 (2015)

11. Shi, H., Li, X.: A sentiment analysis model for hotel reviews based on supervised learning. In: International Conference on Machine Learning and Cybernetics, ICMLC 2011, Guilin, China, July 10–13, 2011, Proceedings, pp. 950–954 (2011)
12. Stantic, B., Pokorný, J.: Opportunities in big data management and processing. Front. Artif. Intell. Appl. **270**, 15–26 (2014). IOS Press

EUStress: A Human Behaviour Analysis System for Monitoring and Assessing Stress During Exams

Filipe Gonçalves, Davide Carneiro, Paulo Novais and José Pêgo

Abstract In today's society, there is a compelling need for innovative approaches for the solution of many pressing problems, such as understanding the fluctuations in the performance of an individual when involved in complex and high-stake tasks. In these cases, individuals are under an increasing demand for performance, driving them to be under constant pressure, and consequently to present variations in their levels of stress. Human stress can be viewed as an agent, circumstance, situation, or variable that disturbs the normal functioning of an individual, that when not managed can bring mental problems, such as chronic stress or depression. In this paper, we propose a different approach for this problem. The EUStress application is a non-intrusive and non-invasive performance monitoring environment based on behavioural biometrics and real time analysis, used to quantify the level of stress of individuals during online exams.

Keywords Stress · Human-computer interaction · Mouse Dynamics · Decision making · Big data mining

F. Gonçalves (✉) · D. Carneiro · P. Novais
Algoritmi Research Centre/Department of Informatics, University of Minho, Braga, Portugal
e-mail: fgoncalves@algoritmi.uminho.pt

D. Carneiro
e-mail: dcarneiro@di.uminho.pt

P. Novais
e-mail: pjon@di.uminho.pt

D. Carneiro
CIICESI, ESTG, Polytechnic Institute of Porto, Felgueiras, Portugal

J. Pêgo
School of Health Sciences, Life and Health Sciences Research Institute (ICVS) University of Minho, Braga, Portugal
e-mail: jmpego@ecsaude.uminho.pt

J. Pêgo
ICVS/3B's—PT Government Associate Laboratory, Braga, Guimarães, Portugal

M. Ivanović et al. (eds.), *Intelligent Distributed Computing XI*,
Studies in Computational Intelligence 737, https://doi.org/10.1007/978-3-319-66379-1_13

1 Introduction

The term *stress* is commonly used to describe a set of physical and physiological responses that emerge as a reaction to a challenging stimulus that alter an organism's environment [17]. Perceiving an individual's physiological stress level is nowadays viewed as an important factor to manage individual performance, in a time when individual and team limits are pushed further.

Indeed, prolonged exposure to stress-inducing factors is a growing concern, specially in complex activities, which demands great responsibility and reliability. This can lead to states of emotional exhaustion (burnout) and other mental disorders [12] (e.g. depression, chronic stress, chronic diseases), with potential consequences at the personal, professional, family, social and economic levels [10].

Some good examples of stressful environments can be seen everywhere in our life, from workplaces [5] to academia [19]. In this paper we focus on academic stressors, which include the student's perception of the extensive knowledge base required and the perception of an inadequate time to develop it [6]. Students report experiencing academic stress at predictable times each semester with the greatest sources of stress resulting from taking and studying for exams, grade competition, and the large amount of content to master in a small amount of time [1]. This results in a high prevalence of anxiety disorders among higher education students.

Existing stress monitoring approaches rely on the use of complicated or expensive hardware, or in the collection of biological variables, all of which alter the routine of the student in some way, which is not desirable [9]. In this paper we propose a distributed system for monitoring human behaviour with the aim of measuring stress. Specifically, the paper discusses how groups of medical students can be monitored through their interaction with the computer [16], in order to study the effect of stress on performance during high-demand tasks. Students are monitored in terms of the efficiency of their interaction patterns with the computer and their decision making during the execution of an online exam.

The aim is to provide a stress marker that can be effectively used in large numbers of students, without inconvenience. Information on how stress affects each student will make it possible to improve individualized teaching strategies as well as to empower these students with better coping strategies. All this will result in the development of better future professionals.

The present article is organized as follows. Section 2 provides related technological works about the existing projects that revolve around the use of behavioural biometrics analysis to solve different problems. The architecture of the system is disclosed in Sect. 3. Section 4 describes the set of steps/precautions taken for obtaining the input from the selected population, taking into account their environment and routines. Section 5 defines the preparation data collection processes required for future data analysis and big data mining. Finally, Sect. 6 presents conclusions about the work developed so far and future work considerations.

2 Background

Behavioural biometrics is a relatively new form of analysis, which defines a field that extracts users behavioral features from the use of the mouse and the keyboard [21]. These methods are mostly used for user identification and authentication (intelligent security systems) [2], which use multiple techniques for automatic recognition of individuals based on their physiological and/or behavioural characteristics. By using biometrics, it is possible to confirm or establish an individual's identity based on who the individual is, rather than by what the individual possesses (an ID card) or what the individual remembers (a password) [15].

Yet, behavioural biometrics is also applied in the monitoring of human performance analysis or human healthcare. For example, such techniques are used to assess distraction and fatigue in scenarios of classrooms or workplaces [18], quantify the stress level of an individual [4] or targeted as an assisted living applications, regarding the outbreak of the disease for adverse event prediction, and patient profiling in order to improve a patients everyday life [22].

The Eustress project focuses in creating a monitoring distributed system of stress assessment, in which the software correlates data collected through different devices involving biometric and behavioural analysis to assess the stress of an individual. In this work we present the part of the project that use the mouse performance behaviour analysis (developed by our own research team in the past [4]) and the decision making performance analysis of individuals under pressure in workplace/study environments.

3 Architecture

As explained before, in order for the EUStress application to assess the stress level of an individual, it requires the analysis of two types of information: mouse behaviour and decision-making behaviour. As such, the EUStress application can be decomposed into the following components:

- *MedQuizz:* An e-assessment management system that enables trainers, educators and testing professionals to author, schedule, deliver, and report on surveys, quizzes, tests and exams, and an useful tool to create item banks. It allows the management of information about the quality of the items supporting the individual in the decision to design his/her assessments. It also has fail-safe features in the case of network failure and has functions capable of creating a log of the individuals actions.[1] This component is used to study the cognitive performance and the behaviour decision making patterns of the individual.

[1]The website of MedQuizz can be assessed at http://www.medquizz.com/.

- *MouseDynamics:* A framework that includes not only the sheer acquisition and classification of the mouse input data, based on the biometric behaviour, but also a presentation tier that supports the human-based or autonomous decision-making mechanisms [3, 4].

Another important factor is the use of saliva as a biomarker of exam stress and a predictor of exam performance. In the study published in the "journal of psychosomatic research", done by Miri Cohen and Rabia Khalaila, it is shown that pH levels of saliva may serve as a reliable, assessible and inexpensive means by which to assess the degree of physiological reactions to exams and other naturalistic stressors [8]. Also, salivary cortisol is routinely used as a biomarker of psychological stress and related mental or physical diseases [14].

In our case study, a sample of the individual's saliva is taken before and after each exam, as a mean to analyse his/her levels of cortisol in their biological system, and to predict the levels of stress of the individual. The variance of the levels of cortisol is calculated, using the following formula:

$$DeltaCortisol_{\alpha,\beta} = PreCortisol_{\alpha,\beta} - PosCortisol_{\alpha,\beta},$$

where α represents the identification of the individual, and β represents the identification of the exam.

Since the study of behavioural features are mostly related to the individual's conduct habits, the calculation of stress level considers the individual's ID, actions of mouse and decisions made during the exams (CyberPsychological computation methods), which are monitored and acquired by the computation system [23]. After pre-processing, those data are stored in a database for machine learning. With these techniques, it allows us to study the relationships between the ubiquitous individual's psychological reactions and their behavioral patterns in cyber space for the psychological assessment of their situations in the learning process [11]. The explained architecture is shown in Fig. 1.

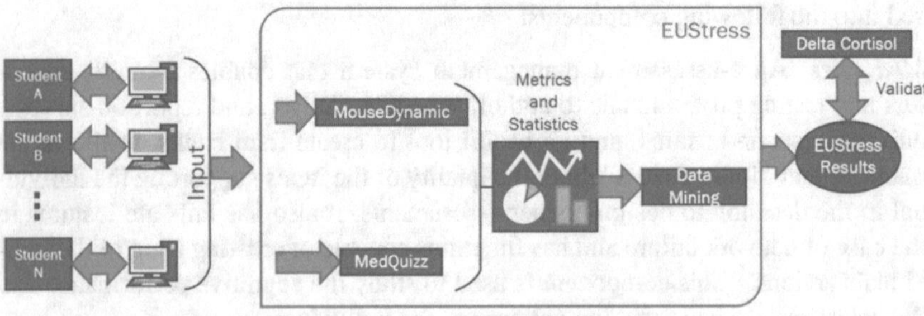

Fig. 1 EUStress Architecture

3.1 MouseDynamics Performance Analysis

MouseDynamics data collection output analyses the individual's mouse behaviour, and calculates his/her behavioural biometrics. These features aim at quantifying the individual mouse performance. Taking as example the movement of the mouse, one never moves it in a straight line between two points, there is always some degree of curve. The larger the curve, the less efficient the movement is [3, 4]. Some of the most important calculated metrics are:

- *Absolute Sum of Degrees (ASD):* This feature seeks to find how much the mouse turned, independently of the direction to which it turned (in degree units).
- *Average Distance of the Mouse to the Straight Line (ADMSL):* This feature quantifies the average sum of the successive distances of the mouse to the straight line defined by two consecutive clicks (in pixels).
- *Average Excess of Distance Between Clicks (AED):* This feature measures the average excess of distance that the mouse travelled between each two consecutive mouse clicks (in pixels).
- *Click Duration (CD):* Measures the time span between two consecutive mouse clicks. The longer the clicks, the less efficient the interaction is (in milliseconds).
- *Distance Between Clicks (DBC):* Represents the total distance travelled by the mouse between two consecutive mouse clicks (in pixels).
- *Mouse Velocity:* The distance travelled by the mouse (in pixels) over the time (in milliseconds).
- *Mouse Acceleration:* The velocity of the mouse (in pixels/milliseconds) over the time (in milliseconds).
- *Time Between Clicks:* The time span between two consecutive mouse clicks (in milliseconds).

3.2 MedQuizz Decision Making Analysis

MedQuizz Data Output: MedQuizz data collection output shows the individuals actions recorded during the exam into a log file. This file shows a list of actions, sorted by student and action time, making it possible to study and analyse the decision making behaviour of an individual. Some of the most important variables of this log are:

- *Exam Code:* Unique identification of the exam.
- *Student Code:* Unique identification of the individual/student.
- *Exam Question No:* Identifies the number of the question where the decision was made.
- *Action Time:* Defines the action time of the decision making, written in date and hour format.

- *Action Description:* Defines the type of the decision made by the individual. For example, action description can show when the individual entered or left a question, an answer was inserted or removed, among other actions.
- *Action Result:* When an answer is inserted or changed, shows if the decision made was right or wrong, based on the correct answer for the question.

MedQuizz Analysed Features: By monitoring the actions log file of each individual during the execution of an exam, it is possible to analyse his/her decision making behaviour. Everything an individual does, consciously or unconsciously, is the result of some decision. The information we gather is to help us understand occurrences, in order to develop good judgements to make decisions about these occurrences. To make a decision, an individual needs to know the problem, the need and purpose of the decision, the criteria of the decision, and the alternative actions to take. Then there's the need to determine the best alternative [20]. Decision making, for which we gather most of our information, has become a mathematical science today. It formalises the thinking we use so that, what we have to do to make better decisions is transparent in all its aspects.

With that in mind, decision making analysis aims to evaluate the performance of the individual, based on the time between decisions, the correctiveness of the selected decisions, if those decisions serve the objectives of the decision maker, number of times a question was visualized, among other features. Some of the most important calculated behavioural features are:

- *Average Time Between Decision (ATBD):* This feature seeks to find the average time it takes each individual to take a decision. All decisions are taken into account. The decisions analysed vary from entering or leaving a question, inserting, changing or removing answers, marking or unmarking questions for review, among others (in milliseconds).
- *Median Time Between Decision (MTBD):* This feature quantifies the median of the time it takes each individual to take a decision (in milliseconds). The average is a measure greatly influenced by large or small number of values, even if these values appear in small numbers in the sample. These values are responsible for the misuse of the average in many situations where it would be more meaningful to use the median.
- *Standard Deviation/Variance Time Between Decision:* This feature measures the standard deviation/variance of the time between decisions for each individual (in milliseconds).
- *Average Time Between Questions (ATBQ):* This feature measures the average time the individual spent between all visualized question (in milliseconds).
- *Decision Making Ratio (DMR):* Verifies the ratio between the number of answers inserted, changed and removed and the total number of actions recorded (in percentage).

- *Correct Decision Making Ratio (CDMR):* Verifies the ratio between the number of decisions considered correct and the number of answers inserted, changed and removed (in percentage). A decision is considered correct when an individual inserts or changes an answer into a correct option or when he/she removes an incorrect answer from the question.

4 Study Design

In order to determine the levels of stress of a group of individuals, based on their mouse behaviour performance and decision making behaviour performance, data was collected from the participation of a group of medical students in computer-based high stake exams. Through the evaluation of exams, these students test their academic knowledge in a monthly period. In these exams, students are indicated to their seats, and at the designated time they log in the exam platform using their personal credentials and the exam begins. The participation in the data-collection process does not imply any change in the student's routine, and all monitored metrics are calculated through background processes (using MedQuizz and MouseDynamics), making the collection data process completely transparent from the student's point of view, just like a normal routine exam. These exams consists mostly of single-best-answer multiple choice questions, where the students only use the mouse as an interaction means. When the exams end, students are allowed to leave the room. As explained in Sect. 3, a sample of the individual's saliva is taken before and after the execution of each exam. From these saliva samples will be later used to compare with the predicted stress levels computed by the EUStress application, as a mean to validate its results and conclusions.

Specifically, the case study considers a group of 105 medical students, which are monitored to study the effect of stress/anxiety in the performance of high demand tasks, such as the execution of exams. The methods used for data collection were taken into account, since any external factor can influence variations in the decision making and mouse performance behaviour of the individuals. As such, it was important to include non-intrusive and non-invasive measures as essential requirements during the execution of the exams.

5 Data Collection and Preparation

In order to analyse the data received from both applications (MedQuizz and MouseDynamics), some conditions were required in the study.

5.1 Temporal Approaches

The first step of the preparation process was the verification of the variations in the decision making of the individual, during an exam. In order to study these variations, the set of events of each individual are ordered by action time, cloned and split into three different datasets:

- *Chronological Time:* The data collected is divided in intervals of five minutes.
- *Percentage Time:* Data is divided in intervals, each one comprising 5% of the total duration of the exam of the individual.
- *Quarter Time:* Data collected is divided in four intervals, each containing twenty five percentage of the total duration of the exam taken by the individual.

This preparation process is used as a way of presenting different approaches for our case study, and consequently to take advantage of the different conclusions for future data mining.

5.2 Data Dimension Reduction

The second step in the preparation of the data is the implementation of data transformation processes that can provide additional insights. Moreover, when dealing with real-world data, it is often necessary to reduce dimensionality to improve big data management. Dimensionality reduction is the transformation of high-dimensional data into a meaningful representation of reduced dimensionality [7]. Ideally, the reduced representation should have a dimensionality that corresponds to the intrinsic dimensionality of the data. The intrinsic dimensionality of data is the minimum number of parameters needed to account for the observed properties of the data [13]. As a result, dimensionality reduction facilitates, among others, classification, visualization, prediction, and compression of high-dimensional data.

The data collected from from Mouse Dynamics comprises a significantly large amount. To reduce its dimensionality and given the usual shape of the data series (exemplified in Fig. 2), we construct a linear fit and use the resulting quadratic function to represent the raw data. The model build of the linear and the quadratic model are represented as $f(x) = \alpha + \beta x$ for the linear model and as $g(x) = \alpha + \beta x + \gamma x^2$ for the quadratic model.

After calculating the coefficients of both models, a mean squared error (MSE) calculation is performed. This mathematical formula measures the quality of an estimator, that is, the difference between the estimator and what is estimated. In other words, values closer to zero are better. By comparing the MSE of both functions, we can choose which model is more accurate for the set of values. The MSE of the predictor can be estimated by:

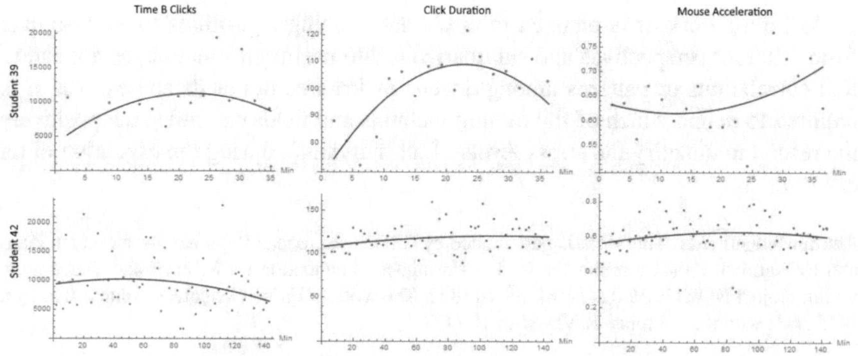

Fig. 2 Example of dimensionality reduction for three features in two students

$$MSE = \frac{1}{n} \sum_{i=1}^{n} (\hat{Y}_i - Y_i)^2$$

in which $\hat{Y}_i$ is a vector of n predictions, and Y is the vector of observed values corresponding to the inputs to the function which generated the predictions.

Figure 2 partially depicts the outcome of this process. In this figure, the dots represent the aggregated raw data at regular intervals while the lines represent the resulting quadratic model. After this reduction process, it is possible to use the parameters of the quadratic function instead of the raw data, which simplifies the posterior use of machine learning techniques.

6 Conclusions and Future Work

In this paper we present a technological approach for a non-intrusive analysis of performance in groups of people. This approach is implemented in the form of a distributed system, that constantly collects, processes, stores and monitors data describing the behaviour of multiple individuals simultaneously, during the execution of online high-stakes exams. Through the metrics monitored during the execution of a set of exams, the platform will be able to correlate data between the mouse performance metrics (using MouseDynamics output) and decision making performance metrics (analysing MedQuizz features), in order to quantify the stress levels of an individual.

By monitoring human behaviour, our aim is to study the effect of stress in the performance of high demand tasks and to point out how each individual is affected by stress. This will allow the educational institution to act on each student, through personalized teaching and coping strategies, and thus improve the quality of the future professionals that are being trained.

As future work, it is planned to apply data mining algorithms to analyse data from different perspectives and summarize it into useful information, as a mean to find correlations or patterns among dozens of features in our databases. Yet, it is required to define which of the mining methods and fields can influence positively the results to quantify the stress levels of an individual, during the execution of an exam.

Acknowledgements This work is part-funded by ERDF—European Regional Development Fund and by National Funds through the FCT – Portuguese Foundation for Science and Technology within project NORTE-01-0247-FEDER-017832. The work of Filipe Gonçalves is supported by a FCT grant with the reference ICVS-BI-2016-005.

References

1. Abouserie, R.: Sources and levels of stress in relation to locus of control and self esteem in university students. Educ. Psychol. **14**(3), 323–330 (1994)
2. Bailey, K.O., Okolica, J.S., Peterson, G.L.: User identification and authentication using multi-modal behavioral biometrics. Comput. Secur. **43**, 77–89 (2014)
3. Carneiro, D., Novais, P.: Quantifying the effects of external factors on individual performance. Future Gen. Comput. Syst. **66**, 171–186 (2017)
4. Carneiro, D., Novais, P., Pêgo, J.M., Sousa, N., Neves, J.: Using mouse dynamics to assess stress during online exams. In: International Conference on Hybrid Artificial Intelligence Systems, pp. 345–356. Springer (2015)
5. Cartwright, S., Cooper, C.L.: Managing workplace stress, vol. 1. Sage (1997)
6. Carveth, J.A., Gesse, T., Moss, N.: Survival strategies for nurse-midwifery students. J. Nurse-Midwifery **41**(1), 50–54 (1996)
7. Chen, H., Chiang, R.H., Storey, V.C.: Business intelligence and analytics: from big data to big impact. MIS Q. **36**(4), 1165–1188 (2012)
8. Cohen, M., Khalaila, R.: Saliva ph as a biomarker of exam stress and a predictor of exam performance. J. Psychosom. Res. **77**(5), 420–425 (2014)
9. Cohen, S., Kessler, R.C., Gordon, L.U.: Measuring Stress: A Guide for Health and Social Scientists. Oxford University Press on Demand (1997)
10. Colligan, T.W., Higgins, E.M.: Workplace stress: etiology and consequences. J. Workplace Behav. Health **21**(2), 89–97 (2006)
11. Dai, W., Duch, W., Abdullah, A.H., Xu, D., Chen, Y.S.: Recent advances in learning theory. Comput. Intell. Neurosci. **2015**, 14 (2015)
12. Fink, G.: Stress: Definition and History. Stress Science: Neuroendocrinology, 3–9 (2010)
13. Fukunaga, K.: Introduction to Statistical Pattern Recognition. Academic press (2013)
14. Hellhammer, D.H., Wüst, S., Kudielka, B.M.: Salivary cortisol as a biomarker in stress research. Psychoneuroendocrinology **34**(2), 163–171 (2009)
15. Jain, A.K., Ross, A., Prabhakar, S.: An introduction to biometric recognition. IEEE Trans. Circuits Syst. Video Technol. **14**(1), 4–20 (2004)
16. Novais, P., Carneiro, D.: The role of non-intrusive approaches in the development of people-aware systems. Progress in Artificial Intelligence **5**(3), 215–220 (2016). http://dx.doi.org/10.1007/s13748-016-0085-1
17. OSullivan, G.: The relationship between hope, eustress, self-efficacy, and life satisfaction among undergraduates. Soc. Indic. Res. **101**(1), 155–172 (2011)
18. Pimenta, A., Carneiro, D., Novais, P., Neves, J.: Detection of distraction and fatigue in groups through the analysis of interaction patterns with computers. In: Intelligent Distributed Computing VIII, pp. 29–39. Springer (2015)

19. Ross, S.E., Niebling, B.C., Heckert, T.M.: Sources of stress among college students. Soc. Psychol. **61**(5), 841–846 (1999)
20. Saaty, T.L.: Decision making with the analytic hierarchy process. Int. J. Servi. Sci. **1**(1), 83–98 (2008)
21. Wang, L.: Behavioral Biometrics for Human Identification: Intelligent Applications: Intelligent Applications. IGI Global (2009)
22. Xefteris, S., Andronikou, V., Tserpes, K., Varvarigou, T.: Case-based approach using behavioural biometrics aimed at assisted living. J. Am. Intell. Huma. Comput. **2**(2), 73–80 (2011)
23. Zhou, X., Dai, G., Huang, S., Sun, X., Hu, F., Hu, H., Ivanović, M.: Cyberpsychological computation on social community of ubiquitous learning. Comput. Intell. Neurosci. **2015**, 12 (2015)

Post Sharing-Based Credibility Network for Social Network

V. Carchiolo, A. Longheu, M. Malgeri, G. Mangioni and M. Previti

Abstract Social networks are intensively and extensively used to exchange news and contents in real time. The lack of a global authority for assessing posts truthfulness however allows malicious to exhibit unfair behaviours; identifying methodologies to detect hoaxes and defamatory content automatically is therefore more and more required. Social networks as Facebook and Twitter provided specific solutions and general approaches were also developed; in this paper we present a general model that takes into account both post as well as users' credibility, using a duplex network of acquaintances and credibility among users. First experiments show that it is possible to distinguish individuals who post non-truthful content through a combined analysis of both the news content and the reposts they get from their contacts.

Keywords Credibility · Social network · Social contagion

1 Introduction

Social networks have become popular over last decade, because people use it to keep in touch with friends, family and colleagues or, more generally, with interesting people, chat with them and exchange news and contents in real time. For a part of population they become substitutes of traditional mass media, because posts spread faster

V. Carchiolo · A. Longheu · M. Malgeri · G. Mangioni · M. Previti (✉)
Dipartimento di Ingegneria Elettrica, Elettronica e Informatica (DIEEI),
Università degli Studi di Catania, Catania, Italy
e-mail: marialaura.previti@unict.it

V. Carchiolo
e-mail: vincenza.carchiolo@dieei.unict.it

A. Longheu
e-mail: alessandro.longheu@dieei.unict.it

M. Malgeri
e-mail: michele.malgeri@dieei.unict.it

G. Mangioni
e-mail: giuseppe.mangioni@dieei.unict.it

M. Ivanović et al. (eds.), *Intelligent Distributed Computing XI*,
Studies in Computational Intelligence 737, https://doi.org/10.1007/978-3-319-66379-1_14

than printed paper and television news, so also traditional mass media recently use it to release instant news and government agencies publish on it official announcements.

In addition to these authoritative sources, there are many others that enter contents of questionable truthfulness. The reasons may be the most disparate: harmless users attempt to attract new followers or contacts on their pages, advertising companies receive remuneration based on the number of clicks received on their pages (also known as clickbait phenomenon), malicious redirect users on pages containing computer viruses, commercial companies, through fake accounts, aim at discrediting the rivals' products or promote their owns and users aim at defaming famous personalities to indirectly support the counterparties.

The absence of an authority that is responsible for assessing the truthfulness of the post allows all these improper behaviors, somewhat daunting task when you consider the millions of posts daily placed on large social networks such as Facebook and Twitter. Therefore it is necessary to identify methodologies that allow the detection of hoaxes and defamatory content automatically.

In the last three years, Facebook repeatly changed his news feed update algorithm [1] in order to stem the abovementioned phenomena and currently penalizes all links that lead out from the social platform only for a short time, because they are suspected of having unimportant or untruthful contents with respect to proposed title. Instead Twitter designed and implemented BotMaker [2], a system both detects spam in real time preventing spam content from getting into the system and continuously classifies users and their content to learn spammers' behaviors and self-improves the detector.

These approaches only reduce the presence of untruthful post, but they do not completely solve the problem, so recently the researchers have developed several solutions some of which are presented in Sect. 2. In Sect. 3 we present our model that aims to evaluate not only the content of posts posted on a social network but also the interest that every post gets in order to determine the credibility of who generates or shares the news. In Sect. 4, we describe the simulator that we have implemented to verify the correct behavior of the model, including a toy example to facilitate the reader in understanding the mechanisms and, in Sect. 5, we report the results of the first tests that were performed with the simulator.

2 Related Works

Recently some works have been proposed to evaluate the credibility of posts and users in social media and this section show some of the most relevant.

Jin et al. [3] proposed a hierarchical propagation model called News Credibility Propagation (NewsCP). For each news event, it generates a three-layer credibility network consisting of event, sub-event and message, calculating the credibility value for each entity and linking them with their semantic and social associations. After this step, it processes the credibility propagation as a graph optimization problem and provides a globally optimal solution with an iterative algorithm. Kwon et al. [4, 5] identified characteristics of rumors examining temporal, structural and linguistic aspects of rumor diffusion, by respectively proposing a model, called Periodic External Shocks (PES), that considers daily and external shock cycles, extracting properties related to the propagation process and examining the textual contents using text classifiers. Conversion of the continuous time series of tweet volume into only three fitting parameters for capturing the temporal fluctuations of features may cause significant information loss and make it difficult to extend the number of fitting parameters in the model, hence to overcome this issues. Ma et al. [6] proposed a different time series model, called Dynamic Series-Time Structure (DSTS), to capture content-based, user-based and diffusion-based features of social context and use them to train a Support Vector Machine model(SVM). Zhao et al. [7] developed a technique based on searching for the enquiry phrases, clustering similar posts together, collecting related posts that do not contain these phrases and ranking the clusters by their likelihood of really containing rumors. Egele et al. [8] proposed COMPA, an approach that uses a composition of statistical modeling and anomaly detection to identify compromised accounts that wrote spam posts on Twitter or Facebook. Martinez-Romo and Araujo [9] presented a methodology based on the detection of spam tweets in isolation and without previous information of the user and the application of a statistical analysis of language to detect spam in trending topic. Wang [10] crawled Twitter's data and performed content analysis to categorize tweets in legitimates and spams and the social graph analysis for detecting abnormal behaviours. O'Callaghan et al. [11] used neighborhood subnetwork to detect comment spammers in Youtube and demonstrated how discriminating motifs can be used as part of a network motif profiling process that tracks the activity of spam user accounts over time. The problem of credibility is also know as trusting and repution, [12–14] and [15] discuss the problem and their impact in social networks.

3 Building a Post Sharing-Based Credibility Network

A social network is a structure made up of social actors and links among them that represent social interactions. It could be represented by a graph and modelled by a social contagion model. In this section we propose a model to represent this behaviour and an additional structure that allows us to evaluate the credibility of its components. Table 1 shows the symbols used throughout the paper and their definitions.

In our model, we consider a population consisting of N individuals which, with respect to the news, are divided into:

- ignorants, i.e. people who are unaware about the news;
- spreaders, i.e. people who are already aware about the news and intend to share it with others;
- stiflers, i.e. people who are already aware about the news, but have no interest in spreading it.

Table 1 Table of symbols

A	Acquaintance network	C	Credibility network
N	Total number of individuals	E	Total number of interactions
S	Slices of multilevel network A	S_x	Slice related to the diffusion of news x
N_x	Number of individuals in slice S_x	E_x	Number of interactions in slice S_x
M_x	Number of repost for news x	H	Total number of news
t_0	Initial instant of net observation	ΔT	Duration of network observation
$U(i)$	Neighborghood of i	$C_{D_{j\to i}}$	Confidence of node j toward i
$d^{S_x}_{news}$	Dose of news exposition due to news contents	$d^{S_x}_{U(j)}$	Fraction of spreaders neighbors credibility respect to credibility of neighborhood of j
$P^{S_x}_j$	Fraction of exposition to the news x of node j	pt_j	Attitude to repost news with credible contents of node j
A_{ij}	Adjacency matrix	a_{ij}	Adjacency matrix element
$X^{S_x}_i$	State of node i in slice S_x	n_i	Number of news spread from node i

We assume that the news are spread by directed contact of the spreaders with ignorants, hence ignorants become spreaders at a rate p or stifler at rate $1 - p$. The decision to spread a news depends on both the news credibility and the proposer credibility. Therefore, our network is a duplex network, where a layer is the acquaintance network A and the other is the credibility network C.

The acquaintance network A is a multislice network [16] $A = (N, E, S)$, where $S = \{S_1, S_2, ..., S_H\}$ is a family of subgraphs (or slices) of A and each slice $S_x = (N_x, E_x)$ represents the diffusion that a news x had on the acquaintance network. In each slice, not all acquaintances nodes are active at the same time, but everyone interacts with someone at least once in observation time window $[t_0, t_0 + \Delta T]$, hence:

$$N = \bigcup_{x=1}^{H} N_x \quad \text{and} \quad E = \bigcup_{x=1}^{H} E_x \tag{1}$$

where H is the number of different news inserted in the network during observing time window.

The interactions among nodes occur in the following manner: if an individual i behaves as spreader, inserting a new post at time t, he makes it visible to his neighborhood $U(i)$ up to a time $t + \Delta t$, so if an individual $j \in U(i)$ is active in this time window, he established a contact e_x, hence $E_x = \{e_1, e_2, ..., e_{M_x}\}$ where M_x is number of repost for news x.

The credibility network C is a directed network, because the end nodes of a link do not trust each other in the same way. This network, at time t_0, consists of all nodes and all arcs of the acquaintance network. Precisely at each undirected arc of the acquaintance network correspond two directed arcs of the credibility network, one for each direction. Each arc is weighed and the weight represents the direct credibility $C_{D_{j\to i}} \in [0, 1]$, i.e., the confidence that node j has toward node i. We assume that the starting credibility is 0.5, modelling there is no information about reputation of the nodes yet.

The direct credibility $C_{D_{j \to i}}$ is closely linked to the credibility of post published, specifically if a spreader i contacts an ignorant j and he becomes spreader, it is likely he considers the news is true and/or the spreader is reliable, but if he decides not to propagate it, it is likely he considers it false and/or the spreader is unreliable, hence it is influenced by the news content $d^{S_x}_{news} \in [0, 1]$ and by the neighborhood activity $d^{S_x}_{U(j)}$. Another important factor affecting the decision to repost a news is the propensity of each ignorant individual to repost true news $pt_j \in [0, 1]$, in fact some individuals are more careful in news content evaluation, while others repost news without checking their sources or generate false news deliberately for the reasons mentioned above. Therefore, the reposting is influenced by the news content $d^{S_x}_{news} \in [0, 1]$ and by the propensity of each ignorant individual to repost true news $pt_j \in [0, 1]$. Moreover, the greater the exposure to the news, the greater the propensity to repost it, so also the neighborhood activity $d^{S_x}_{U(j)}$ affects the node decision of share contents. The fraction that is exposed to the news x at time t of node j is:

$$P^{S_x}_j(t) = \frac{\alpha d^{S_x}_{news} + \beta d^{S_x}_{U(j)}(t)}{\alpha + \beta} \tag{2}$$

where α and β are weighted constants.

The fraction of credibility in slice S_x of spreader nodes in neighborhood of j respect to the totality of his neighbors is:

$$d^{S_x}_{U(j)}(t) = \frac{\sum_{i=1}^{N} a^{S_x}_{ij} X^{S_x}_i(t) C_{D_{j \to i}}(t)}{\sum_{i=1}^{N} a^{S_x}_{ij} C_{D_{j \to i}}(t)} \tag{3}$$

where $a^{S_x}_{ij}$ are the elements of adjacency matrix of slice S_x, $A^{S_x}_{ij}$.

$X^{S_x}_i(t)$ is a variable representing the state of node i at time t. Each node become spreader if his personal propensity to repost true news $pt_j \in [0, 1]$, fixed at time t_0, is overcome, hence:

$$X^{S_x}_i(t) = \begin{cases} 1, & P^{S_x}_i(t-1) > pt_i \\ 0, & P^{S_x}_i(t-1) \leq pt_i \end{cases} \tag{4}$$

The node i's direct credibility is the following:

$$C_{D_{j \to i}}(t) = \frac{1}{n_i} \sum_{x=1}^{n_i} P^{S_x}_j(t) \tag{5}$$

where $n_i \in [0, H]$ is number of news spread from node i.

The local credibility of each node i is calculate considering the local reputations that he has in the neighborhood, because only they can see, and hence repost, the contents inserted by node i. The local credibility of node i is an average of local credibility assigned to incoming edges.

$$C_L(i) = \frac{1}{k}\sum_{n=1}^{k} C_{D_{i\to j}}(n) \tag{6}$$

where $n \in [0, H]$.

4 Simulator

To evaluate the proposed model, we have implemented a simulator, the purpose of which is to verify whether a individual's credibility is directly proportional to his ability to repost true news. In this section, we propose a toy example to facilitate the reader in understanding the mechanisms.

Before starting the simulation, we set some constants: the number of nodes in the network (N), the initial (t_0) and final ($t_0 + \Delta T$) instants of the observing time window and $\alpha = \beta = 1$ (we give the same importance to $d^{S_x}_{news}$ and $d^{S_x}_{U(j)}(t)$ in eq. (2)). The simulator randomly assigns to each node a value of pt and, before starting time evolution, generates a scale-free acquaintance network with the request number of nodes, declares empty list of slices and clones all the nodes belonging to the acquaintance network to the credibility network. Moreover the simulator generates two links for each link of acquaintance network with an initial neutral direct credibility of 0.5. At last, the time of observation window starts running and for each time slot are performed a series of steps. At the beginning of each time slot, each node can be arbitrarily turned on or off, in order to simulate individuals who enter and leave the social network. Each active node can generate one new news. In this case, a new slice is generated where the seed node is the only spreader and the other nodes are ignorant by default.

Afterwards, if a node has become a spreader during the previous timeslot (or it is the seed in the first time step), it tries the contagion of the neighbors who have not spread that news yet. In this phase, Eqs. (2) and (3) are performed exploiting status variables and the local credibilities calculated in the previous time step (or the default value in the initial instant).

During contagion cycle all the status variables and the local credibilities are updated using Eqs. (4) and (5). At the end of the observation window, the local credibility of each node is calculated using Eq. (6) to be compared with the attitude of repost true news. The final results allow us to compare the credibilities C_L with initial values of pt. Figures 1, 2a–c show a simple example.

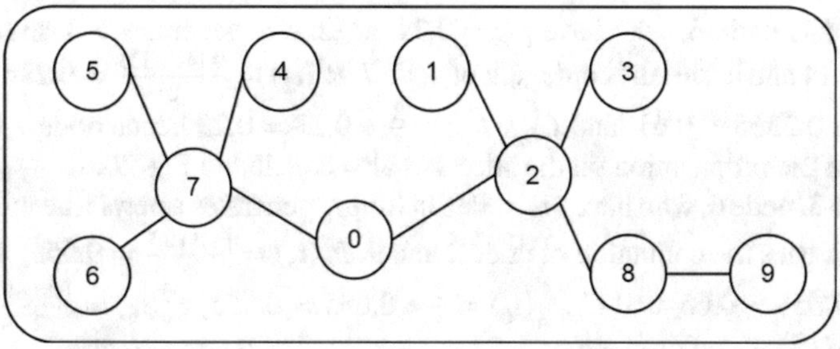

Fig. 1 Acquaintance network

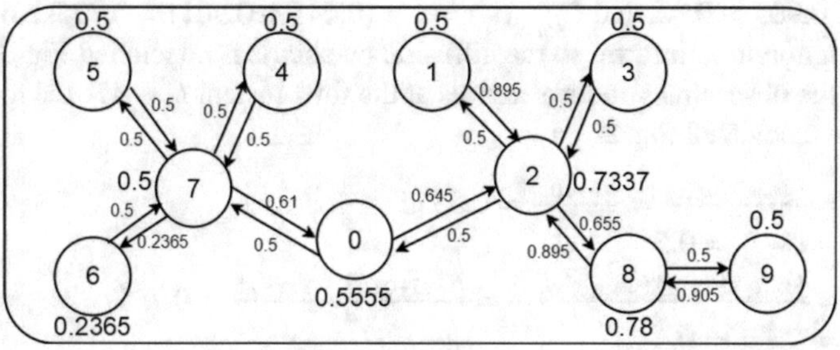

Fig. 2 Credibility network

We considered a network of 10 nodes and 10 time slots. The *pt* randomly assigned are $pt_0 = 0.53$, $pt_1 = 0.65$, $pt_2 = 0.82$, $pt_3 = 0.1$, $pt_4 = 0.6$, $pt_5 = 0.9$, $pt_6 = 0.24$, $pt_7 = 0.61$, $pt_8 = 0.9$ and $pt_9 = 0.92$.

In time slot t_0, node 2 generates a new news. It has a high *pt* value, so it has a high probability of generating true news, in fact d^0_{news} generated by the simulator is 0.79. Slice 0 is created and all the active nodes at time t_0 are assigned to it.

Afterward, node 2 attempts propagation to active neighbors 1, 0 and 8.

Using eq. (3): $d^0_{U(1)} = d^0_{U(8)} = \frac{0.5}{0.5} = 1$ and $d^0_{U(0)} = \frac{0.5}{1} = 0.5$.

Using eq. (2): $P^0_1(t_0) = P^0_8(t_0) = \frac{0.79+1}{2} = 0.895$ and $P^0_0(t_0) = \frac{0.79+0.5}{2} = 0.645$

Using eq. (4): $X^0_1(t_1) = 1$ because $P^0_1(t_0) = 0.895 > pt_1 = 0.65$, $X^0_8(t_1) = 0$ because $P^0_8(t_0) = 0.895 < pt_8 = 0.9$ and $X^0_0(t_1) = 1$ because $P^0_0(t_0) = 0.645 > pt_0 = 0.53$.

Using eq. (5): $C_{D_{1\to 2}}(t_0) = C_{D_{8\to 2}}(t_0) = \frac{1}{1} * 0.895 = 0.895$ and $C_{D_{0\to 2}}(t_0) = \frac{1}{1} * 0.645 = 0.645$.

In time slot t_1, node 1 wants to propagate the news but all neighbors already know it, so it does not propagate it, while node 0 attempts to transmit to node 7. With the same operations of the previous time slot we obtain: $P^0_7(t_1) = \frac{0.79+0.333}{2} = 0.56$ and $X^0_7(t_1) = 0$ because $0.56 < 0.61$, $C_{D_{7\to 0}}(t_1) = \frac{1}{1} * 0.56 = 0.56$, hence node 7 become stifler, and the propagation on the slice 0 is concluded Fig. 2a.

Analogous procedures are performed for slice 1 and 2.

In slice 1, node 6, who have $pt_6 = 0.24$, at time t_2 generates a false news with $d^1_{news} = 0.14$ and it tries the contagion of node 7. $P^1_7(t_2) = \frac{0.14+0.333}{2} = 0.2365, X^1_7(t_3) = 0$ because $0.2365 < 0.61$, and $C_{D_{6\to7}}(t_2) = \frac{1}{1} * 0.23 = 0.23$ hence node 7 become a stifler, and the propagation on the slice 1 is also concluded Fig. 2b.

In slice 2, node 6, who have $pt_8 = 0.9$, at time t_6 generates a news true with $d^2_{news} = 0.81$ and it tries the contagion of node 2 and 9. $P^2_2(t_6) = \frac{0.81+0.5}{2} = 0.655, X^2_2(t_7) = 1$ because $0.655 > 0.65$, and $C_{D_{2\to8}}(t_6) = \frac{1}{1} * 0.655 = 0.655$, $P^2_9(t_6) = \frac{0.81+1}{2} = 0.905$, $X^2_9(t_7) = 0$ because $0.905 < 0.92$, and $C_{D_{9\to8}}(t_6) = \frac{1}{1} * 0.905 = 0.905$.

At time t_7 node 2 tries the contagion of node 0. $P^2_0(t_7) = \frac{0.81+1}{2} = 0.905, X^2_2(t_8) = 1$ because $0.905 > 0.82$, and $C_{D_{0\to2}}(t_7) = \frac{1}{2} * (0.645 + 0.905) = 0.775$. Node 0 have not other ignorant to inform, so the diffusion on slice 2 is concluded Fig. 2c.

When the observing windows arrives at the final instant $t_0 + \Delta T$, the local credibilities are calculated Fig. 2.

$$C_L(0) = \frac{C_{D_{7\to0}} + C_{D_{2\to0}}}{2} = \frac{0.61+0.5}{2} = 0.5555$$
$$C_L(1) = C_{D_{2\to1}} = 0.5$$
$$C_L(2) = \frac{C_{D_{1\to2}} + C_{D_{0\to2}} + C_{D_{8\to2}} + C_{D_{3\to2}}}{4} = \frac{0.895+0.645+0.895+0.5}{4} = 0.7337$$
$$C_L(3) = C_{D_{2\to3}} = 0.5$$
$$C_L(4) = C_{D_{7\to4}} = 0.5$$
$$C_L(5) = C_{D_{7\to5}} = 0.5$$
$$C_L(6) = C_{D_{7\to6}} = 0.2365$$
$$C_L(7) = \frac{C_{D_{6\to7}} + C_{D_{5\to7}} + C_{D_{4\to7}} + C_{D_{0\to7}}}{4} = \frac{0.5+0.5+0.5+0.5}{4} = 0.5$$
$$C_L(8) = \frac{C_{D_{2\to8}} + C_{D_{9\to8}}}{2} = \frac{0.655+0.905}{2} = 0.78$$
$$C_L(9) = C_{D_{8\to9}} = 0.5$$

5 Simulation Results

In a good model, it would be desirable that the individual's propensity to post true news (pt) is as close as possible to the credibility of the corresponding node (C_L). To verify this, five simulations were carried out with 100 nodes and a time window of 100 time slots. We calculated the difference of the two abovementioned values for each node and we counted the corresponding nodes by inserting them in band of 10% of the difference. The nodes that did not share and repost any news during the observation window were excluded from counting (0-post nodes).

As shown in Table 2, in each simulation over 60% of the nodes have the aforementioned difference between credibility and propensity to share and repost true news less than 30%.

Table 2 Simulation results

	Sim 1	Sim 2	Sim 3	Sim 4	Sim 5
No. news	532	476	498	512	458
0-post nodes	2	2	3	4	7
0–10%	26	24	31	25	22
10–20%	25	20	21	24	24
20–30%	14	20	16	17	17
30–40%	16	21	13	14	16
40–50%	12	12	14	9	9
50–60%	5	1	1	6	4
60–70%	0	0	0	1	1
70–80%	0	0	0	0	0
80–90%	0	0	0	0	0
90–100%	0	0	0	0	0

6 Conclusions

In this paper, we proposed a model for the assessment of posts truthfulness in social networks. To this purpose, we created a credibility network starting from an acquaintance network using both the news contents and the fraction of each node neighbor reposts. The latter parameter is, in turn, influenced by credibility acquired in last time slot, hence in our model the history of the network is also taken into account. Some preliminary results have shown that, applying our model, individual with high attitude to repost true news receive high credibility whereas those who post false news receive low credibility, therefore the proposed model agree with real behaviours. In future work, we will validate our model on larger networks and larger time windows in order to apply it on real social network.

References

1. Peysakhovich, H.: News feed FYI: Further reducing clickbait in feed. http://newsroom.fb.com/news/2016/08/news-feed-fyi-further-reducing-clickbait-in-feed/. Accessed Mar 2017
2. Jeyaraman: Fighting spam with botmaker. https://blog.twitter.com/2014/fighting-spam-with-botmaker. Accessed Mar 2017
3. Jin, Z., Cao, J., Jiang, Y.G., Zhang, Y.: News credibility evaluation on microblog with a hierarchical propagation model. In: 2014 IEEE International Conference on Data Mining (ICDM), pp. 230–239. IEEE (2014)
4. Kwon, S., Cha, M.: Modeling bursty temporal pattern of rumors. In: ICWSM. (2014)
5. Kwon, S., Cha, M., Jung, K., Chen, W., Wang, Y.: Prominent features of rumor propagation in online social media. In: 2013 IEEE 13th International Conference on Data Mining (ICDM), pp. 1103–1108. IEEE (2013)

6. Ma, J., Gao, W., Wei, Z., Lu, Y., Wong, K.F.: Detect rumors using time series of social context information on microblogging websites. In: Proceedings of the 24th ACM International on Conference on Information and Knowledge Management, pp. 1751–1754. ACM (2015)
7. Zhao, Z., Resnick, P., Mei, Q.: Enquiring minds: Early detection of rumors in social media from enquiry posts. In: Proceedings of the 24th International Conference on World Wide Web, pp. 1395–1405. ACM (2015)
8. Egele, M., Stringhini, G., Kruegel, C., Vigna, G.: Compa: Detecting compromised accounts on social networks. In: NDSS (2013)
9. Martinez-Romo, J., Araujo, L.: Detecting malicious tweets in trending topics using a statistical analysis of language. Expert Syst. Appl. **40**(8), 2992–3000 (2013)
10. Wang, A.H.: Don't follow me: spam detection in twitter. In: Proceedings of the 2010 International Conference on Security and Cryptography (SECRYPT), pp. 1–10. IEEE (2010)
11. O'Callaghan, D., Harrigan, M., Carthy, J., Cunningham, P.: Identifying discriminating network motifs in youtube spam (2012). arXiv preprint arXiv:1202.5216
12. Carchiolo, V., Longheu, A., Malgeri, M., Mangioni, G.: Users' attachment in trust networks: reputation vs. effort. IJBIC **5**(4), 199–209 (2013)
13. Carchiolo, V., Longheu, A., Malgeri, M., Mangioni, G.: Trust assessment: a personalized, distributed, and secure approach. Concurr. Comput. Pract. Exp. **24**(6), 605–617 (2012)
14. Carchiolo, V., Longheu, A., Malgeri, M., Mangioni, G.: Trusting evaluation by social reputation. In Badica, C., Mangioni, G., Carchiolo, V., Burdescu, D.D. (eds.): IDC II—Proceedings of the 2th IDC, 2008, Catania, Italy, 2008. vol. 162 of Studies in Computational Intelligence, pp. 75–84. Springer (2008)
15. Carchiolo, V., Longheu, A., Malgeri, M., Mangioni, G.: Gain the best reputation in trust networks. In: IDC V—Proceedings of the 5th IDC 2011, Delft, The Netherlands, pp. 213–218 (2011)
16. Criado, R., Flores, J., García del Amo, A., Gómez-Gardeñes, J., Romance, M.: A mathematical model for networks with structures in the mesoscale. Int. J. Comput. Math. **89**(3), 291–309 (2012)

Ontological Hybrid Storage for Security Data

Igor Kotenko, Andrey Chechulin, Elena Doynikova and Andrey Fedorchenko

Abstract The paper investigates different security data sources and analyzes the possibility of their sharing in a uniform data storage on the basis of the ontological approach. An ontological model of the uniform hybrid storage is suggested. A common technique for security data inference based on this approach is developed. The results of experiments with the suggested ontology to construct the security data storage are discussed.

Keywords Ontology · Security data · Data sources · Data analysis · Hybrid storage

1 Introduction

The process of gathering and accumulation of the security related data was started at the end of 1980s. At that date computer technologies started to be penetrated by intruders in different application domains (data processing tools automation, network communications, remote maintenance, etc.). The task of security becomes more important while level of evolution of such domains is growing. For instance,

I. Kotenko (✉) · A. Chechulin · E. Doynikova · A. Fedorchenko
St. Petersburg Institute for Informatics and Automation of the Russian Academy of Sciences, 14-Th Liniya, 39, 199178 St. Petersburg, Russia
e-mail: ivkote@comsec.spb.ru

A. Chechulin
e-mail: chechulin@comsec.spb.ru

E. Doynikova
e-mail: doynikova@comsec.spb.ru

A. Fedorchenko
e-mail: fedorchenko@comsec.spb.ru

I. Kotenko · A. Chechulin · E. Doynikova · A. Fedorchenko
St. Petersburg National Research University of Information Technologies, Mechanics and Optics, ITMO University, 49, Kronverkskiy Prospekt, Saint-Petersburg, Russia

M. Ivanović et al. (eds.), *Intelligent Distributed Computing XI*,
Studies in Computational Intelligence 737, https://doi.org/10.1007/978-3-319-66379-1_15

the number of new detected vulnerabilities of the firmware stays rather big during the last ten years compared to the previous years. Currently, in addition to the vulnerability databases that store specific types of security related data, there are different databases of exploits, platforms, attack patterns and many others.

The global problem with application of this information for state, military, industrial and other security areas consists in a big number of heterogeneous data sources which should be collected and processed. Consequently it prevents to efficient usage of this security information. In its turn it affects the common security level of computer infrastructures. One way to address diversity and absence of links in security information sources is their integration. Investigations and development in this direction is under way for many years, but so far there is no generally accepted solutions. The current status is defined by the many factors: commercial companies and research organizations usually use only proprietary formats to describe the security data; the lack of techniques for linking of disparate security data; inconsistent filing of security data storages, etc. In this paper we analyze several kinds of security information sources and suggest their integration in one storage by applying an ontological approach. We suppose that the resulting ontological storage will allow not only to look through the information on the detected by the security scanners vulnerabilities, attacks etc., but also to get new knowledge on the system using analysis of the current security situation and relationships between information objects in the ontological storage. The novelty of the paper is in a new technique of the ontological inference which is based on the proposed ontology and the hybrid security data storage. This technique will enable us to create an intelligent system for security information and event management. Currently this research is especially relevant due to continuous filling of various security data sources, expansion of the areas of security related applications (for instance, "Internet of Things", social and cyber-physical systems), relations between security related data, etc.

The paper is organized as follows. Section 2 reviews related work on security data integration and analyzes various types of security data sources. These types of sources are used further to construct an integrated security data storage. Section 3 considers the suggested ontological approach for security data integration and the technique for ontological inference. Section 4 discusses the experiments investigating the links between different information sources and case studies for application of the security storage. Section 5 describes the obtained results and the future research plans.

2 Related Work and Security Information Sources

The challenge of security data integration has been investigated for more than 10 years. The most valuable for the community are those security databases that have links with other databases or information objects because it helps security specialists to create an interconnected understanding of the security information.

Vulnerability databases, as databases of a particular type of security data, are used in vulnerability scanners [1], Web application firewalls [2], and also applied in conjunction with other security information to evaluate network infrastructure security by attack modeling [3–5] and risk assessment [6, 7].

The primary task which arises during application of several different vulnerability databases is the integration of their records [8]. The process of vulnerability databases integration is also outlined in [9, 10]. These papers consider the data fields that uniquely define the vulnerabilities, but did not take into account the belonging of vulnerabilities to specific software and hardware components.

Exploit databases are less widely available, but they are more applicable for making real attack actions. One can say that these databases are the most practical among the all types of security information sources. These databases can be used for attack detection and security evaluation and as a part of penetration testing means which can be used not only for research purposes. As well as vulnerability databases these databases should be integrated to maximize the efficiency of their application. The most popular and comprehensive project in this area is Metasploit [11].

The application of the ontological approach for security information integration is not new. For example, in [12–14] the vulnerability-centric ontologies for security analysis are presented. In [15] the common approach to the hybrid ontological storage generation was introduced that we evolve in this paper. In [16] a security metrics ontology for security assessment is suggested. We should also mention the paper [17] which deals with the construction of a common ontology for the SCAP protocol [18]. The SCAP protocol includes the following types of security data: vulnerabilities [19], configurations [20], software and hardware [21], etc. Based on this, we can say that the goal of this protocol is in the integration of security data.

The application of a specific data type from one source in the heterogeneous integrated storage considerably simplifies the process of generation of this integrated storage. In this case there is no need to integrate data of one type. In its turn it is connected with lack of alternatives for some common data sources. Also it is determined by the completeness of data in one source. But when a few sources of data of the same type (for example, different vulnerability databases) are used, there is a challenge of data inconsistency. In this case a preprocessing is needed to transform data on the same objects from different sources to one format.

Currently a multitude of data, characterizing different security aspects, is used for security monitoring. This data includes: vulnerabilities; weak places; (vulnerable) configurations; exploits; platforms; attack patterns; remediation; malware; black and white lists; software and hardware updates; network traffic; security events and many others. This paper analyzes and assesses the ability to integrate the first seven aforementioned data types used in different databases.

Vulnerability databases are one of the oldest sources of security information. To date more than 100,000 vulnerabilities for more than 60,000 hardware and software products have been found. The basis of almost any vulnerability description format consists of the next elements: an identifier of the source database; explicit and implicit identifiers on external databases in relation to the source database; the list or configuration of vulnerable software and hardware; textual description; vulnerability description using own source scoring system or CVSS [28]; references on an additional description, some of these references can be implicit identifiers. Vulnerability records are presented in the following sources: "Common Vulnerabilities and Exposures" (CVE) [22], "National Vulnerabilities Database" (NVD) [23], "Open Source Vulnerabilities Data Base" (OSVDB) [24], "Vulnerability Notes Database" (VND) [25], SecurityFocus project with BugTraq [26], IMB X-Force [27].

Dictionaries of software and hardware (platforms) products are important security data. They enable identifying potentially vulnerable objects. At the moment there is only one open, standardized and accessible dictionary—it is a Common Platform enumeration (CPE) standard developed by MITRE. The format of the product dictionary of the Common Vulnerability Reporting Framework (CVRF) is a successor of the CPE format. The main peculiarity of this format is a hierarchical structure of product names which provides a more understandable representation and the ability to uniquely identify records. The product records are stored in the following dictionaries: "Common Platform enumeration" (CPE) [21], "Common Vulnerability Reporting Framework" (CVRF) [30].

Exploit databases contain descriptions of a software, files, requests, or a sequence of commands that takes advantage of a vulnerability in order to cause unintended or unanticipated behavior. Exploits are often used during the penetration testing, malicious attack performing and malware activity. But they also can be used in attack recognition process—if a security system detects the presence of an exploit or its traces, it usually generates alarms and starts the actions aimed on attack prevention. The main source with information about exploits is "Exploit DataBase" (EDB) [31].

Attacks patterns are also very important security data aimed for monitoring and protection of distributed networks. The basis of the attack pattern specification consists of description of attack implementation methods, attack steps and attack step techniques, as well as fields that refer on the exploited vulnerabilities and weaknesses. This information can also be obtained in other formats from the various sources. For example, it can be attack patterns from specific intrusion detection systems. This kind of security information can be found in the "Common Attack Pattern Enumeration and Classification" (CAPEC) database [32].

Weaknesses of the software and hardware are represented as vulnerabilities classification. Thus, the presence of weakness indicates a potential vulnerability, and the presence of a vulnerability is the direct evidence of a weakness. Information about weaknesses available in the "Common Weakness Enumeration" (CWE) database [29].

Remediation databases are valuable sources of security data. One of the formats to represent countermeasures is "Common remediation enumeration" (CRE) [33].

Configuration databases contain descriptions of recommended secure settings for specific software platforms. Usually these settings are defined by software developers on the basis of their experience and best practices. But currently application of this opportunity is limited due absence of the CRE database. The main source of information about configurations is the "Common Configuration Enumeration" (CCE) database [20].

3 Ontological Security Data Storage and Inference Technique

As it was shown there is a multitude of standards and databases that describe different security aspects. In the same time the main disadvantage of existing databases consists in impossibility to form the common picture, because of disunity of these databases. Thus, despite the fact that entries of some databases refer on entries of other databases (for example attack patterns refer on the description of exploited vulnerabilities). It should be noticed, that these references represent the top level of relations and can be extended with additional data.

One of the solutions to represent the interconnected data to process the data of complex structure is an ontological approach. It allows to express complex relations between entities using description logic. The approach consists in definition of the set of concepts in the selected subject area. Then the connections between the concepts are generated considering their relations and interaction. Though the papers listed in the previous section suggest the ontologies for the security related data, there is no ontological storage that incorporates data from different databases considering the nature of the relationships for these data.

The paper presents the hybrid structure of interconnections between such information objects as vulnerabilities, software, software weaknesses, exploits, attack patterns, software and hardware configurations and remediations. To identify these interconnections we reviewed main open databases and outlined relations between them. The common representation of these relations is represented in Fig. 1. Relations between the presented databases are defined by fields with

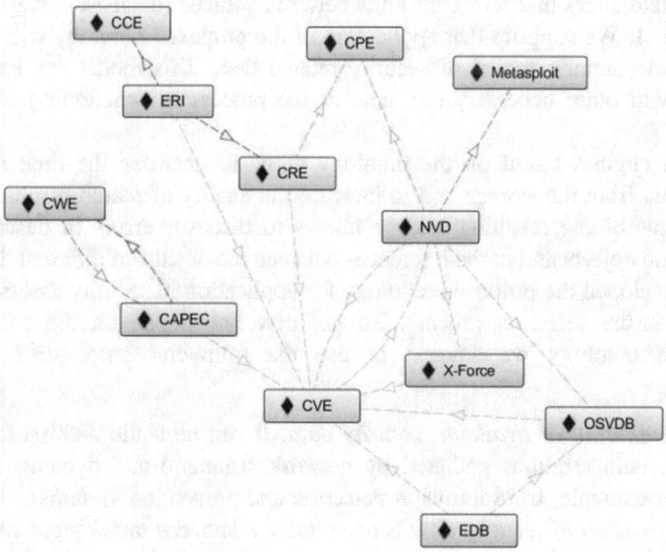

Fig. 1 Relations between databases

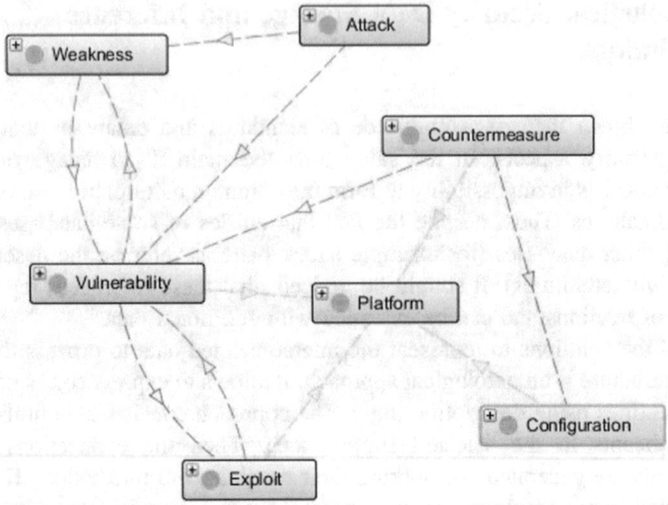

Fig. 2 Relations between security data

identifiers of internal and external databases. It should be noted that currently some databases are not available in the open access though appropriate data formats exist. For example, remediation database (ERI and CRE formats) should be additionally generated. Results of analysis of relations between the databases allowed to generate ontological representation of the security related data and relations between them manually using Protégé 5.1.0. The ontological hybrid storage is filled in semi automatic mode. The top level representation of this result is in Fig. 2. This figure contains references that represent links between sources (databases) that presented in the Fig. 1. We suppose that application of the proposed ontology will permit to construct a common model of security related data. This model can be complemented with other necessary concepts in the process of functioning of security systems.

Search engines based on the ontology allow to decrease the time needed to extract data from the storage and to increase the quality of search results. Besides, the analysis of the resulting storage allows to discover errors in description of information objects and in the references between the objects of different data types.

We developed the proposed ontology for application in security assessment and countermeasure selection process. To get new knowledge on the basis of the developed ontology we suggest to use the following *ontological inference technique*.

Step 1. Gathering of available security data. It can be static data (software and hardware, vulnerabilities gathered by network scanners) and dynamic data (obtained, for example, from intrusion detection and prevention systems).
Step 2. Extension of security data using relations between ontological concepts. It is performed bypassing security data instances using links. We outline "strong" and

Table 1 Types of the relations between ontological concepts

	Platform	Vulnerability	Attack	Weakness	Exploit	Configuration
Platform		Strong				
Vulnerability	Strong			Strong	Strong	
Attack		Weak		Weak		
Weakness		Weak	Strong			
Configuration	Weak					
Countermeasure	Weak	Weak				Weak
Exploit	Weak	Strong				

"weak" relations. Link is "strong" if the ancestor concept instance uniquely indicates existence of the descendant instance. Link is "weak" if the ancestor concept instance does not guarantee existence of the descendant concept instance. If link is "strong" (Table 1, the leftmost column contains ancestor concepts, and the top row contains descendant concepts) we proceed bypassing and put the appropriate concept instance to the output dataset; if link is "weak" (Table 1) we stop bypassing by this link and put the appropriate concept instance to the dataset for additional analysis.

Step 3. Data analysis. On this stage we analyze output dataset (obtained via "strong" links) and dataset for the additional analysis (obtained via "weak" links). Processing of the dataset for additional analysis includes analysis of particular fields of the ontological concept instances. For example, the link between a software in the CPE format and vulnerabilities in the CVE format is "strong". It means that existence of the software instance in the system uniquely indicates existence of the vulnerabilities of this software. On the other hand, the link between a CAPEC attack pattern and vulnerabilities in the CVE format is "weak". It means that detection of the appropriate attack pattern does not determine existence of the linked vulnerabilities. But we can additionally analyze CAPEC attack pattern fields related to vulnerable software and make more accurate conclusions.

4 Experiments and Discussion

Currently we implemented the integrated vulnerability storage using PostgreSQL and Java [8] and are developing a hybrid security storage that incorporate data of different types on the basis of the proposed ontology using Virtuoso Server from OpenLink. Let us describe two stages of the conducted experiments with the hybrid security storage. The goal of the *first stage* consisted in analysis of existing security related databases to approve the opportunity to extend security knowledge using relationships between the concepts. The *second stage* was devoted to the specific case studies of derivation of new knowledge using relations between information objects.

On the first stage to construct the hybrid security storage we obtained statistical characteristics of links between databases (Table 2). Table 2 contains the next characteristics: (1) name of the relevant database (B2) which is referenced in the analyzed (targeted) database (B1); (2) total number of references to B2 in B1; (3) number of B1 entries that have references to B2; (4) number of references to unique B2 elements in B1; (5) percentage of B1 elements that have references to B2; (6) percentage of unique references to B2 in B1 (relation of value in column (2) to value in column (4)); (7) average number of references to B2 from one B1 element; (8) exploitability of B2 database in B1 database (i.e. relation of number of references to B2 elements in B1 to the total number of elements in B2). Note that reference from the analyzed database to itself in the table (NVD-CVE, CAPEC, EXPLOIT-DB) specifies the total number of entries in the appropriate database.

We used the last versions of the security databases (01.04.17) to analyze target and relevant data sources (01.04.17) except exploits database (Jan.15). Low level of the exploitation (column 8) was observed only for the CVE database from the CAPEC database. This is related with the fact that vulnerabilities in CAPEC database are provided only as examples of vulnerabilities that can be used for the attack implementation. In its turn extra-exploitability of references to CPE from NVD is related with: (1) significant extension of the unique entries of the CPE dictionary through the NVD; (2) features of the CPE specification that lead to the impossibility of the unique identification of the vulnerable products in some cases. Also the high level of connectivity (column 5) should be noted for: (1) NVD(CVE) database with CPE dictionary and CWE database, (2) CAPEC database with CWE database, and (3) ExploitDB database with NVD(CVE) database. Obtained results of analysis of the relations between the data sources approve possibility of extension of the security knowledge using logical inference on the basis of the developed ontology.

On the second stage we demonstrated the usage of the developed ontological hybrid storage on two case studies.

Example 1 This example shows how the storage can be used for automated detection of network weak places and selection of countermeasures (Fig. 3). The relations between CPE, NVD, CWE and CAPEC are used. he security monitoring

Table 2 Statistical characteristics of links between data sources

(1)	(2)	(3)	(4)	(5)	(6)	(7)	(8)
NVD (84,557)							
EXPLOITS	2752	2599	2409	3,1%	87,5%	0,03	7,21%
CPE	2,048,178	82,740	199,160	97,9%	9,7%	24,22	169,97%
CWE	51,131	50,519	88	59,7%	0,2%	0,60	12,45%
CAPEC (528)							
CVE	69	36	57	6,8%	82,6%	0,13	0,07%
CWE	964	234	241	44,3%	25,0%	1,83	34,09%
EXPLOIT-DB (33,394)							
CVE	33,993	33,993	16,879	100,0%	49,7%	0,99	19,96%

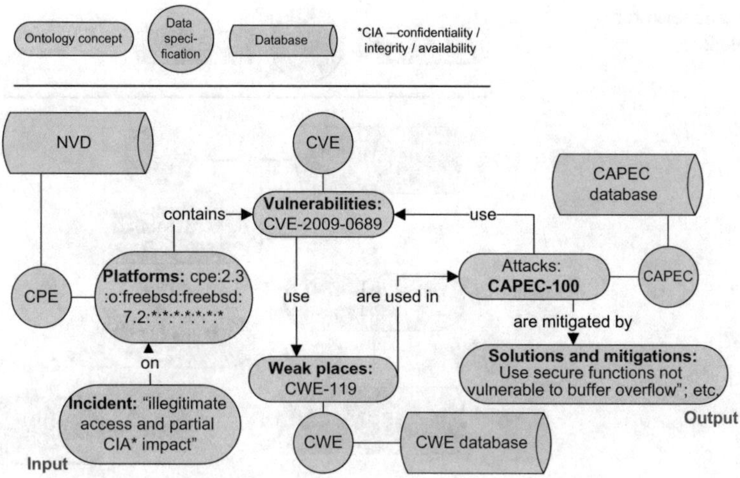

Fig. 3 Inference for example 1

system reports on a security incident. Information on this incident includes information on the compromised host and the level of its compromise (it can be illegitimate privileges and/or confidentiality, integrity or availability violation). Using this information it is possible to define host vulnerabilities in the CVE format that lead to the fixed attack consequences. In this example we detected illegitimate access to the web server and partial confidentiality, integrity and availability violation on it.

On the basis of the analysis of hardware and software on the host, it was defined that exploitation of the vulnerability CVE-2009-0689 (large precision value in a format string triggers overflow) in FreeBSD 7.2 software (cpe:2.3:o:freebsd: freebsd:7.2:*:*:*:*:*:*:*) leads to these consequences. In NVD [23] this vulnerability is connected with weakness from CWE database: CWE-119 "Improper Restriction of Operations within the Bounds of a Memory Buffer". In its turn the weakness can be connected with the possible attack patterns: CAPEC-100 "Overflow Buffers", and others (CAPECs 8, 9, 10, 14, 24, 42, 44, 45, 46 and 47).

All these attack patterns are children of CAPEC-100 "Overflow Buffers". To avoid the attacks of this type in the future the solutions that are proposed for attacks of this type in CAPEC database ("Solutions and Mitigations" field) can be implemented, namely: "Use a language or compiler that performs automatic bounds checking"; "Use secure functions not vulnerable to buffer overflow"; "If you have to use dangerous functions, make sure that you do boundary checking"; etc. It should be noticed that this approach is not intended to eliminate the specific vulnerability or to respond to the specific attack, but to avoid the problems of this type in the future.

Example 2 This example demonstrates ontological inference for identification of weak places in case of exploit detection (Fig. 4). On the host with Apple Mac OS X 10.12.1 operation system an execution of the exploit "macOS HelpViewer 10.12.1 -

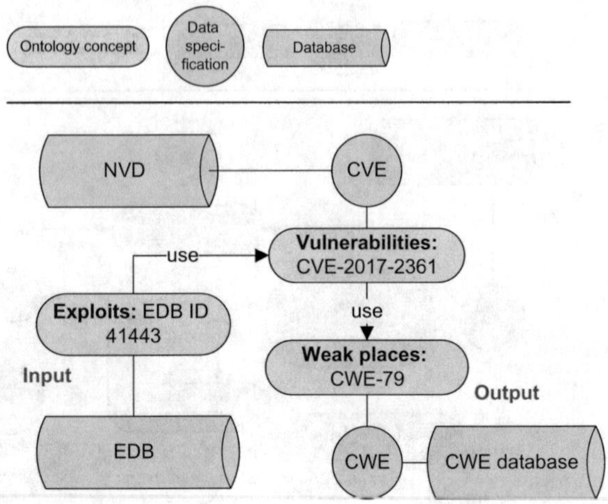

Fig. 4 Inference for Example 2

XSS Leads to Arbitrary File Execution and Arbitrary File Read" was detected. This exploit has identifier "41,443" in the EDB database.

On the basis of connection between EDB database and vulnerability database we can define the exploited vulnerability—CVE-2017-2361. Using connection between the vulnerability database and the weaknesses (CWE) database it could be defined the exploited weakness—CWE-79 "Cross-Site Scripting (XSS)". This leads to the conclusion about the definite measures to prevent attacks of this type on the host.

Consideration of additional interconnections on the low level allows to increase an accuracy of conclusions. For example, some CAPEC attack patterns refer on the observed vulnerability instances from the CVE database. Consideration of this information helps to connect the CVE instance with the CAPEC attack pattern instance in Example 1 more accurate. But it requires an additional analysis when this field is filled not for all attack patterns and contains not all possible vulnerabilities.

The conducted experiments and description of the case studies show existence of opportunity to extend security knowledge through construction and using hybrid storage on the basis of the ontological approach. To generate connections between the ontological concepts we used explicit references between different security databases. Besides references on each other, the used databases have implicit connections on the basis of the fields of the same type. For example CWE standard has the field "Applicable Platforms" for the weaknesses in the CWE database, the CVE vulnerabilities in NVD reference on the platforms in the CPE format, and the attack patterns in the CAPEC database have the field "Technical Context". Comparative analysis of content of these fields allows to define more accurate connection between vulnerabilities and attack patterns. Another field that requires additional analysis is related to the attack consequences. Namely, NVD contains values of the CVSS "Attack Impact" index for vulnerabilities. This index specifies consequences of vulnerability exploitations. In its turn, the CWE database contains the "Common Consequences" field for weaknesses that defines consequences of weakness exploitations. Finally, the attack patterns in the CAPEC database have the "CIA

Impact" field. Comparative analysis of content of these fields makes it possible to define a more accurate connection between vulnerabilities and attack patterns.

To form the proposed hybrid ontological storage, two things are needed: the ontological data model (described in Sect. 3) and access to the sources of security information. It should be noticed, that because of its structure the resulted storage can be updated both by adding new information as well as by modifying old records.

Experimental results shown that application of the ontological approach for the security situation analysis is important. One of its advantages consists in opportunity to generate a common (not overloaded) data model. This model should be extended for each specific application area. One of the significant advantages of the ontological approach compared with relational approach is low resource intensity of the meta scheme modifications. The disadvantage consists in dependency of quality of the ontological data representation from the quality of the input data. Thus existing security databases frequently contain errors and inaccuracies, and extraction of the interconnections between the objects is obstructed by the absence of the unified reference format, especially for the databases from the different producers.

5 Conclusion

The paper proposes an ontology based approach for creating the integrated security storage. Existing standards for representation of security related data and sources of these data were reviewed. Their characteristics and interconnections between them were analyzed. On the basis of these interconnections we developed and implemented (using Protégé 5.1.0) the ontology for generation of the integrated security data storage. On the basis of the suggested ontology and the hybrid security data storage we developed the technique of ontological inference to get new security knowledge. We assume that this technique will allow us in the future to create an intelligent system for security monitoring. We conducted the experiments on the interconnections between data instances in the different security sources to get statistics on the number of existing links and to prove the possibility to fill the suggested storage. Examples of application of the developed ontology and technique for derivation of new security related knowledge were demonstrated and discussed. The further research will be connected with the development of the technique for detection and correction of errors in the source databases. Also we intend to extend the suggested ontological model with security events to increase the quality of security assessment and countermeasure selection.

Acknowledgements The work is performed by the grant of RSF #15-11-30029 in SPIIRAS.

References

1. OPENVAS. Web. http://www.openvas.org/
2. PT Application Firewall. Web. https://www.ptsecurity.com/ww-en/products/af/
3. Kotenko, I., Chechulin, A.: A cyber attack modeling and impact assessment framework. In: 5th International Conference on Cyber Conflict 2013 (CyCon 2013), pp. 119–142 (2013)
4. Kotenko, I., Chechulin, A.: Computer attack modeling and security evaluation based on attack graphs. In: 7th International Conference on Intelligent Data Acquisition and Advanced Computing Systems: Technology and Applications, pp. 614–619 (2013)
5. Chechulin, A., Kotenko, I.: Attack tree-based approach for real-time security event processing. Automatic Control Comput. Sci. **49**(8), 701–704 (2015). Allerton Press Inc
6. Kotenko, I., Doynikova, E.: Dynamical calculation of security metrics for countermeasure selection in computer networks. In: 24th Euromicro International Conference on Parallel, Distributed and Network-Based Processing, pp. 558–565 (2016)
7. Doynikova, E., Kotenko, I.: Countermeasure selection based on the attack and service dependency graphs for security incident management. In: Lecture Notes in Computer Science (LNCS), vol. 9572, Springer, pp. 107–124 (2016)
8. Fedorchenko, A., Kotenko, I., Chechulin, A.: Design of integrated vulnerabilities database for computer networks security analysis. In: 23th Euromicro International Conference on Parallel, Distributed, and Network-Based Processing (PDP 2015), pp. 559–566 (2015)
9. Sufatrio, Yap, R.H.C., Zhong, L.: A machine-oriented integrated vulnerability database for automated vulnerability detection and processing. In: LISA XVIII, pp. 47–58 (2004)
10. Tierney, S.: Knowledge discovery in cyber vulnerability databases. A project report submitted in partial fulfillment of the requirements for the degree of Master of Science (2005)
11. Metasploit official website. Web. https://www.metasploit.com/
12. Elahi, G., Yu, E., Zannone, N.: A modeling ontology for integrating vulnerabilities into security requirements conceptual foundations. In: ER-2009, pp. 99–114. Springer-Verlag (2009)
13. Guo, M., Wang, J.: An ontology-based approach to model common vulnerabilities and exposures in information security. In: 2009 ASEE SE Section Conference, 10 p. (2009)
14. Guo, M., Wang, J.: Security data mining in an Ontology for vulnerability management. In: Conference on Bioinformatics, Systems Biology and Intelligent Computing, pp. 597–603 (2009)
15. Kotenko, I., Saenko, I., Polubelova, O., Chechulin, A.: Design and implementation of a hybrid ontological-relational data repository for SIEM systems. Future Int. **5**(3) (2013)
16. Kotenko, I., Saenko, I., Polubelova, O., Doynikova, E.: The ontology of metrics for security evaluation and decision support in SIEM systems. In: 8th International Conference on Availability, Reliability and Security (ARES 2013), pp. 638–645 (2013)
17. Parmelee, M.C.: Toward an ontology architecture for cyber-security standards. In: 2010 Semantic Technology for Intelligence, Defense, and Security Conference, 8 p. (2010)
18. Waltermire, D., Quinn, S., Scarfone, K., Halbardier, A.: The technical specification for the security content automation protocol (SCAP): SCAP version 1.2. 66 p. (2011)
19. Common Vulnerabilities and Exposures (CVE). Web. http://cve.mitre.org
20. Common Configuration Enumeration (CCE). Web. https://nvd.nist.gov/cce/index.cfm
21. Common Platform Enumeration (CPE) official website. Web. https://nvd.nist.gov/cpe.cfm
22. Common Vulnerabilities and Exposures (CVE). Web. https://cve.mitre.org/
23. National Vulnerability Database (NVD) official website. Web. https://nvd.nist.gov
24. Open Source Vulnerability Database (OSVDB) blog. Web. https://blog.osvdb.org/
25. US Computer Emergency Readiness Team (US-CERT). Web. http://www.us-cert.gov/
26. SecurityFocus (BugTraq database) official website. Web. http://securityfocus.com/
27. IBM X-Force exchange project official website. Web. http://xforce.iss.net
28. Common Vulnerability Scoring System (CVSS) official website. Web. https://www.first.org/cvss

29. Common Weakness Enumeration (CWE) official website. Web. https://cwe.mitre.org/
30. ICASI Common Vulnerability Reporting Framework (CVRF) official website. Web. http://www.icasi.org/cvrf/
31. Offensive security's exploit database archive. Web. https://www.exploit-db.com/
32. Common Attack Pattern Enumeration and Classification (CAPEC) official website. Web. https://capec.mitre.org/
33. Common Remediation Enumeration (CRE) official website. Web. https://scap.nist.gov/specifications/cre/

Part V
Machine Learning

Wind Power Production Forecasting Using Ant Colony Optimization and Extreme Learning Machines

Maria Carrillo, Javier Del Ser, Miren Nekane Bilbao, Cristina Perfecto and David Camacho

Abstract Nowadays the energy generation strategy of almost every nation around the world relies on a strong contribution from renewable energy sources. In certain countries the relevance taken by wind energy is particularly high within its national production share, mainly due to its large-scale wind flow patterns. This noted potentiality of wind energy has so far attracted public and private funds to support the development of advanced wind energy technologies. However, the proliferation of wind farms makes it challenging to achieve a proper electricity balance of the grid, a problem that becomes further involved due to the fluctuations of wind generation that occur at different time scales. Therefore, acquiring a predictive insight on the variability of this renewable energy source becomes essential in order to optimally inject the produced wind energy into the electricity grid. To this end the present work elaborates on a hybrid predictive model for wind power production forecasting based on meteorological data collected at different locations over the area where a wind farm is located. The proposed method hybridizes Extreme Learning Machines with a feature selection wrapper that models the discovery of the optimum subset of predictors as a metric-based search for the optimum path through a solution graph efficiently tackled via Ant Colony Optimization. Results obtained by our approach for two real wind farms in Zamora and Galicia (Spain) are presented and discussed, from which we conclude that the proposed hybrid model is able to efficiently reduce the number of input features and enhance the overall model performance.

M. Carrillo · J. Del Ser (✉) · M. Nekane Bilbao · C. Perfecto
University of the Basque Country UPV/EHU,48013 Bilbao, Spain
e-mail: javier.delser@ehu.eus

M. Carrillo
e-mail: mcarrillo007@ikasle.ehu.eus

J. Del Ser
TECNALIA, 48160 Derio, Spain

J. Del Ser
Basque Center for Applied Mathematics (BCAM), 48009 Bilbao, Spain

D. Camacho
Universidad Autónoma de Madrid, 28049 Madrid, Spain
e-mail: david.camacho@uam.es

M. Ivanović et al. (eds.), *Intelligent Distributed Computing XI*,
Studies in Computational Intelligence 737, https://doi.org/10.1007/978-3-319-66379-1_16

Keywords Wind production · Supervised learning · Feature selection · Ant Colony Optimization · Extreme Learning Machines

1 Introduction

The pace at which the energy grid is becoming fully digital has accelerated in the last few years as a result of the advent and massive deployment of ICT-powered infrastructure over this large-scale network [1]. Such a progressive digitalization of the grid finds its roots not only in the need for a more fine-grained supervision of the energy delivery process along its lines, but also in the technical advantages of ICT technologies allowing for bidirectional information flows from the grid to the operator, supervisor or customer, e.g. demand side management, fraud detection or improved energy efficiency in buildings [2].

A particular byproduct of the aforementioned digitalization is the fact that technicians supervising and managing different levels of the grid are provided with a rich data substrate from which to infer valuable, timely insights on the current status and operation of the grid. Examples abound in this matter: following the above cases, by properly sampling and transmitting information about the energy consumption of end customers—by means of smart meters—operators can trigger actions to match the overall generation to consumption along time or detect abnormal patterns in the consumed energy traces that could be symptomatic of a non technical loss (e.g. tampering). In the context of renewable energy sources profitable benefits also arise from the digitalization of the equipment required to collect and convey the captured energy flow, with a strong emphasis on the crucial role taken by this technology when the focus is placed on the maximization of the installation productivity or the pattern characterization of the produced energy towards its injection upstream.

The wind energy sector has particularly leveraged the plethora of technical advantages and possibilities unchained by the digitalization of equipments over the grid [3]. A significant share of such advantages rely on the application of predictive models for the estimation of the produced energy within a certain time horizon. Such a prediction can be inferred not only from the past produced energy up to the time when the prediction is made, but also from other parameters that impact on the wind flow patterns of the geographical area where the wind is located and, ultimately, on its predicted generation. This close link between wind dynamics and the energy produced by a wind farm has hitherto steered research efforts towards the prediction of wind-related physical characteristics, on the assumption that the generated power can be estimated therefrom.

From the technical point of view a real plethora of predictive models have been applied to this problem, from naïve Machine Learning approaches such as Neural Networks [4–6], Support Vector Machines [7, 8] or Decision Tree Regressors [9] to more elaborated schemes such as model ensembles [10] or Deep Learners [11, 12]. Among them a research grain that has lately gained momentum focuses on the hybridization of nature-inspired heuristics and machine learning models as a computationally efficient workaround to deal with the usually high dimensionality of datasets processed in this application scenario. To cite a few, evolutionary solvers and nature-inspired heuristics have been often utilized as efficient wrappers to configure the underlying predictive model [13, 14] and/or select a subset of features [15–17] under a maximal generalization performance criterion. Comprehensive surveys can be found in [18–21].

This work joins the latter research trend by exploring the practical performance of a hybrid wind power generation forecasting model based on Extreme Learning Machines (ELM), a low-complexity variant of neural networks characterized by a fast training process [22]. The novel ingredient with respect to the state of the art in this topic is the conception of the feature selection process as a search for the optimal path through a graph, which is algorithmically tackled by Ant Colony Optimization (ACO, [23]), a bioinspired solver that has been applied in many graph-related problems such as scheduling [24, 25] and network analysis [26]. If the input features to the ELM model are conceived as nodes of a fully connected graph, a colony of ants can be used to find a *good* path through this feature space efficiently by virtue of the collaborative behavior of this multi-agent solver. Results obtained for two different wind farms located in Spain will be discussed, from where we will conclude that the proposed hybrid scheme excels at constructing datasets of reduced dimensionality and improved generalization performance.

The rest of the manuscript is structured as follows: the notation used throughout the paper and a formal statement of the feature selection problem on which this research gravitates is given in Sect. 2, whereas Sect. 3 delves into the proposed hybrid model, stressing on how ELM and ACO are combined in a single algorithmic flow. Results are presented and discussed in Sect. 4, and finally Sect. 5 concludes the paper and outlines future lines of related research.

2 Notation and Problem Formulation

Following the schematic diagram in Fig. 1 we assume a wind farm comprising M wind turbines, producing a total instantaneous power P_t [W] at time t which we aim to predict at time $t - \Delta_t$, with Δ_t denoting the prediction horizon. We further consider that V meteorological variables of interest (e.g. wind speed modulus, wind direction, U/V components and temperature) for the target variable are obtained over the geographical location where wind turbines are located by a numerical weather prediction (NWP) model. Let $\mathbf{X}_t^{\Diamond,\mathbf{p}}$ denote the vector of meteorological variables obtained for position $\mathbf{p} \in \{\mathbf{p}_1, \ldots, \mathbf{p}_P\}$, where $\mathbf{p}_i \in \mathbb{R}^2$ denotes the geographical coordinates (latitude/longitude) of point i and $|\mathbf{X}_t^{\Diamond,\mathbf{p}}| = V$. Therefore, the entire set of meteorological features registered over all P locations at time t will be denoted as

$$\mathbf{X}_t^{\Diamond} \doteq \{\mathbf{X}_t^{\Diamond,\mathbf{p}_1}, \mathbf{X}_t^{\Diamond,\mathbf{p}_2}, \ldots, \mathbf{X}_t^{\Diamond,\mathbf{p}_P}\}, \tag{1}$$

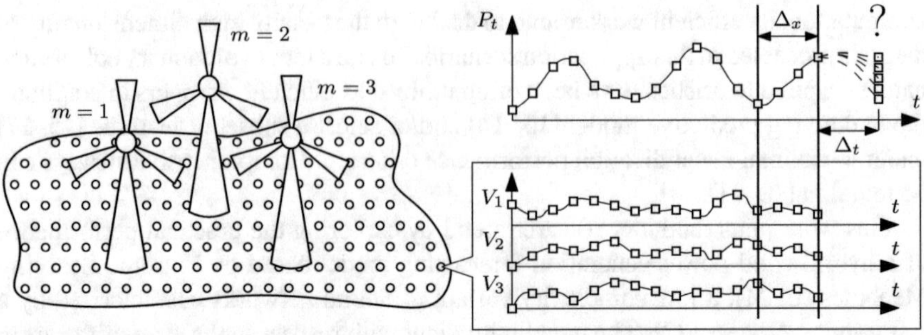

Fig. 1 Diagram of the system model addressed in this work with $M = 3$ turbines, $V = 3$ meteorological variables and $\Delta_t = \Delta_X = 3$ time steps

comprising a total of PV variables. In order to properly capture short-term time correlations that could prevail beneath the data we will extend the above partial feature vector $\mathbf{X}_t^{\diamondsuit}$ with a Δ_X-sized window of both produced power values and meteorological variables before the instant at hand, namely,

$$\mathbf{X}_t \doteq \{\mathbf{X}_t^{\diamondsuit}, \mathbf{X}_{t-1}^{\diamondsuit}, \ldots, \mathbf{X}_{t-\Delta_X}^{\diamondsuit}, P_t, P_{t-1}, \ldots, P_{t-\Delta_X}\}. \quad (2)$$

which is used as an input to a predictive model $M_{\theta}(\cdot)$ controlled by parameters θ used to predict $P_{t+\Delta_t}$ as $P_{t+\Delta_t} = M_{\theta}(\mathbf{X}_t)$. To this end the model is trained over a set of supervised training examples $\{(\mathbf{X}_t, P_{t+\Delta_t})\}_{t\in\mathcal{T}_{trn}}$ and evaluated in terms of generalization performance over a test set $\{\mathbf{X}_t\}_{t\in\mathcal{T}_{tst}}$. Here, $\mathcal{T}_{trn}$ and $\mathcal{T}_{tst}$ denote the time instants corresponding to supervised training instances and unsupervised test samples, respectively.

Following the same rationale as in prior work on feature selection in wind prediction, by filtering out irrelevant or redundant features not only the model learning process is less time consuming, but also the model itself becomes less prone to overfitting (lower variance) as per a more compact learned knowledge. This is specially important when dealing with supervised learning problems with a relatively high number of features. In this context, wind power prediction problems based on multi-parametric, multi-site meteorological variables with an additional window to account for autoregressive components undoubtedly calls for the adoption of feature selection schemes: the total number of features contained in $\mathbf{X}_t$ is $|\mathbf{X}_t| = P(V+1)\Delta_x$ which for a minimal setup of e.g. $P = 20$ points, $V = 8$ variables, and a window of $\Delta_x = 3$ time instants, amounts up to $|\mathbf{X}_t| = 540$ input features.

Having said this, feature selection can be formulated as an optimization problem guided by a fitness metric that reflects the predictive performance of the model when processing any given subset of features $\mathcal{X}' \subseteq \mathcal{X}$, with $\mathcal{X}$ denoting the set of all variables included in the feature vector $\mathbf{X}_t$ as per (2). Such a metric should quantify the generalization performance of the model when processing unseen data instances. To this end, cross-validation (CV) methods help estimating how the predictive model will generalize to an independent test set. In its simplest form (k-fold CV), the training set is divided into k disjoint subsets, and the estimation of the generalization of the model is achieved by averaging the partial performance scores attained by the model when trained over $k-1$ folds and tested with the remaining one.

Mathematically, if we denote as $\mathcal{T}_{trn}^{k}$ the time instants corresponding to the k-th fold in which the training set is split (i.e. $\mathcal{T}_{trn}^{1} \cup \mathcal{T}_{trn}^{2} \cup \ldots \cup \mathcal{T}_{trn}^{K} = \mathcal{T}_{trn}$), a model can be learned from $\{(\mathbf{X}_t, P_{t+\Delta_t})\}_{t\in\mathcal{T}_{trn}-\mathcal{T}_{trn}^{k}}$, whose output $M_\theta(\{\mathbf{X}_t\}_{t\in\mathcal{T}_{trn}^{k}})$ can be compared to $\{P_{t+\Delta_t}\}_{t\in\mathcal{T}_{trn}^{k}}$ to yield a score $\varphi(k) \in \mathbb{R}^{+}$ associated to fold k. By averaging such scores over k an estimation of the expected performance $\widehat{\varphi}$ of the model when facing the test dataset $\{\mathbf{X}_t\}_{t\in\mathcal{T}_{tst}}$ can be obtained. It is important to note that $\widehat{\varphi}$ depends on the candidate feature subset $\mathcal{X}'$, as it determines the dimensions of the datasets involved in the model construction. For coherence through subsequent formulation we will explicitly indicate this dependence in the performance scores $\varphi(k)$ and $\widehat{\varphi}$ as $\varphi(k, \mathcal{X})$ and $\widehat{\varphi}(\mathcal{X})$, respectively.

With this notation in mind, the optimization problem considered in this manuscript aims at finding an optimal feature subset such that the expected generalization performance of a model M_θ is maximized, i.e.

$$\underset{\mathcal{X}'\subseteq\mathcal{X}}{\text{Maximize}}\ \widehat{\varphi}(\mathcal{X}') \doteq \frac{1}{K}\sum_{k=1}^{K}\varphi(k, \mathcal{X}'), \tag{3}$$

where the search is done over the space of all possible combinations of features. Evaluating exhaustively all such possibilities for $\mathcal{X}'$ would require a total of $\sum_{x=1}^{|\mathcal{X}|}\binom{|\mathcal{X}|}{x}$ cross-validation processes with the model $M_\theta(\cdot)$ and dataset at hand. The exponential complexity of this search space motivates the adoption of heuristic wrappers capable of exploring it efficiently. This is indeed the rationale of the hybrid ACO-ELM model explained in the following section.

3 Proposed Hybrid Scheme

As described in Algorithm 1, the proposed predictive model leverages the low complexity featured by ELMs and the efficient search over graphs provided by ACO. The main idea is to model the search space as a fully connected graph $\mathcal{G} = \{\mathcal{U}, \mathcal{V}\}$, where each node $u \in \mathcal{U}$ represents a feature in $\mathcal{X}$.

When sent through this graph, ants construct the subset of features proposed as a candidate solution by starting at a randomly selected node and moving along the edges connecting every node to each other. As explained in [23], Ant System (AS), when an ant finds a food source, it deposits a pheromone on its way back to the nest. This pheromone trace can be detected by other ants that prefer to follow trails where more pheromone was deposited. However, pheromones evaporate over time, which implies that either more ants deposit pheromone on a trail or the pheromone on this trail disappears.

Algorithm 1: Proposed ACO-EML model for wind power forecasting.

Input : Historical meteorological information $\mathbf{X}_t^{\diamondsuit,\mathbf{p}_1}, \ldots, \mathbf{X}_t^{\diamondsuit,\mathbf{p}_P}$ at positions $\{\mathbf{p}_1, \ldots, \mathbf{p}_P$, test times $\mathcal{T}_{tst}$, number of folds K, prediction horizon Δ_t, feature window size Δ_x, ELM model $M_\theta^{ELM}(\cdot)$, number of ACO generations I, number of ants A.

Output : Predicted wind power $P_{t+\Delta_t}$ of every test instance.

Construct training instances $\{(\mathbf{X}_t,)\}_{t\in\mathcal{T}_{trn}}$ as per (1) and (2).
Construct test instances $\mathbf{X}_t$ by proceeding accordingly for $t \in \mathcal{T}_{tst}$.
Set all edge probabilities of the solution graph to $1/(|\mathcal{X}| - 1)$ (i.e. equally likely).
foreach $i = 1$ *to* I **do**
  foreach $a = 1$ *to* A **do**
    Deploy ant a in the solution graph $\mathcal{G}$ based on the average sum of edges connected to every node in the graph.
    Let ant a move along the graph based on the existing edge probabilities, assigning the visited nodes to the components of the solution $\mathcal{X}_a(i) \subseteq \mathcal{X}$.
    Evaluate the quality $\widehat{\varphi}(\mathcal{X}_a(i))$ of the path $\mathcal{X}_a(i)$ as the average predictive performance of $M_\theta^{ELM}(\cdot)$ computed via K-fold cross-validation over $\{(\mathbf{X}_t, P_{t+\Delta_t})\}_{t\in\mathcal{T}_{trn}}$ using the reduced feature subset $\mathcal{X}_a(i)$.
    Update edge probabilities $p_v(i)$ using Expression (4) with the pheromone $\tau_v(i)$ equal to $\varphi(\mathcal{X}_a(i))$.
  end
  Evaporate the quantity of deployed pheromones on edge v as per (5).
end
Let the path with highest quality $\widehat{\varphi}(\mathcal{X}_a(I))$ among all ants denote the optimal feature subset $\mathcal{X}_a^{best}$ produced by the ACO wrapper.
Learn a model $M_\theta^{ELM}(\cdot)$ from the training set $\{(\mathbf{X}_t, P_{t+\Delta_t})\}_{t\in\mathcal{T}_{trn}}$ using $\mathcal{X}_a^{best}$.
Predict wind power for each test instance as $P_{t+\Delta_t} = M_\theta^{ELM}(\mathbf{X}_t)$ with $t \in \mathcal{T}_{tst}$.

This nature-inspired principle is indeed embraced in our proposed approach: the movement of ants through the solution graph is guided probabilistically by the pheromone deployed by other ants through their paths, whose intensity is driven by the quality of the solution found by every ant. The fitness of a given path is computed by computing a K-fold cross-validated measure of predictive performance of an ELM model $M_\theta^{ELM}(\cdot)$ when learning the features of the training set discriminated by nodes that compose the path of the ant at hand. When arranged in colonies, a specific number of artificial ants A build a solution step by step selecting the next edge considering the quantity of pheromone deposited by the previous ants. If more pheromone is deposited, the probability is higher that the ant leads to this node. In this regard, the probability that in step $i \in \{1, \ldots, I\}$ ant $a \in \{1, \ldots, A\}$ goes from node $u \in \mathcal{V}$ to node $w \in \mathcal{V}$ in the solution graph is given by

$$p_{uw}^a(i) = \frac{\tau_u(i) + \tau_w(i)}{\sum_{w'\in\mathcal{V}/u,w} \tau_{uw'}(i)}, \tag{4}$$

where $\tau_{uw}(i)$ is the total quantity of pheromones on the edge connecting nodes (features) u and w at generation i, and $\mathcal{V}/u, w$ denotes the set of all nodes in the graph except u and w. If pheromones become more important at two any given nodes (features), then a promising area of found solutions is deeper explored.

Pheromone evaporation is also included in the model as a form of forgetting the pheromone deployed on traversed edges if it is not reinforced by new ants passing along them. At the end of a generation i, that is when all ants have built a solution, the amount of pheromones on each node is updated as

$$\tau_u(i+1) = \rho \cdot \tau_u(i) + \sum_{a=1}^{A} \mathbb{I}(u \in \mathcal{X}_a(i))\varphi(\mathcal{X}_a(i)), \quad (5)$$

where ρ is the evaporation rate aimed at avoiding the convergence to a local optimal solution, and $\mathbb{I}(\cdot)$ is an auxiliary indicator function taking value 1 if its argument is true and 0 otherwise. Every ant also maintains a memory of its visited nodes so as to avoid loops along its path. Once all generations have been completed, the path characterized by the highest quality $\widehat{\varphi}(\mathcal{X}_a(I))$ among all ants in the colony is declared as the optimal feature subset, based on which the ELM model is trained and the predicted wind power for the test set is produced.

To end with the description of the proposed hybrid model, the ELM model used in this paper is selected due to its fast learning procedure over a similar topological neural structure to that of multi-layer perceptrons. The most significant characteristic of the ELM training procedure is that it can be carried out by randomly setting the weights of the underlying neural network, and then taking the inverse of the hidden-layer output matrix [22]. This yields an extremely agile learning procedure which makes this learner very suitable for wrapping-based feature selection problems.

4 Experimental Results and Discussion

In order to assess the performance of the proposed hybrid model several Monte Carlo experiments have been carried out by using data from two different wind farms in Spain, namely, Peña Roldana (Zamora, hereafter labeled as ROLDANA) and Faro Farelo (Galicia, FARELO). The farm corresponding to the ROLDANA dataset comprises $M = 22$ turbines with a total nominal power of 36,740 KW, whereas FARELO comprises $M = 18$ turbines with a total nominal power of 30,060 KW. The collected data span from January 2013 to October 2015, with a time step between wind power measurements of 1 h. A NWP model was used to interpolate temperature (V_1), wind module (V_2) and wind U/V components (V_3 and V_4) over a rectangular grid of $P = 45$ points located in the surroundings of the wind farm. The prediction horizon was set to $\Delta_t = 1$ time steps (i.e. short-time forecasting), while a total of $\Delta_X = 2$ past values of every feature were accumulated as potential input predictors, giving rise to T$45 \cdot 5 \cdot 2 = 450$ possible features per scenario.

The scope of the simulations discussed in this section is to validate the predictive performance gain achieved when using the proposed ACO wrapper with respect to the case when no feature selection is made. Methodologically speaking several simulation aspects are worth to highlight: to begin with, 20 independent experiments per every simulated scenario have been run in order to account for the stochasticity of the ACO algorithm as per (4). Consequently, results must be assessed statistically. Furthermore, folds in which the training dataset is split towards evaluation (line 8 in Algorithm 1) are not the same for all produced candidate solutions; otherwise there is a risk for the overall feature selection process to overfit the model to the specific distribution of folds computed from the beginning of the algorithm. To end with the specifications of the simulation benchmark, a colony of $A = 10$ ants, an EML with 30 hidden neurons and $I = 50$ iterations have been configured. The measure of predictive performance will be the so-called coefficient of determination or R^2 score, whose best value (i.e. perfect prediction) is 1 while $R^2 = 0$ corresponds to the case where the model always output the expected value of the target variable.

Figure 2a, b summarize the results obtained by the ELM-ACO model over ROLDANA and FARELO datasets, respectively. The plots depict the convergence of the cross-

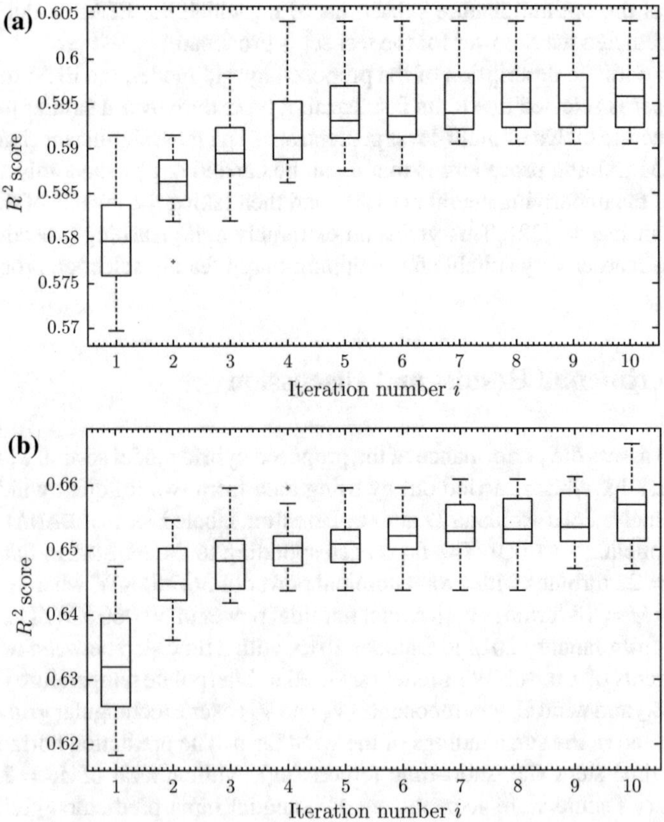

Fig. 2 In blue, R^2 convergence plots of the proposed ACO-ELM model for **a** ROLDANA **b** FARELO datasets. Cross-validated R^2 scores of 0.32 (ROLDANA) and 0.302 (FARELO) are obtained by the ELM model when no feature selection is made

validated score R^2 used as the fitness of the ACO wrapper during its search process. It is important to see that in both simulated scenarios the feature selection process provides a predictive gain with respect to the case when no feature selection is made (in the order of 0.3 in the R^2 scale), which is of interest due to the 10 features selected on average by the ACO wrapper for both scenarios (less than 3% of the original feature set).

5 Concluding Remarks and Future Research Lines

This paper has elaborated on a hybrid predictive model that combines Extreme Learning Machines and Ant Colony Optimization for wind power production forecasting. Its main design principle is to represent the space of possible input features to the model as nodes of a solution graph, which is efficiently explored by using ant colonies guided by a fitness equal to the cross-validated prediction score of the underlying model. The adoption of Extreme Learning Machines ensures an light optimization procedure of the overall model due to the renowned low-complexity training process of this particular class of supervised learners. The performance of the proposed regression model has been put to practice with real data recorded in two different wind farms located in Spain characterized by very distinct wind patterns. The performance enhancement obtained by the proposed hybrid approach is promising, with R^2 increases of near 0.3 in terms of R^2 with respect to the case where no feature selection is made.

Future research efforts will be invested towards accelerating the convergence properties of the ACO wrapper by adding heuristic information to the pheromone calculation in Expression (4). Among other ideas, we will concentrate on how to reflect the collinearity between nodes u and v in this expression so as to avoid transitions between nodes (features) when they are strongly correlated to each other. Furthermore, other swarm heuristics will be also under active investigation as alternative feature selection wrappers.

Acknowledgements This work has been co-funded by the following research projects: EphemeCH (TIN2014-56494-C4-4-P) by the Spanish Ministry of Economy and Competitivity, CIBERDINE (S2013/ICE-3095), both under the European Regional Development Fund FEDER, by Airbus Defence & Space (FUAM-076914 and FUAM-076915), and by the Basque Government under its ELKARTEK program (KK-2016/00096, BID3ABI project).

References

1. Farhangi, H.: The path of the smart grid. IEEE Power Energy Mag. **8**(1) (2010)
2. Wissner, M.: The smart grid-a saucerful of secrets? Appl. Energy **88**(7), 2509–2518 (2011)
3. Murthy, K.S.R., Rahi, O.P.: A comprehensive review of wind resource assessment. Renew. Sustain. Energy Rev. **72**, 1320–1342 (2017)

4. Kaur, T., Kumar, S., Segal, R.: Application of artificial neural network for short term wind speed forecasting. In: International Conference on Power and Energy Systems: Towards Sustainable Energy (PESTSE), pp. 1–5 (2016)
5. Ata, R.: Artificial neural networks applications in wind energy systems: a review. Renew. Sustain. Energy Rev. **49**, 534–562 (2015)
6. Alexiadis, M.C., Dokopoulos, P.S., Sahsamanoglou, H.S., Manousaridis, I.M.: Short-term forecasting of wind speed and related electrical power. Solar Energy **63**(1), 61–68 (1998)
7. Mohandes, M.A., Halawani, T.O., Rehman, S., Hussain, A.A.: Support vector machines for wind speed prediction. Renew. Energy **29**(6), 939–947 (2004)
8. Zhao, P., Xia, J., Dai, Y., He, J.: Wind speed prediction using support vector regression. In: 5th IEEE Conference on Industrial Electronics and Applications (ICIEA), pp. 882–886 (2010)
9. Troncoso, A., Salcedo-Sanz, S., Casanova-Mateo, C., Riquelme, J.C., Prieto, L.: Local models-based regression trees for very short-term wind speed prediction. Renew. Energy **81**, 589–598 (2015)
10. Heinermann, J., Kramer, O.: Machine learning ensembles for wind power prediction. Renew. Energy **89**, 671–679 (2016)
11. Hu, Q., Zhang, R., Zhou, Y.: Transfer learning for short-term wind speed prediction with deep neural networks. Renew. Energy **85**, 83–95 (2016)
12. Dalto, M., Matuško, J., Vašak, M.: Deep neural networks for ultra-short-term wind forecasting. In: IEEE International Conference on Industrial Technology (ICIT), pp. 1657–1663 (2015)
13. Salcedo-Sanz, S., Ortiz-Garcia, E.G., Perez-Bellido, A.M., Portilla-Figueras, A., Prieto, L.: Short term wind speed prediction based on evolutionary support vector regression algorithms. Expert Syst. Appl. **38**(4), 4052–4057 (2011)
14. Liu, D., Niu, D., Wang, H., Fan, L.: Short-term wind speed forecasting using wavelet transform and support vector machines optimized by genetic algorithm. Renew. Energy **62**, 592–597 (2014)
15. Jursa, R., Rohrig, K.: Short-term wind power forecasting using evolutionary algorithms for the automated specification of artificial intelligence models. Int. J. Forecast. **24**(4), 694–709 (2008)
16. Salcedo-Sanz, S., Pastor-Sanchez, A., Prieto, L., Blanco-Aguilera, A., Garcia-Herrera, R.: Feature selection in wind speed prediction systems based on a hybrid coral reefs optimization Extreme learning machine approach. Energy Convers. Manag. **87**, 10–18 (2014)
17. Salcedo-Sanz, S., Pastor-Sanchez, A., Del Ser, J., Prieto, L., Geem, Z.W.: A Coral Reefs Optimization algorithm with Harmony Search operators for accurate wind speed prediction. Renew. Energy **75**, 93–101 (2015)
18. Foley, A.M., Leahy, P.G., Marvuglia, A., McKeogh, E.J.: Current methods and advances in forecasting of wind power generation. Renew. Energy **37**(1), 1–8 (2012)
19. Colak, I., Sagiroglu, S., Yesilbudak, M.: Data mining and wind power prediction: A literature review. Renew. Energy **46**, 241–247 (2012)
20. Giebel, G., Brownsword, R., Kariniotakis, G., Denhard, M., Draxl, C.: The state-of-the-art in short-term prediction of wind power: A literature overview. ANEMOS. plus (2011)
21. Lei, M., Shiyan, L., Chuanwen, J., Hongling, L., Yan, Z.: A review on the forecasting of wind speed and generated power. Renew. Sustain. Energy Rev. **13**(4), 915–920 (2009)
22. Huang, G.B., Zhu, Q.Y., Siew, C.K.: Extreme learning machine: theory and applications. Neurocomputing **70**(1), 489–501 (2006)
23. Dorigo, M., Birattari, M., Stutzle, T.: Ant Colony Optimization. IEEE Comput. Intell. Mag. **1**(4), 28–39 (2006)
24. Gonzalez-Pardo, A., Camacho, D.: A new csp graph-based representation for ant colony optimization. In: IEEE Congress on Evolutionary Computation, pp. 689–696 (2013)
25. Gonzalez-Pardo, A., Camacho, D.: A new csp graph-based representation to resource-constrained project scheduling problem. In: IEEE Congress on Evolutionary Computation (CEC), pp. 344–351 (2014)
26. Gonzalez-Pardo, A., Jung, J.J., Camacho, D.: ACO-based clustering for Ego Network analysis. Future Gener. Comput. Syst. **66**, 160–170 (2017)

Convolutional Neural Networks for Four-Class Motor Imagery Data Classification

Tomas Uktveris and Vacius Jusas

Abstract In this paper the use of convolutional neural networks (CNN) is discussed in order to solve four class motor imagery classification problem. Analysis of viable CNN architectures and their influence on the obtained accuracy for the given task is argued. Furthermore, selection of optimal feature map image dimension, filter sizes and other CNN parameters used for network training is investigated. Methods for generating 2D feature maps from 1D feature vectors are presented for commonly used feature types. Initial results show that CNN can achieve high 68% classification accuracy for the four class motor imagery problem with less complex feature extraction techniques. It is shown that optimal accuracy highly depends on feature map dimensions, filter sizes, epoch count and other tunable factors, therefore various fine-tuning techniques must be employed. Experiments show that simple FFT energy map generation techniques are enough to reach the state-of-the-art classification accuracy for common CNN feature map sizes. This work also confirms that CNNs are able to learn a descriptive set of information needed for optimal electroencephalogram (EEG) signal classification.

Keywords Convolutional neural network · Motor imagery · Feature map · Image classification, FFT energy map

1 Introduction

Motor imagery classification is one of many widespread machine-learning problems of brain-computer interface (BCI) systems. With the need for human mind controlled applications the recording of electroencephalograms (EEG) has emerged as

T. Uktveris (✉) · V. Jusas
Software Engineering Department, Kaunas University of Technology,
Studentu St. 50, Kaunas, Lithuania
e-mail: tomas.uktveris@ktu.lt

V. Jusas
e-mail: vacius.jusas@ktu.lt

M. Ivanović et al. (eds.), *Intelligent Distributed Computing XI*,
Studies in Computational Intelligence 737, https://doi.org/10.1007/978-3-319-66379-1_17

an optimal solution for non-interventional brain activity analysis. The ability to fully understand this brain induced electrical signal would greatly simplify the life for people with disabilities or break the barrier for natural interaction in entertainment industry.

This work focuses on four-class motor imagery problem where the recorded EEG signal is classified into four different classes that correspond to four different human subject imagined motoric actions (left hand, right hand, feet and tongue movement). Even if a simpler two-class (binary) problem achieves good classification performance, the four-class still struggles to reach same results and requires more scientific investigation.

A relatively new and perspective approach to EEG data classification was found in deep learning branch of machine learning. Convolutional neural network (CNN) is a novel animal visual cortex inspired method for image based classification that has not been widely used with EEG, let alone motor-imagery task. With the abilities to generalize/pool and self-learn the needed features in non-linear ways it can benefit EEG classification. Since EEG motor imagery task lacks accurate solutions the CNN could be the new perspective way to look deeper into the same problem. Regarding its novelty and success in other fields it was chosen as the main tool for four-class EEG motor imagery problem analysis in this paper.

By using CNN for classification subtle fine tuning is required to receive best results. This involves selecting proper neural network architecture, feature method and feature map size. These nuances and their effect on classification performance are further analyzed and discussed in this paper.

2 Related Work

In recent years an increasing number of papers that use CNN for EEG classification task have been published. Multiple approaches have been proposed for solving motor imagery and other related problems. A short review of the common techniques is presented in the next paragraphs.

Two-class motor imagery problem was presented in paper [1] and a technique was proposed to analyze the EEG in frequency domain. A time-frequency distribution (TFD) images were constructed based on complex Morlet wavelet decomposition for electrode pairs. The TFDs were subtracted from symmetrical channels to form weight matrices that were used to compute weighted energy for classification. A Laplacian filter was used for signal preprocessing. Average classification rate of 78% was achieved for this method. Another approach based on energy entropy preprocessing and Fisher class separability criteria was proposed in literature [2]. The paper analyzed a two-class motor imagery problem in time-frequency domain. Similar TFD distributions (spectrograms) were constructed from EEG short-term Fourier transform (STFT) data. Three different classification methods were compared. Classification accuracy for the two class problem was 85%. A more complicated approach for 3-class motor imagery analysis was given in [3]. The study proposed a new method to extract the MRICs (movement related independent components) and utilized ICA (Independent Component Analysis) spatial distribution patterns for such a task. Different ICA filter designs were tested. ICA filter design was confirmed to be subject invariant. Classification accuracy ~62% received.

A more recent study [4] on 4-class motor imagery proposed a novel Wavelet-CSP (Common Spatial Patterns) with ICA-filter method. The EEG artifacts were removed using negative entropy-based ICA. Mean accuracy 76% was achieved using SVM (Support Vector Machine) classifier.

One of the latest works in the field of CNN and 4-class motor imagery is [5]. The authors of the paper proposed a frequency complementary feature map selection (FCMS) method. ACSP (Augmented CSP) feature filtering was used in the work. Two other feature selection methods—random map selection (RMS) and selection of all feature maps (SFM) were analyzed. FCMS was the best performing method due to its ability to limit the ACSP feature redundancy in different frequency bands. The CNN used 5 layer architecture with 5 × 5 filters (kernels). The work also demonstrated that CNNs are capable of learning discriminant, deep structure features for EEG classification without relying on the handcrafted features. Average classification accuracy achieved –69%.

3 Methods of Analysis

3.1 Common Spatial Patterns (CSP)

A widely adopted signal pre-processing method that decomposes the raw EEG into subcomponents (spatial patterns) having maximum differences in variance [6]. Technique allows better feature separation [7] in feature space and thus more accurate signal classification. Also the property of CSP to decrease feature dimensionality is very suitable for EEG data complexity reduction. It has been shown in [8] and other works that this method gives a substantial EEG signal classification performance increase, thus is a highly recommended filtering method.

The filter is a spatial coefficient matrix W:

$$S = W^T E$$

where S—the filtered signal matrix, E—original EEG signal. Columns of W denote spatial filters, while inverse, i.e. W^{-1}, are spatial patterns of EEG signal. The criterion of CSP for a two C1, C2 class problem is given by:

$$maximize\colon tr\left(W^T \textstyle\sum_1 W\right) \tag{1}$$

$$subject\ to\colon W^T\left(\textstyle\sum_1 + \sum_2\right)W = I \tag{2}$$

where $\sum_1$ and $\sum_2$ are the class covariance matrices.

Solution can be acquired by solving a generalized eigenvalue problem. Since CSP was designed for a binary problem multiclass solutions are combined of multiple spatial filters.

Due to the broad and positive acknowledgement of CSP, the method was used in the work to filter EEG data before commencing feature extraction.

3.2 Feature Extraction Methods

A multitude of EEG feature extraction methods have been studied in [8] and other literature. Their output usually is a one dimensional feature vector that can be used for classification. The ability to adapt the algorithms for two-dimensional CNN has not been thoroughly analyzed. It is also important to know if the adapted methods can give similar or better results when applied in 2D for CNN. Thus, a review of the most common feature extraction techniques and their implementations for CNN is presented in this work. A short list of the EEG feature methods that were tested and analyzed in this paper is given further in Table 1.

Table 1 Feature extraction methods

Method	Feature generation function
Mean channel energy (MCE)	$y_i = \log\left(\frac{1}{N}\sum_{k=1}^{N} x_i[k]^2\right),\ i=\overline{1,n}$ (3)
	here x_i is *i-th* channel EEG signal, y_i is feature *i-th* component
Channel variance (CV)	$y_i = \log\left(\frac{1}{N}\sum_{k=1}^{N} (x_i[k]-\overline{x_i})^2\right),\ i=\overline{1,n}$ (4)
	here $\overline{x_i}$ is *i-th* channel EEG signal mean
Mean window energy (MWE)	$H_{i,j} = \log\left[\frac{1}{N}\sum_{j=1}^{N}\left(\frac{1}{W}\sum_{k=0}^{W} x_i[k]^2\right)\right], N=\overline{1,p}$ (5)
	here $H_{i,j}$ is *2D* feature map component, *W* is window size, *p* is window count , $p\in[1;22]$
Principal component analysis (PCA)	$y_i = \log\left(\frac{1}{N}\sum_{k=1}^{N} \lvert PCA(x_i)\rvert_k^2\right), i=\overline{1,n}$ (6)
Mean band power (BP)	$y_i = \frac{1}{N}\sum_{j=1}^{N}\left[ln\left(\frac{1}{w}\sum_{k=0}^{w} x[j-k]^2\right)\right], i=\overline{1,n}$ (7)
	here w is smoothing window size
Channel FFT energy (CFFT)	$y_i = \log\left(\sum_{k=1}^{N} FFT(x_i)^2\right), i=\overline{1,n}$ (8)
Discrete cosine transform (DCT)	$y_i = \log\left(\sum_{k=1}^{N} DCT(x_i)^2\right), i=\overline{1,n}$ (9)
Time domain parameters (TDP)	$p_j(t) = \frac{d^j x(t)}{dt^j}, j=0,1,\ldots,m$ (10)
	$y_i = \frac{1}{N}\sum_{k=1}^{N} \ln\left(u\cdot p_j[k]-(1-u)\cdot p_j[k-1]\right), i=\overline{1,n}$ (11)
	here u is the moving average parameter, $u\in[0;1]$

(continued)

Table 1 (continued)

Method	Feature generation function
Teager-Kaiser energy operator (TKEO)	$y_i = \log\left[\frac{1}{N}\sum_{k=1}^{N}(x^2[k] - x[k-1]x[k+1])\right], i = \overline{1,n}$ (12)
FFT energy map (FFTEM)	$H_i = \lvert FFT(x_i)\rvert, \quad i = \overline{1,n}$ (13) *here* H_i is *2D* feature map *i-th* row
Complex wavelet transform (CWT)	$W_x(\tau,f) = \int_{-\infty}^{+\infty} x(u)\left(\sqrt{f}e^{i2\pi f(u-\tau)}e^{-\frac{(u-\tau)^2}{2\sigma^2}}\right)du$ (14) $H_i = \frac{1}{N}\sum_{k=1}^{N}\lvert W_x(\tau,f_k)\rvert, \quad i = \overline{1,n}$ (15) *here* τ, σ is complex wavelet parameters, $f \in [0;30]$
Raw signal features (RAW)	$H_i = x_i, \quad i = \overline{1,n}$ (16)
Signal energy map (SEM)	$H_i = \log x_i^2, \quad i = \overline{1,n}$ (17)

Table 2 Different evaluated CNN architectures

#	CNN configuration	Notes
1	IC(4)RPFSO	4 filters
2	IC(4)RP(4)FSO	Stride 4
3	IC(8)RPFSO	8 filters
4	ICRPFSO	
5	IC(32)RPFSO	32 filters
6	IC(64)RPFSO	64 filters
7	ICRPCRPFSO	
8	ICRFSO	
9	ICFSO	
10	IC(7 × 1)RC(1 × 7)RPFSO	Non-rect filters
11	IC(1 × 7)RPC(7 × 1)RPFSO	Non-rect filters

3.3 CNN Architecture Selection

Choosing the correct network architecture for the problem gives a greater probability of getting better classification results. CNN supports serially connected layers. Due to the large number of different layer types it is not trivial to find the optimal chain that closely matches the given problem.

Tests for 11 different CNN architectures were completed. The architecture configurations in a simplified notation are given in Table 2 further. Used notation is explained in Table 3.

Evaluation results are shown in Fig. 1. It can be seen, that testing accuracy is similar ~65% between most of the configurations. However, training accuracy displays a more dynamic profile from 50% to 80%. In this case, the CNN configuration with the least amount of computational-processing resources (i.e. simplest) should be selected as optimal –1, 2, 4 or 10.

Table 3 CNN layer symbolic notation

#	Description (Default parameters)
I	Input layer of size (44 × 44 x 1)
C	Convolutional layer (7 × 7, 16 filters)
R	ReLU layer
P	Max pooling layer (2 × 2, stride 2)
F	Fully connected layer (4 classes)
S	Softmax layer
O	Classification (output) layer

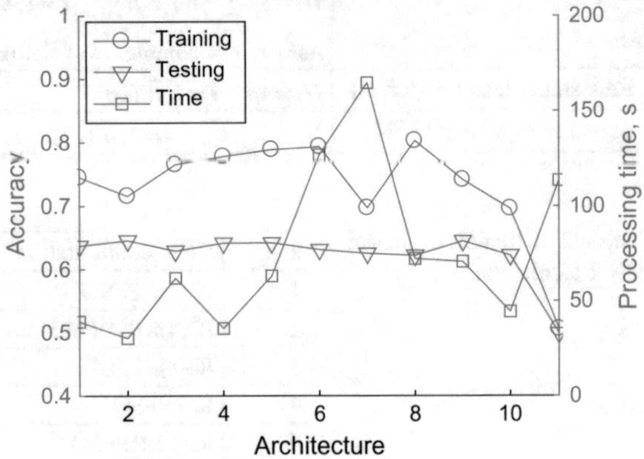

Fig. 1 CNN architecture evaluation

3.4 CNN Parameter Tuning

CNNs are more complex since they have more hyper-parameters than a standard MLP. However, the usual learning rates and regularization constants still apply. CNN training parameters: initial learning rate, momentum, batch size and number of epochs must be tuned for best performance. Since a 4D parameter grid based search is too resource intensive, a parameter range scanning approach was carried out to find sub-optimal values.

The momentum value denotes the contribution for the next gradient value from previous iteration in Stochastic Gradient Descent (SGD) method. Larger parameter values decrease the effectiveness of faster learning as shown in Fig. 2. In tests, values above 0.6 push the CNN to overfitting and thus decrease generalization and testing accuracy. Value of zero for momentum is not recommended since that invokes loss of historical gradient learning information.

Optimal number of training epochs ensures that the network learns and generalizes the provided features. Excessive epochs deteriorate the testing accuracy since the network is overfitting. Figure 3 shows that the optimal count for training is 400–500 epochs.

Batch size is the image count that is used for single epoch training. This size has direct effect on the network learning quality as show in Fig. 4. The maximum batch

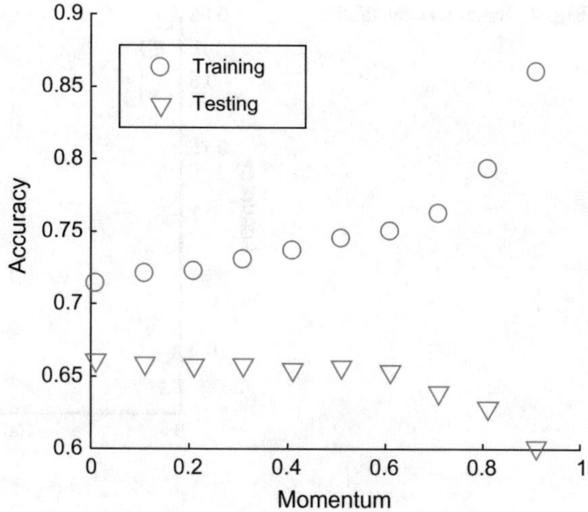

Fig. 2 Momentum evaluation

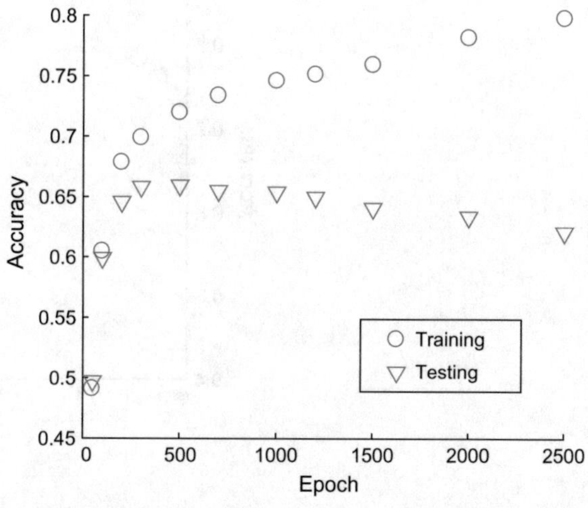

Fig. 3 Epoch count evaluation

size is the number of total images, e.g. N = 288 in experiments. Lower values than N/4 prevent the network from fully maximizing learning efficiency, greater values only increase computational costs at the price of no change in testing accuracy.

Initial learning rate must be adopted for each problem. Experiments show that the value should be no bigger than 0.1, while the network testing accuracy 0.66 optimum is achieved with values close to 0.01 as shown in Fig. 5. Lower values allow to learn fine grained features, while large have the tendency to overfit the network.

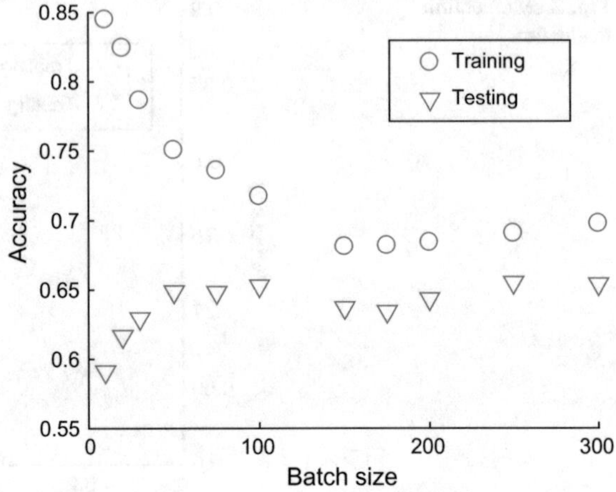

Fig. 4 Batch size evaluation

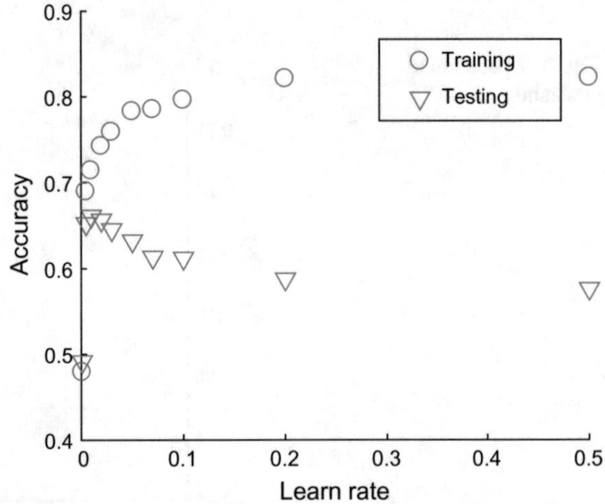

Fig. 5 Initial learning rate evaluation

3.5 *Feature Map and Filter Dimensions*

The problem is to find the right level of granularity in order to create data abstractions at the proper scale, given a particular dataset. Different feature map and filter sizes were analyzed for the motor imagery problem. Dimensions from 8 × 8 to 64 × 64 of feature maps were tested. Test results are given in Fig. 6.

The plot shows that the optimal feature map size is 24 × 24 with accuracy 65%, even though a more accurate solution of 66% exists at size 44 × 44. Choosing a smaller size feature map ensures faster computation and processing speeds. Also note that the accuracy convergence is reached when the feature map size is at least twice (15 × 15) the size as the convolution layer filter size (7 × 7 in the

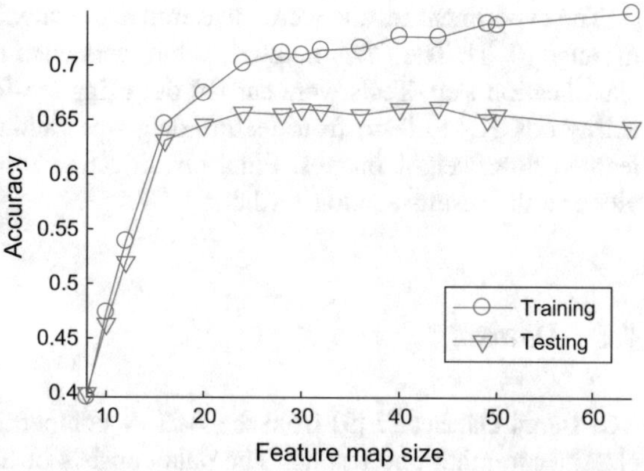

Fig. 6 Feature map size evaluation

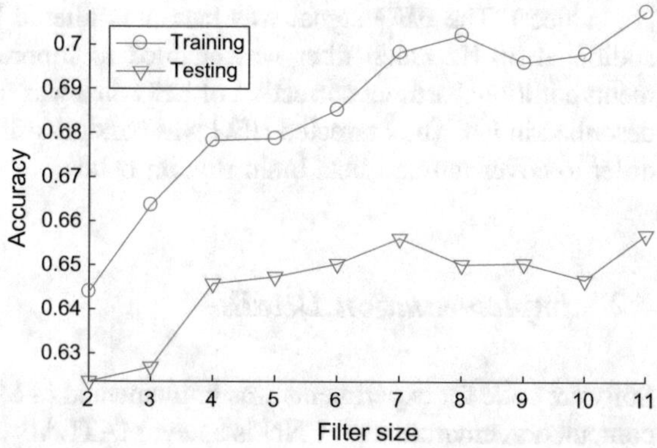

Fig. 7 Convolution layer filter size evaluation

experiment). When the optimal size is reached the further increase in dimension only introduces extra computational costs.

Convolution layer filter size limits the learning granularity by encompassing fixed size feature map regions. Ten different filter sizes were tested in range [2, 11] for 22 × 22 feature maps. Test results are displayed in Fig. 7.

The optimal filter size, that gives highest accuracy, is 7 × 7 and 11 × 11. Choosing the smaller filter size ensures faster processing speeds. Filters of size 2 × 2 and 3 × 3 exhibit too few weights to fully learn the details of the provided data.

4 Experiments

The main purpose of the experimentation activities was to investigate the capabilities of CNN classifier for four-class motor imagery classification problem. Also to analyze various CNN architectures, feature maps, filter size optimization and other parameter dependency with respect to classification accuracy. Experiments were conducted in the analysis step (tuning the CNN network parameters) and also in the main motor imagery classification step (for each subject).

The experiment results were measured and evaluated using normalized accuracy in range [0; 1]. The CNN network parameters were tuned and verified before final classification step. Tests were carried out using ten-fold cross validation. Also, the ability of CNN to learn from feature data was validated visually by inspecting the learned filter/weight images. Final classification results for each subject are provided in the results section further.

4.1 Dataset

BCI signal Dataset 2a [9] from the BCI IV competition held in 2008 was used for classifier training and testing. The data consists of 22 channels of 250 Hz sample rate recorded EEG signal for 9 healthy test subjects (total 288 motor imagery trials per subject). The EEG signal was bandpass-filtered between 0.5 and 100 Hz and additional 50 Hz notch filter was enabled to suppress line noise. Before experiments additional artifact correction of EEG data was done to discard invalid trials as described in [9]. The corrected EEG was bandpass-filtered between 7 and 30 Hz in order to cover mu and beta brain rhythm bands.

4.2 Implementation Details

Software code for experiments was implemented in MATLAB 2016b/9.1 numerical computation environment. CNN is a new MATLAB functionality (starting from the 2016a/9.0 version), which uses GPU processor for parallel computations. Other alternatives for convolutional neural networks exist such as the open source MatConvNet library [10], however the library was left as an option for future CNN evaluations. Parts of the open source BioSig library for biomedical signal processing and imaging were used in EEG signal analysis.

CNN convolution layer initial filter weights in all tests were set to have a Gaussian distribution with a mean of 0 and standard deviation of 0.01. The default for the initial bias was 0.

4.3 Results

Final classification results were obtained after analysis and CNN parameter fine-tuning step. A CNN with initial learning rate –0.01, momentum –0.1, batch size –128, epochs –200 and architecture I(22 × 22)C(4 × 4,16) RPFSO was trained and tested for final evaluation on all subjects. Results were verified by using 10-fold cross-validation scheme. The accuracies with their standard deviation values are displayed in Table 4. From the results it can be seen that the best performing

Table 4 Classification results for feature extraction methods

Method	MCE	CV	MWP	PCA	BP	CFFT	DCT	TDP	TKEO	FFTEM	CWT	RAW	SEM
Training	0.66 ± 0.19	**0.68 ± 0.18**	0.66 ± 0.19	0.61 ± 0.16	0.52 ± 0.18	0.66 ± 0.19	0.5 ± 0.17	0.4 ± 0.11	0.43 ± 0.12	**0.70 ± 0.18**	0.46 ± 0.10	0.48 ± 0.14	**0.67 ± 0.18**
Testing	0.5 ± 0.20	**0.61 ± 0.22**	0.58 ± 0.20	0.55 ± 0.20	0.39 ± 0.11	0.58 ± 0.20	0.42 ± 0.11	0.31 ± 0.07	0.34 ± 0.05	**0.68 ± 0.20**	0.43 ± 0.13	0.37 ± 0.11	**0.61 ± 0.20**

(70% in training) and (68% in testing) is the FFT energy map method. Second and third best methods in tests were the Channel variance (68%/61%) and Signal energy map (67%/61%) features. The lowest accuracy (41%/31%) was achieved by the TDP feature method.

5 Conclusions

This work analyzed Convolutional Neural Networks and their application to four class motor-imagery based problem. After an in-depth CNN analysis and parameter fine-tuning promising results were achieved for the selected problem. The FFT energy map method demonstrated the best feature determination abilities and achieved 68% mean testing accuracy for all the BCI IV competition 2a dataset subjects. The gained accuracy is slighltly better than in new techniques [11] and similar to more complex state of the art [5] EEG analysis techniques. The use of simpler feature extraction methods as FFT energy maps shows a high CNN method potential for motor imagery EEG analysis.

Further work will continue in order to provide more efficient feature extraction methods favoring processing speed and accuracy.

References

1. Qin, L., He, B.: A wavelet-based time-frequency analysis approach for classification of motor imagery for brain-computer interface applications. J. Neural Eng. **2**, 65–72 (2005)
2. Xiao, D., Mu, Z.D., Hu, J. F.: Classification of motor imagery EEG signals based on energy entropy. In: International Symposium on Intelligent Ubiquitous Computing and Education, 61–64 (2009)
3. Zhou, B., Wu, X., Zhang, L., Lv, Z., Guo, X.: Robust spatial filters on three-class motor imagery EEG data using independent component analysis. J. Biosci. Med. **2**, 43–49 (2014)
4. Bai, X., Wang, X., Zheng, S., Yu, M.: The offline feature extraction of four-class motor imagery EEG based on ICA and Wavelet-CSP. In: Control Conference *(CCC)*, pp. 7189–7194 (2014)
5. Yang, H., Sakhavi, S., Ang, K.K., Guan, C.: On the use of convolutional neural networks and augmented CSP features for multi-class motor imagery of EEG signals classification. In: 37th Annual International Conference of the IEEE Engineering in Medicine and Biology Society (EMBC), pp. 2620–2623 (2015)
6. Naeem, M., Brunner, C., Pfurtscheller, G.: Dimensionality reduction and channel selection of motor imagery electroencephalographic data. Comput. Intell. Neurosci. (2009)
7. Wang, Y., Gao, S., Gao, X.: Common spatial pattern method for channel selection in motor imagery based brain-computer interface. In: IEEE Engineering in Medicine and Biology 27th Annual Conference, pp. 5392–5395 (2005)
8. Uktveris, T., Jusas, V.: Comparison of Feature Extraction Methods for EEG BCI Classification, Information and Software Technologies: 21st International Conference, pp. 81–92 (2015)

9. Brunner, C. et al.: BCI Competition 2008—Graz data set A (2008)
10. Vedaldi, A., Lenc, K.: MatConvNet—convolutional neural networks for MATLAB. In: Proceedings of the ACM International Conference on Multimedia (2015)
11. Tabar, Y. R., Halici, U.: A novel deep learning approach for classification of EEG motor imagery signals. J. Neural Eng. **14**(1) (2016)

Binary Classification of Images for Applications in Intelligent 3D Scanning

Branislav Vezilić, Dušan B. Gajić, Dinu Dragan, Veljko Petrović, Srđan Mihić, Zoran Anišić and Vladimir Puhalac

Abstract Three-dimensional (3D) scanning techniques based on photogrammetry, also known as Structure-from-Motion (SfM), require many two-dimensional (2D) images of an object, obtained from different viewpoints, in order to create its 3D reconstruction. When these images are acquired using closed-space 3D scanning rigs, which are composed of large number of cameras fitted on multiple pods, flash photography is required and image acquisition must be well synchronized to avoid the problem of 'misfired' cameras. This paper presents an approach to binary classification (as 'good' or 'misfired') of images obtained during the 3D scanning process, using four machine learning methods—support vector machines, artificial neural networks, *k*-nearest neighbors algorithm, and random forests. Input to the algorithms are histograms of regions determined to be of interest in the detection of image misfires. The considered algorithms are evaluated based on the prediction accuracy that they achieved on our dataset. The average prediction accuracy of 94.19% is obtained using the random forests approach under cross-validation.

B. Vezilić · S. Mihić
Doob Innovation Studio, Bulevar oslobođenja 127/V, 21000 Novi Sad, Serbia
e-mail: b.vezilic@doobinnovation.com

S. Mihić
e-mail: s.mihic@doobinnovation.com

D.B. Gajić (✉) · D. Dragan · V. Petrović · Z. Anišić
Faculty of Technical Sciences, University of Novi Sad, Trg Dositeja Obradovića 6, 21000 Novi Sad, Serbia
e-mail: dusan.gajic@uns.ac.rs

D. Dragan
e-mail: dinud@uns.ac.rs

V. Petrović
e-mail: pveljko@uns.ac.rs

Z. Anišić
e-mail: anisic@uns.ac.rs

V. Puhalac
Doob Group AG, Speditionstraße 13, 40221 Düsseldorf, Germany
e-mail: v.puhalac@doobgroup.com

M. Ivanović et al. (eds.), *Intelligent Distributed Computing XI*,
Studies in Computational Intelligence 737, https://doi.org/10.1007/978-3-319-66379-1_18

Therefore, the application of the proposed approach allows the development of an 'intelligent' 3D scanning system which can automatically detect camera misfiring and repeat the scanning process without the need for human intervention.

Keywords Machine learning · Image processing · Image classification · Binary classification · Decision trees · Random forest · 3D scanning · Photogrammetry · Structure-from-motion

1 Introduction

Various methods for 3D (three-dimensional) scanning have gained significant attention of researchers in recent years. Photogrammetry-based methods for 3D scanning, known as Structure-from-Motion (SfM), require a large number of images of an object obtained from different camera viewpoints [1, 2]. Using the SfM, a 3D model of the object—its structure—is created from the images obtained using triangulation, from what is referred to as camera motion [1, 2]. While it is possible to perform the SfM with one camera in motion, it is equally possible to form the dataset instantly by using multiple identical cameras whose position and orientation vary. This has its own limitations, naturally, but allows for much faster capture and for the capture of ephemeral events—at the very least at the 100 ms scale. These images are most often acquired using closed-space rigs, with a large number of cameras (typically 50–70) fitted on multiple pods (typically 8–12), which are positioned around the object in order to cover the full 360° view.

In this setting, artificial flash lightning is required and image acquisition must be well synchronized to avoid the problem of 'misfired' cameras. The images from the cameras that weren't triggered synchronously with the flash have their light levels impacted. Those fired completely out of synchronization are dark and easily detected, but in practice the critical window is just missed leading to an area of the image being darker, due to the 'rolling shutter' design of modern cameras. We call these anisotropically darkened images 'misfired'.

This paper discusses the problem of binary classification (as 'good' or 'misfired') of images obtained during the 3D scanning process in a closed-space rig, using four different machine learning approaches. The algorithms under consideration are support vector machines (SVMs), artificial neural networks (ANNs), the *k*-nearest neighbors algorithm (*k*-NN), and random forest (RF) [3–10]. The evaluation is performed on a data set of 26027 images of 500 objects and is conducted in terms of the prediction accuracy. The highest prediction accuracy of 94.70% is obtained using the random forest approach. Therefore, the application of the proposed approach allows the construction of 'intelligent' 3D scanning rigs which can automatically detect whether some of the cameras have misfired and repeat the scanning process without the need for human intervention.

The use of machine learning for image classification has been a topic of intense interest of researchers for several decades. In [7] and [8], the use of SVMs for image classification is shown to produce high accuracy. Results in [9] show that the SVMs outperform the ANNs in classification tasks. A strong case for the application of the nearest neighbor algorithm for image classification has been made in [10]. More recently, the RF algorithm has been used with considerable success for feature selection and classification [4]. Based on the literature review, we have selected the four classification machine learning methods considered in this paper. The results presented in this paper show that the RF algorithm outperforms all the other considered machine learning approaches, including SVMs, for the specific task of binary image classification. To the best of our knowledge, there are no other papers discussing the problem of automatic detection of camera misfiring.

The remainder of the paper is split into four sections. The approach, methodology, and dataset are described in Sect. 2. Section 3 presents the experimental setup and the results. The final section outlines the principal conclusions reached, as well as possible directions for future work.

2 Methodology

In this section, we describe the used dataset, as well as our approach and final implementation.

2.1 Data

The dataset used in the experiments is a collection of 3D scan data in the form of multiple images of the same object obtained from different viewpoints simultaneously using multiple cameras in a closed-space 3D scanning rig. The dataset contains 26027 images which are courtesy of Doob Group AG. The images are distributed among 500 distinct scans with each image being 3456 × 5184 pixels and compressed using lossy JPEG.

Since the dataset was drawn directly from the industry, the incidence rate for misfired images was low—naturally the rig is designed to minimize image errors of this type. Only 2729 images out of a total of 26027 (10.49%) were designated as misfired by a human operator. Because of this, the first step includes balancing the two classes considered, so that bias may be reduced. 2729 'good' images are sampled at random from the remaining data set, producing a balanced training set.

The final data set, after outliers and edge-cases were removed, contains 2589 'misfired' images and 2589 'good' images.

2.2 Approach

The key step in solving the before-mentioned problem is determining how to process our images into a form which is tractable for further analysis using machine-learning methods we have already selected. The first problem is performance. The images in our dataset are of size 3456 × 5184 pixels, which makes them difficult to analyze in reasonable time. Given that the misfired nature of an image is a global feature and not a per-pixel one, we considered it reasonable to work with smaller, simpler images. A reduction of roughly 13 times on the horizontal and the vertical dimension proved to be satisfactory during exploratory testing, as was the reduction of the image to grayscale. Grayscale proved adequate to our needs as the phenomenon we were trying to identify affected only the illumination level of the image.

The most important step is determining how to reduce our image to a few indicators which can then be fed into a machine learning algorithm. The key insight is in noticing the anisotropy of the phenomenon: the misfire effect only affects one part of the image causing a marked contrast between the left and the right or top and bottom as can be seen in Fig. 1. This is due to the 'rolling shutter' technology employed on cameras with MOS (metal-oxide-semiconductor) sensors. In this type of camera instead of the exposure being instant, the image is formed by reading out pixel by pixel of the image (either left-to-right top-to-bottom, or top-to-bottom left-to-right), while the rest of the individual photo-sensors keep collecting photons. This means that the end image is composed of pixels which are not contemporaneous but instead represent a sort of 'smear' during a short but nontrivial length of

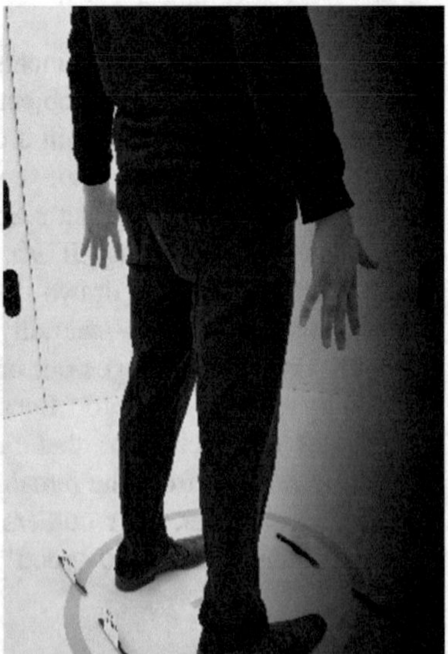

Fig. 1 'Misfired' image example (courtesy of Doob Group AG)

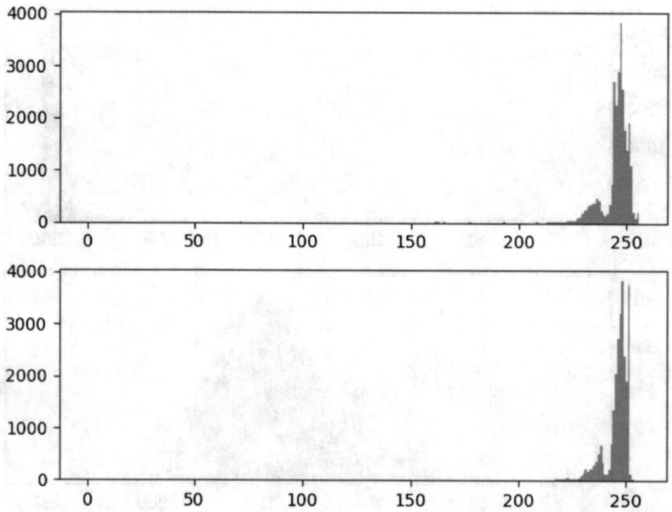

Fig. 2 Image histograms for 'good' images—left (*top*) and right (*bottom*) part of the image

time. In swiftly moving images this produces distortions, but in the case of swiftly changing *lighting* it produces anisotropic changes to the illumination of the scene.

This meant that we could (given our control of the scanning rig and its background) use as a powerful indicator of partial exposure the relative illumination of the left and right portions of the screen. Therefore, we cropped out the leftmost and the rightmost (the approach is trivial to adapt to rolling shutters of cameras that go from top to bottom first) part of the image and computed its illuminance histogram to use as a global feature descriptor. The optimal cropping line was determined empirically. Exploratory testing using this approach produced promising results. The histograms of an image coded as 'good' produced results as in Fig. 2 showing that we could expect a certain amount of histogram similarity. They are not identical and a fuzzy approach to their comparison is necessary but the human eye finds them very similar indeed.

Figure 3 shows a histogram coded as 'misfired' by a human operator. As can plainly be seen, it shows *very* prominent differences between left and right. The only thing required is to quantify precisely the nature of this difference and train some artificially intelligent system to distinguish between the two.

We attempted to create four such systems, based on SVMs, ANNs—specifically a multi-layer perceptron with a single hidden layer, the *k*-NN, and the RF algorithm [3], by training them on the set described in Sect. 2.1 and then testing them using a standard 11-fold cross-validation approach.

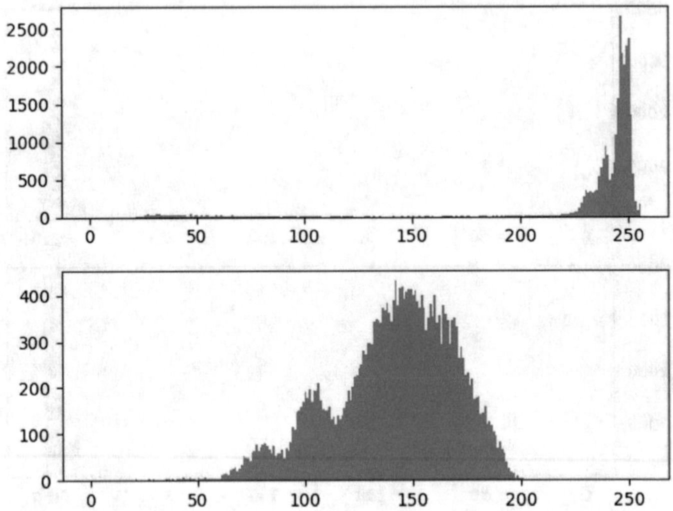

Fig. 3 Image histograms for 'misfired' images—left (*top*) and right (*bottom*) part of the image

2.3 Implementation

The implementation of our binary classifier system has four phases:

1. Image processing,
2. Histogram creation,
3. Classifier definition and
4. Classifier training.

Image processing. For each image, we reduce its resolution from 3456 × 5184 pixels to 266 × 400 pixels (using the cv2.INNER_AREA interpolation method, available as part of the OpenCV 3.2 library [11]) for easier processing and reading. We also reduced the complexity of the image by converting it to 8-bit grayscale.

Histogram creation. Given the properties of the rolling shutter technology, and visual inspection of all infected images during the coding procedure, we elected to take 25% of total width from both sides and compute a histogram of the illuminance using a histogram resolution of 256 bins, corresponding to the possible quantized illumination levels in an 8-bit grayscale image. The resulting two vectors representing histogram values are then concatenated into single vector with 512 elements. This vector then represents the chief predictor of partial exposure, and is the input to our classifier.

Classifier definition. We defined an SVM classifier, an ANN classifier, a *k*-NN classifier, and an RF classifier as part of our experiment. The implementations of the relevant classifiers are provided by scikit-learn for SVM, *k*-NN, and RF [12], while the ANN implementation was provided by Keras [13]. The implementations used were standard but were parameterized based partially on domain knowledge and partially on iterated exploratory phase tests. With the SVM classifier the main parameter to consider was the kernel. RBF (radial basis function) produced overfitting, and so a fourth-order polynomial kernel was used with the degree of the polynomial being determined empirically. The ANN classifier was designed with 512 input layer neurons, 128 hidden layer neurons, and 2 output layer neurons, with the input and hidden layer neurons having a rectifier activation function, and the output layer having a softmax activation function. The *k*-NN classifier [3] was only parameterized with *k* which was empirically set to 5. The RF classifier was parameterized with 512 decision trees and the Gini impurity index as the criterion for tree splitting [5].

Classifier training. The training and parameter optimization was done as part of a continuous test using an 80/20 training/testing data set split based on random sampling as provided by Scikit-learn's split_data_set function. Most training was henceforth trivial once the classifiers were properly defined, but the ANN required further parameterization by establishing a testing regimen of 200 epochs with 32 samples in each batch and dropout functionality ensuring that 50% of the ANN's neurons were disabled at any one time during training in order for the remainder to better adjust their weights, leading to better control over overfitting. The network optimizer was Nesterov-Adam [13].

3 Experiments

3.1 Experimental Setup

We performed all tests on an Intel Core™ i7-4710HQ central processing unit (CPU), with 8 GBs of RAM, and an NVIDIA GeForce GTX 850M graphics processing unit (GPU). The system was running a Windows 10 Pro 64-bit operating system, and using Scikit-learn 0.18.1, Keras 2.0.3, Theano 0.9.0, OpenCV 3.2, and Python 2.7.

3.2 Experimental Results

Having trained our classifier with the 80% of the data, we tested them with the remainder and got estimates of prediction accuracy presented in Table 1. To

Table 1 Experimental results obtained using the test set

Classifier	Prediction accuracy (%)
ANN	86.42
SVM	93.00
KNN	93.43
RF	94.70

Table 2 Experimental results obtained using 11-fold cross validation

Classifier	Prediction accuracy (%)
ANN	69.34 (±15.85)
SVM	91.12 (±1.90)
KNN	92.91 (±1.20)
RF	94.19 (±1.29)

Table 3 Training time for classifier

Classifier	Time (s)
KNN	0.18
SVM	2.73
RF	15.8
ANN	126.21

confirm that we ought to be confident in the results we got, we then performed 11-fold cross-validation (Table 2), whose results did not vary significantly from our initial measurements.

From the results, we can conclude that the RF method performed best, followed closely by the *k*-NN and the SVM, with the ANN being last. The poor performance of the ANN can be attributed to the relatively simple structure of the used network, as described in Sect. 2.3. Plans for future work include the development of a more complex ANN, with a deep hidden layer structure, and testing its performance on the considered image classification task.

Since our data set is not large (only 5178 images before split), and input for every classifier is an array with only 512 elements, most classifiers can be trained quickly. The outlier, when it comes to performance, is the ANN which required 126.21 s to fully train. The training time results in Table 3 do not include the image processing and histogram computation steps which took around 20 s on their own.

Confusion matrix is used to display which classes are most often misclassified, and what classes they are mistaken for. From the confusion matrix for the Random Forest algorithm (see Fig. 4), we can conclude that the model achieves good generalization across both classes of considered images. Further, we can see that the model has a slight tendency towards 'misfired' images.

We have established that the most important features for classification are brightly colored pixels on the left and right side (corresponding to positions in the mid-200s to high-400s, respectively) of the image. This shouldn't come as a surprise because if we look at the histograms, shown in Figs. 5 and 6, we can see the asymmetrical deficit in high-brightness pixels that misfiring produces.

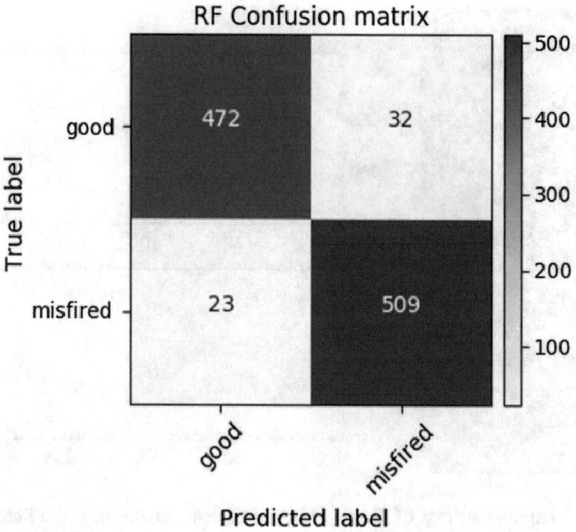

Fig. 4 The confusion matrix

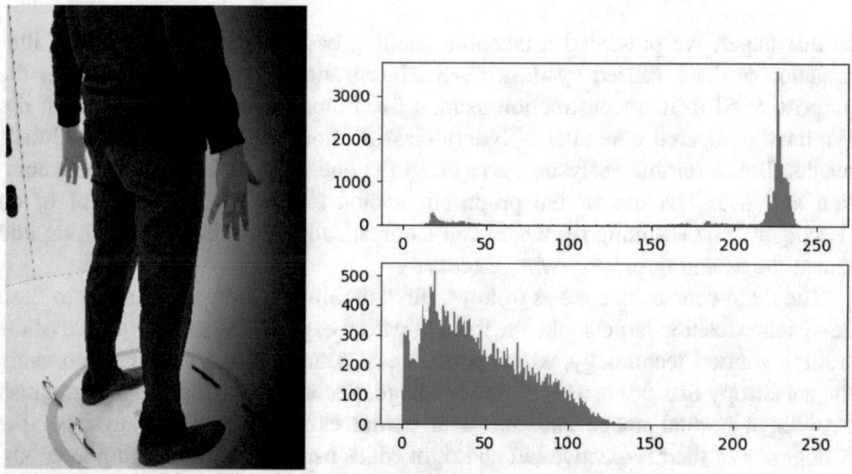

Fig. 5 Example image (courtesy of Doob Group AG) and histogram. Predicted as 'misfired' with 92.38% certainty

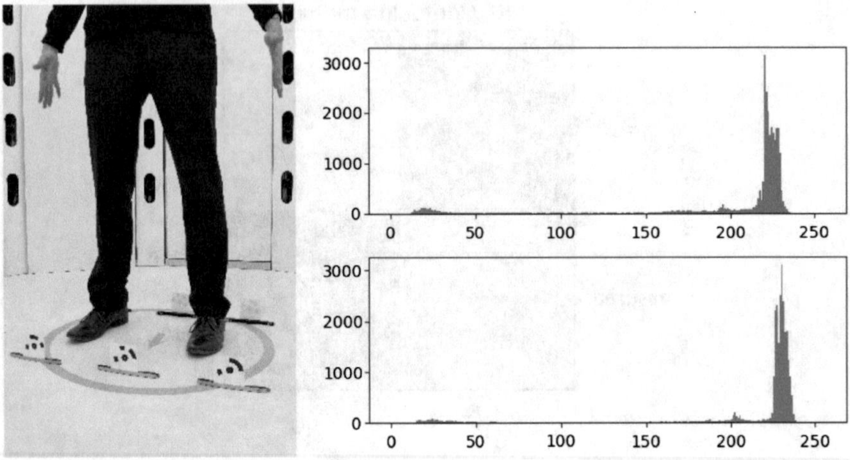

Fig. 6 Example image (courtesy of Doob Group AG) and histogram. Predicted as 'good' with 81.83% certainty

4 Conclusions and Future Work

In this paper, we presented a machine learning-based method for detecting illumination artifacts caused by flash de-synchronization in images acquired for the purpose of SfM 3D reconstruction using a fixed multiple-camera closed-space rig. We have compared a number of binary classification solutions and analyzed their results. Based on this analysis, we selected the one most suitable for implementation and use. The use of the proposed method allows the development of an 'intelligent' 3D scanning rig which can automatically detect camera misfiring and repeat the scanning process when necessary.

The main conclusions are as follows. First, the illumination artifacts due to flash de-synchronization largely take the form of partial exposure as a consequence of the 'rolling shutter' technology, which permits easier analysis and identification using the anisotropy that this causes to its advantage. Second, it is possible to distinguish between a normal image and one with partial exposure by the analysis of the histograms of their respective left and right edges using machine learning methods, with best results achieved using the random forest classifier.

Possible directions for future work include increasing accuracy, especially in edge cases, i.e., eliminating as much as possible both false positive and false negative rates. Methods which may accomplish this can potentially be developed by combining multiple classifiers, making use of camera position and rotation information, and adding additional image descriptors.

Acknowledgements The authors would like to thank Doob Group AG for the support and dataset provided for this research. The reported research is also partly supported by the Ministry of Education, Science, and Technological Development of the Republic of Serbia, projects TR32044 (2011–2017), ON174026 (2011–2017), and III44006 (2011–2017).

References

1. Hartley, R., Zisserman, A.: Multiple View Geometry in Computer Vision, 2nd edn. Cambridge University Press (2004)
2. Luhmann, T., Robson, S., Kyle, S., Boehm, J.: Close-Range Photogrammetry and 3D Imaging, revised 1st edn. De Gruyter (2013)
3. Theodoridis, S., Koutroumbas, K.: Pattern Recognition, 4th edn. Elsevier (2008)
4. Menze, B.H., Kelm, B.M., Masuch, R., Himmelreich, U., Bachert, P., Petrich, W., Hamprecht, F.A.: A comparison of random forest and its Gini importance with standard chemometric methods for the feature selection and classification of spectral data. BMC Bioinformatics 10:213 (2009). doi:10.1186/1471-2105-10-213
5. Random forest. https://en.wikipedia.org/wiki/Random_forest. Last accessed 13 Apr 2017
6. Basten, K.: Classifying Landsat Terrain Images via Random Forests. Radboud University, Nijmegen, Netherlands (2016)
7. Chapelle, O., Haffner, P., Vapnik, V.N.: Support vector machines for histogram-based image classification. IEEE Trans. Neural Netw. **10**(5), 1055–1064 (1999)
8. Melgani, F., Bruzzone, L.: Classification of hyperspectral remote sensing images with support vector machines. IEEE Trans. Geosci. Remote Sens. **42**(8), 1778–1790 (2004)
9. Wong, W., Hsu, S.: Application of SVM and ANN for image retrieval. Eur. J. Oper. Res. **173** (3), 938–950 (2006)
10. Boiman, O., Shechtman, E., Irani, M.: In defense of Nearest-Neighbor based image classification. In: Proceedings IEEE Conference on Computer Vision and Pattern Recognition (CVPR), Anchorage, USA, pp. 1–8 (2008)
11. OpenCV library, www.opencv.org. Last accessed 13 Apr 2017
12. Scikit-learn, http://scikit-learn.org/stable/. Last accessed 26 Apr 2017
13. Keras, https://keras.io/. Last accessed 26 Apr 2017

Part VI
Internet of Things and Cloud Computing

Context Aware Resource and Service Provisioning Management in Fog Computing Systems

Saša Pešić, Milenko Tošić, Ognjen Iković, Mirjana Ivanović, Miloš Radovanović and Dragan Bošković

Abstract Complexity of IoT systems and tasks that are put before them require shifts in the way resources and service provisioning are managed. The concept of fog computing is introduced so as to enhance IoT systems scalability, reactivity, efficiency and privacy. In this paper we present fog computing solution with context aware decision-making procedures distributed between IoT cloud platform and IoT gateways. The solution performs decision-making for smart actuation, based on analysis of sensory data streams, and context informed fog computing resource and service provisioning management based on topology changes. The state-of-the-art mainly focuses either on smart actuation enabled through insightful data analysis and machine learning, or on managing fog system itself in order to improve performance and efficiency. Our solution showcases how one software framework can be used to achieve both. Proof of concept experiments executed on a fog computing testbed validate our solutions performance in improving resilience and responsiveness of the fog computing system in context of topology changes.

Keywords Internet of Things · Fog computing · IoT gateway · MQTT · Data analysis · Decision-making

S. Pešić (✉) · M. Tošić · O. Iković · D. Bošković
Foundation VizLore Labs, Novi Sad, Serbia
e-mail: sasha.pesic@vizlore.com; sasa.pesic@dmi.uns.ac.rs

M. Tošić
e-mail: milenko.tosic@vizlore.com

O. Iković
e-mail: ognjen.ikovic@vizlore.com

D. Bošković
e-mail: dragan.boscovic@vizlore.com

S. Pešić · M. Ivanović · M. Radovanović
Faculty of Sciences Department of Mathematics and Informatics,
University of Novi Sad, Novi Sad, Serbia
e-mail: mira@dmi.uns.ac.rs

M. Radovanović
e-mail: radacha@dmi.uns.ac.rs

M. Ivanović et al. (eds.), *Intelligent Distributed Computing XI*,
Studies in Computational Intelligence 737, https://doi.org/10.1007/978-3-319-66379-1_19

1 Introduction

The importance of cloud computing in Internet of Things (IoT) systems cannot be overstated. Scalability, robustness, computing and storage resources provided by the cloud enabled revolution in smart sensing, monitoring and data analysis (DA) services in the IoT domain. Another important segment of IoT systems are IoT gateways (GWs) which provide an interface between the managed system and the cloud. The legacy approach is to format collected sensory and contextual data at the gateways and transport them to the cloud for analysis. Decisions made based on DA, performed at the cloud, are then transferred to the GWs to be translated into execution commands. This approach introduces several major issues limiting wider acceptance of IoT: The Internet is still the best effort service and relying too much on the cloud limits control of the whole process and introduces a single point of failure; Inefficient bandwidth usage; Latency in the decision-making (DM) chain; Privacy and security. To address these challenges, modern IoT systems introduced the concept of fog computing [1, 2].

Fog computing (see Fig. 1) is a decentralized architecture that brings computational resources and application services closer to data sources. It creates an environment for a new type of applications and services that rest on responsiveness, privacy protection, location awareness, with improved quality of service for direct streaming of data [3]. In IoT systems, fog computing aggregates and enables utilization of locally available computing, communication and storage resources. A fog computing system, like any other highly dynamic and distributed system, is vulnerable to node and link failure. Network management approaches like software defined networking and advanced routing protocols provide certain level of robustness. If the network of IoT GWs performs only data formatting and bridging to the cloud platform, then these network management solutions are very effective in overcoming topology disturbances. But what happens when GWs are responsible for DA and reactive DM? Load balancing and handover allow adapting the topology so as to compensate coverage holes as a result of node/link failure.

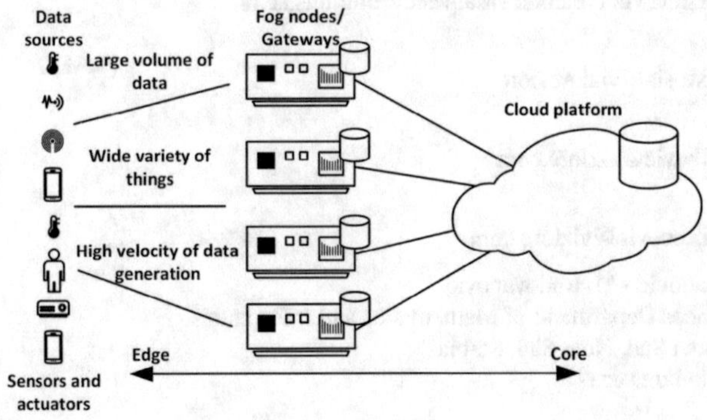

Fig. 1 Fog computing architecture

However, they cannot address the DM processes that go offline. One solution is a failover mechanism towards the cloud platform. This is not always possible nor desirable, especially for fog computing systems that rely on data privacy and low latency. Therefore, the fog computing systems need a solution which allows participating nodes to take over not only data streams from their failing peers, but also DA and DM processes. In this paper we present a software framework which provides a solution to this challenge.

The presented solution employs a novel approach for scalable DA and hierarchically distributed DM for context-aware IoT services ensuring full adaptivity of the fog computing processes to the dynamism of the environment. The IoT GWs represent a distributed fog computing environment and perform protocol translation, data analysis, reactive DM and smart actuation.

The MQTT[1] protocol is used for data and command transportation. The cloud platform hosts the main MQTT broker with the mechanism for orchestration of local brokers deployed on IoT GWs and updating their publisher and subscriber bases in line with topology changes. The introduced solution is based on the requirement that each IoT GW in the fog system can take over sensors, analytical and DM processes from surrounding GWs.

In Sect. 2 we present the analysis of related work. Next, in Sect. 3 we present our software framework. Finally, we validate the presented solution through proof of concept experiments.

2 Related Work

The importance of intelligent distributed computing in the IoT has been promoted by many authors as a key technique to enable near real-time DA, reduce latency and create independent, context-aware devices [4, 5]. The state-of-the-art mainly focuses either on smart actuation enabled through insightful data analysis and machine learning, or on auto-managing a fog system in order to improve performance and efficiency. Our solution showcases how one software framework can be used to achieve both. A mission critical fog computing system must be resilient to constant node failures and redundancy must be leveraged to substantially reduce the bandwidth consumption and latency. While many authors stress out the importance of fog nodes throughput [6, 7] through employing different approaches, we argue that correct node failure management and load balancing strategy, to ensure the stability and resilience of fog systems are equally important, especially for context-aware, mission-critical IoT fog systems.

While authors of [8] tackle node failure management, their attention is rather on enabling computational continuity than on distributing the workload of the failed node. Computational continuity is a much more complex challenge when taking into account the fact that different GWs have different roles in an IoT system. The

[1]Message Queuing Telemetry Transport: http://www.mqtt.org.

system restores optimal performance as soon as the node is back online, however works with noted lowered accuracy while nodes are offline. Our fog computing software framework is used for realization of services and applications for end users and the managed system (i.e. smart building, home automation and microgrid) and for context-aware management of fog computing resources (storage, computing and networking) and processes. The introduced solution focuses on enhancing system capabilities for detecting and acting upon node failure. When a fog computing node fails, other capable nodes will automatically pick up the pace, according to the load-balancing strategy decided on the cloud.

3 Context Aware Resource Management in the Fog

The presented solution is comprised of three main mechanisms (see Fig. 2). There is a MQTT publish/subscribe framework distributed across two hierarchical layers, a Soft sensors approach for scalable and hierarchically distributed DA [9, 10], as well as context informed DM for smart actuation and fog system management.

The main feature of the presented fog computing solution is its' adaptability to the changing context in the controlled IoT system and fog topology. This is achieved through dynamic, context informed management of the fog resources where DM is distributed between the IoT GWs and the IoT cloud platform. The cloud platform manages all of the IoT GWs deployed in the fog system and maintains overall topology tables.

3.1 Publish/Subscribe Mechanism—MQTT

Publish/subscribe mechanisms are a suitable choice for machine-to-machine correspondence due to loose coupling and simple communication architecture [11]. One of the widely accepted standards is the MQTT protocol which passes messages between multiple clients through a central broker. It has a strong foothold in IoT and fog computing systems [12, 13].

In our solution, MQTT is in charge of bridging IoT GWs and ensuring communication with the cloud broker (see Fig. 2). There is a local broker on each GW and every DA and machine learning process is implemented as a software module with MQTT publish and subscribe functionality. Also, physical sensors connected to IoT GWs have virtual representation in the GW system with integrated MQTT publisher, which is achieved with the local FHEM server[2]. Local MQTT bridging enables internal and external exchange of sensory and contextual data, analytical results and actuation commands. DA and DM processes from one IoT GW can access results and physical sensor readouts from another GW through the local network. There is no

[2]FHEM: http://www.fhem.de.

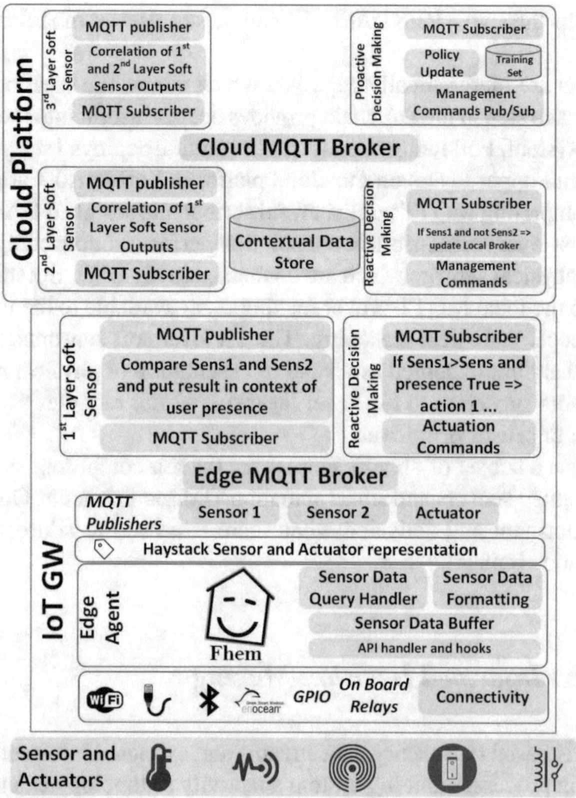

Fig. 2 Proposed solution

need for transmitting data streams to the cloud. Subscriber processes on IoT GWs are listening for MQTT Last Will and Testament messages, enabling detection of peers that have disconnected from the network ungracefully. The cloud based MQTT broker is used for distributing policy updates towards the local brokers (broker lists, handover tables and DM policies). This enables a high level of customization of how a local level decision algorithm operates (i.e. set a new thresholds or statistical reasoning rule). Also, the cloud MQTT broker receives IoT GW status data enabling it to maintain up to date topology tables.

3.2 Soft Sensor Approach for Distributed Data Analysis

Presented solution distributes complex DA models, featured in many IoT systems, across self contained software modules performing simple analytical processes and exposing results through a REST API. We call these software modules ‘soft sensors’ [9, 10].

Soft sensors are hierarchically deployed which allows the creation of analytical chains where each step in the DA chain provides deeper insight into the context of the managed IoT system. Following this logic, our solution deploys 1st layer soft sensors on IoT GWs and upper layers on the cloud platform. Soft sensors are implemented as modules comprising MQTT publisher, subscriber and assigned DA operation (see Fig. 2). The first layer soft sensors in our solution perform simple comparison of data coming from physical sensors which are connected to the GWs. Soft sensors publish their results to the local MQTT broker making them available to the upper layer soft sensors for deeper statistical reasoning. The 1st layer soft sensors subscribe to the MQTT publishers implemented as virtual representation of physical sensors. Upper layer soft sensors subscribe to the lower layer soft sensor modules as well as virtual representation of sensor primitives.

IoT GW runs a subset of all soft sensors for the fog computing system based on the set of appointed sensors and smart actuation/DM requirements. Other soft sensor modules are dormant and activated when there is a need to address the topology changes (i.e. node outage).

3.3 Context Informed Decision-Making

Reactive DM is based on sensory data streams and extracted information. It enables smart actuation processes and fog system adaptivity to topology changes. Our solution employs reactive as well as proactive DM, which is more strategic and policy oriented. Based on system monitoring, the proactive algorithms produce updated actuation policies and fog computing system topology tables and handover opportunities. Reactive DM algorithms have two roles in the implemented fog computing solution. The first role is focused on smart actuation and notifications towards the end users based on outputs of soft sensors. The second group manages the fog computing system itself. Based on status of the IoT GWs and their surroundings, reactive DM algorithms conduct sensor stream and process handover.

Figure 3 shows a flow diagram of the solution for managing topology updates and performing handover of processes between GWs. When MQTT subscriber of a locally implemented process cannot access a topic from the predefined publisher, it will start the round robin procedure for checking if the topic is available on its’ immediate neighbours. The list of immediate neighbours is managed by the cloud. The subscriber will ping the local broker on each node maximum 5 times to trigger a handover on the selected neighbouring node(s). This procedure signalizes the DM process for managing handovers to start local physical sensor virtualization module on a neighbouring IoT GW. Next it will start the soft sensor and DM modules originating from the failed node on neighbouring node(s). Finally, it will update the local brokers. The cloud platform constantly monitors status of IoT GWs and detects failing nodes. It then waits a period of time for fog based failover mechanisms to

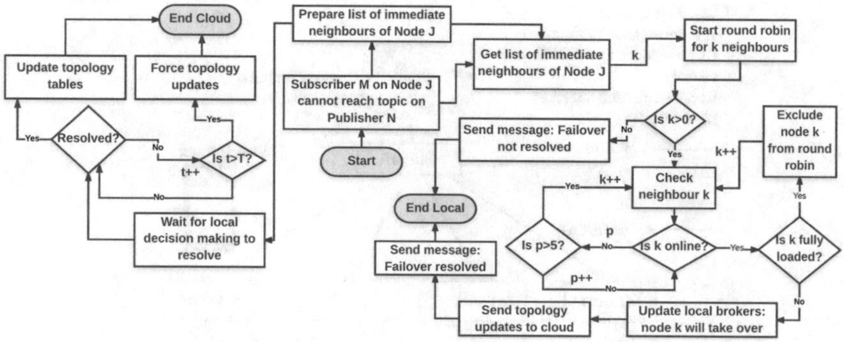

Fig. 3 Failover management and topology update mechanism

converge. If the problem is not locally resolved after the defined period, the cloud system will bypass local DM process and update local MQTT brokers.

We argue that this software framework can be successfully utilized for: 1. performing smart actuation and context informed notification services towards end users and administrators of the underlying system (i.e. smart building, smart home, etc.); and 2. enabling adaptivity in the fog system itself through context informed failover and handover management.

4 Experiments and Results

The proof of concept testbed is set up as a fog computing system required to provide the highest level of privacy (all sensory data stay in the edge) while ensuring 50% robustness, meaning that the fog computing system maintains full operation in case that half of IoT GWs fail. The following features of the presented fog computing solution are validated in the proof of concept experiments: 1. Ability of IoT GWs to take over sensory data streams from failing node; 2. Ability of IoT GWs to take over DA and DM processes from failing nodes; 3. Fog computing systems capability to maintain all analytical and DM processes in case of topology changes ensuring required level of privacy.

Let's assume there are n DA processes and m DM processes in the fog computing system uniformly distributed between g IoT GWs. Each GW operates n/g soft sensors and m/g DM processes. Each DA process is linked with one DM process (actuation, notification or topology update). Because of the MQTT mechanism configuration, they do not need to reside on the same node. Further on, we assume that each soft sensor and DM process require the same amount of CPU and RAM load (c and r respectively) and approximate linear dependency between system CPU/RAM load and the number of processes. The topology of the fog computing system that we analyze here is such that each GW has at least two neighbouring nodes. If one GW fails, the remaining GWs need to take over its' sensory data streams and $(n+m)/g$

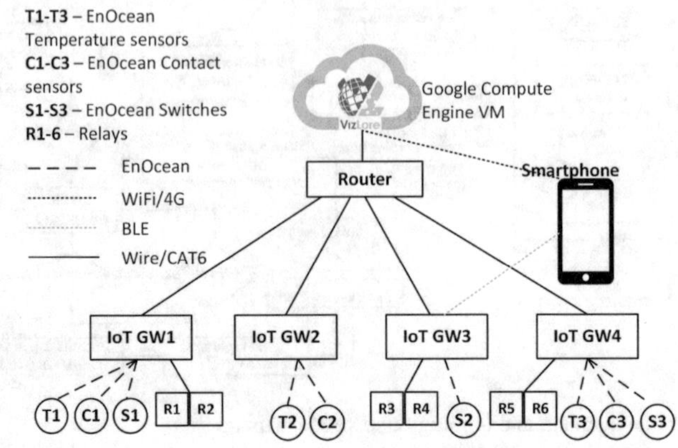

Fig. 4 Deployed fog computing testbed

processes. Our topology update algorithm will first try to distribute these processes to the neighbouring GWs and then increase hop by one until all processes are assigned. Therefore, neighbouring nodes will take on them additional $c \cdot (n + m)/g$ CPU and $r \cdot (n + m)/g$ RAM load.

For a simple simulation we observe a fog computing system with 20 GWs, 50 DA and 50 DM processes. Each GW is burdened with the number of processes set by the context in which the IoT system operates. The topological distribution of GWs reflects the context-awareness feature of the fog system in contrast to pure fault tolerance/redundancy improvement. In order for system to work at full capacity, all GWs should be able to run instances of all processes, which is not the case. Assuming that a GW can run maximum 10 processes, if 2 GWs, with 5 processes each (10%) fail, 10 processes can be distributed to two neighbouring nodes. Following this logic, and with proper topological distribution of failing nodes, we can conclude that 10 GWs (50%) can fail and all processes will remain active, with the remaining 10 GWs working at full capacity, with 0 downtime and no additional cost for maintaining redundancy.

Now, we test the presented logic on the testbed comprised of cloud platform (virtual machine deployed on the Google Compute Engine—GCE VM), four IoT GWs deployed on Raspberry Pi 3 platform, 6 EnOcean[3] sensors (902 MHz, temperature and contact), 3 EnOcean switches, 6 relays, IP router and a smartphone with custom mobile application (see Fig. 4). The testbed integrates MQTT framework (Mosquito library[4]) with hierarchically distributed brokers, the Project Haystack[5] data model and API, low energy bluetooth presence detection soft sensor, and DA and DM modules deployed on GWs and GCE VM.

[3]EnOcean: http://www.enocean.com.

[4]Mosquitto: http://www.mosquitto.org.

[5]Haystack: http://www.haystacktechnologies.com.

To test the presented handover and failover processes, the testbed includes the following redundancy enablers: 1. 802.11n mesh networking between IoT GWs next to the wired connection to the local switch; 2. Three GWs equipped with two relays enabling them to take over actuation process from a failing node with proper local wiring. GCE VM communicates with all testbed devices and runs algorithms for policy-update at the lower levels together with algorithms for GWs failure detection. The smartphone app notifies the end user about network issues and actuation processes success. Thanks to the large coverage area and broadcasting nature of EnOcean technology sensors and configuration of local FHEM servers, IoT GWs are able to automatically pick up sensors originally assigned to nodes that have failed. Our coverage tests indicate that EnOcean sensors have range of up to 20 m in a closed space [14] which makes the proof of concept experiment even more relevant for real world deployments like smart buildings.

The testbed integrates simple DM scripts which combine one or more physical sensors and soft sensors for real time actuation: 1. Comparing temperature reading with a threshold; 2. Comparing temperature readings of two selected sensors; 3. Combination of a temperature sensor reading and status of a contact sensor and/or switch; 4. Combination of a temperature sensor reading and bluetooth presence detection. Different outcomes of these analysis procedures trigger different actuation actions (notification or relay trigger) in DM modules. Development of more complex DM processes will be a part of our future work.

In order to validate ability to adapt the fog computing system to topology changes (node failures) we have tested all possible combinations of one, two and three GWs failing. We have made measurements of CPU/RAM and network load. We have also monitored output of all system soft sensors and DM processes through the smartphone app, which is part of the local MQTT framework. Figure 5 presents results for four experimental steps: step 1 (GW1 fails), step 2 (GW1 and GW2 fail), step 3 (GW2 and GW3 fail) and step 4 (all but GW4 fail). It is evident that the deployed fog computing system is capable of withstanding failure of all but one IoT GW (75% redundancy). The system load is balanced between working IoT GWs in each experimental step. GW4 has enough capacity even when it takes over all processes.

Our research hypothesis, followed throughout this paper, is that a software framework comprising hierarchically distributed MQTT brokers and DA (achieved through soft sensors) improves resilience and responsiveness of the fog computing system. It also enables management of privacy by not requiring fog systems to send all sensory data towards the cloud platform. By packaging DM and DA algorithms as contained software modules with MQTT publish and subscribe procedures we have created a basis for enabling migration of all fog computing processes between nodes participating in the system. Our approach requires that all IoT GWs are configured so as to include all DM processes some being active and other on standby at each time. The presented failover mechanism (Fig. 3) addresses topology changes so as to balance the load and perform sensor and process handover to able nodes. The fog computing system utilizing our resource and process management framework depends on the cloud platform just for policy updates and is capable of performing all reactive DM processes independently even when 50% of participating GWs fail.

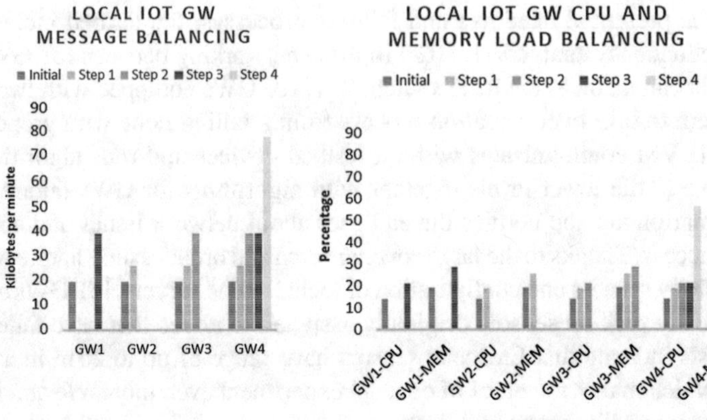

Fig. 5 **a** CPU and RAM load. **b** Message load

We are aware that the proposed fog computing solution seems very situational in a sense that it requires careful system and topology planning. However, the proof of concept experiments show that it provides a solid basis for a generic solution enabling robustness in fog data streams as well as DM processes with one fog computing software framework.

5 Conclusion and Future Work

In this paper we have presented a software framework for management of resources and service provisioning in the fog computing systems. We have deployed the proof of concept testbed and conducted experiments showcasing the solutions ability to perform smart actuation and notification services as well as context informed failover and handover management between IoT GWs. The proof of concept experiments validate the solutions ability to adapt to changes in the fog system topology and perform timely failover and handover. For the future work we will deploy the developed solution in the smart building testbed comprising 150 apartments which will include around 200 IoT GWs. Finally, our strategy is to integrate cloud based management of fog computing topology and resource distribution with software defined networking (SDN) controller and OpenFlow enabled switches (already initiated in [9, 10]).

References

1. Da Xu, L., He, W., Li, S.: Internet of things in industries: a survey. IEEE Trans. Ind. Inf. **10**, 2233–2243 (2014)
2. Sicari, S., Rizzardi, A., Grieco, L.A., Coen-Porisini, A.: Security, privacy and trust in Internet of Things: the road ahead. Comput. Netw. **76**, 146–164 (2015)

3. Luan, T.H., Gao, L., Li, Z., Xiang, Y., Wei, G., Sun, L.: Fog computing: focusing on mobile users at the edge. arXiv preprint arXiv:1502.01815 (2015)
4. Ciccozzi, F., Spalazzese, R.: MDE4IoT: supporting the internet of things with model-driven engineering. In: International Symposium on Intelligent and Distributed Computing, pp. 67–76. Springer (2016)
5. Piette, F., Caval, C., Dinont, C., Seghrouchni, A.E.F., Taillibert, P.: A multi-agent approach for the deployment of distributed applications in smart environments. In: International Symposium on Intelligent and Distributed Computing, pp. 37–46. Springer (2016)
6. Burchard, J., Chemodanov, D., Gillis, J., Calyam, P.: Wireless mesh networking protocol for sustained throughput in edge computing. In: 2017 International Conference on Computing, Networking and Communications (ICNC), pp. 958–962. IEEE (2017)
7. Xu, Y., Mahendran, V., Radhakrishnan, S.: Towards SDN-based fog computing: Mqtt broker virtualization for effective and reliable delivery. In: 2016 8th International Conference on Communication Systems and Networks (COMSNETS), pp. 1–6. IEEE (2016)
8. Wagle, S.: Semantic data extraction over MQTT for IoTcentric wireless sensor networks. In: International Conference on Internet of Things and Applications (IOTA), pp. 227–232. IEEE (2016)
9. Tosic, M., Ikovic, O., Boskovic, D.: SDN based service provisioning management in smart buildings. In: 2016 39th International Convention on Information and Communication Technology, Electronics and Microelectronics (MIPRO), pp. 754–759. IEEE (2016)
10. Tosic, M., Ikovic, O., Boskovic, D.: Soft sensors in wireless networking as enablers for SDN based management of content delivery. In: 2016 39th International Convention on Information and Communication Technology, Electronics and Microelectronics (MIPRO), pp. 559–564. IEEE (2016)
11. Karagiannis, V., Chatzimisios, P., Vazquez-Gallego, F., Alonso-Zarate, J.: A survey on application layer protocols for the Internet of Things. Trans. IoT Cloud Comput. **3**(1), 11–17 (2015)
12. Hakiri, A., Berthou, P., Gokhale, A., Abdellatif, S.: Publish/subscribe-enabled software defined networking for efficient and scalable iot communications. IEEE Commun. Mag. **53**(9), 48–54 (2015)
13. Singh, M., Rajan, M., Shivraj, V., Balamuralidhar, P.: Secure MQTT for Internet of Things (iot). In: 2015 Fifth International Conference on Communication Systems and Network Technologies (CSNT), pp. 746–751. IEEE (2015)
14. Anders, A.: Enocean Wireless Systems—Range Planning Guide. EnOcean Gmbh (2008)

An Argumentative Approach to Smart Home Office Ambient Lighting

Andrei Mocanu

Abstract Numerous studies have linked lighting conditions to how well humans have performed their daily activities. We have designed a system for the purpose of increasing productivity by tailoring the ambient lighting for various tasks, but which at the same time can gain the trust of its users by exposing the way it "thinks" through computational argumentation. With the recent emergence of smart bulbs, we now have the means for creating a software system that is able to understand which task is performed and adapt to it accordingly. While the key role of this system is to make activities more pleasurable and easier to perform, it can also predict future activities and make suitable recommendations.

Keywords Argumentation · Ambient lighting · Internet of Things · Smart home

1 Introduction

Lighting conditions have a significant influence on human behaviour including mental workload [15], focus and concentration [1], mood [12] and even on food choices [4]. Fortunately, recent breakthroughs in technology make lighting easier to control than ever, with smart bulbs being able to control basic parameters such as brightness and colour, while offering even more advanced features such as geofencing, motion control, emergency alerts, syncing with music, movies [5, 6, 17] or even video games.[1]

Perhaps the most widespread smart lighting solution at the time of writing is represented by Philips Hue. While it has its flaws, most notably various security that have been uncovered [14, 16] but ultimately fixed, it remains one of the most versatile smart home lighting solutions available on the market. Essential to development

[1] http://www.philips.com/a-/about/news/archive/standard/news/press/2015/20150715-Worlds-first-video-game-to-be-synched-with-home-lighting-from-Philips.html.

A. Mocanu (✉)
University of Craiova, Blvd. Decebal nr. 107, Craiova, Romania
e-mail: mocanu.andrei@ucv.ro

M. Ivanović et al. (eds.), *Intelligent Distributed Computing XI*,
Studies in Computational Intelligence 737, https://doi.org/10.1007/978-3-319-66379-1_20

is the fact that Philips provides a RESTful API[2] for the Hue range that is well documented and doesn't have a steep learning curve.

While some systems, including [11], strive to create an interface for an entire intelligent living environment, since a significant part of the day is spent working, we chose the office as the main area of focus for our application. Unfortunately, most office work tends to involve extensive time spent in front of the screen which strains the eyes and doesn't include much exercise, which in turn poses potential health problems. It is in this context that we can develop a system that can understand the task that is currently performed, anticipate the next one and provide useful suggestions for the user while maintaining the desired level of privacy.

2 Background

Argumentation [2, 10] is a research field that is concerned with the interaction of parties which contradict or support some conclusion, and which, through this interaction, produces justifications for the selected arguments.

2.1 *Abstract Argumentation*

Abstract argumentation was first introduced by Dung [7] in an attempt to model the engagement in arguments as a graph and the winner of the dispute on the principle that "He who has the last word laughs best!".

Definition 1 ([7]) An argumentation framework is a pair AF = $\langle \mathcal{A}_{rgs}, \mathcal{A}_{tt} \rangle$ where $\mathcal{A}_{rgs}$ is a set of arguments, and $\mathcal{A}_{tt}$ is a binary relation between arguments. We will use the notation (A, B) to denote that argument A attacks argument B.

Definition 2 ([7]) (1) An argument A is **acceptable** wrt. a set S of arguments iff for each argument B: if B attacks A then B is attacked by some argument in S.

(2) A conflict-free set of arguments S $\subseteq \mathcal{A}_{rgs}$ is **admissible** iff each argument in S is acceptable wrt. S.

(3) A set of arguments S $\subseteq \mathcal{A}_{rgs}$ is **preferred** if it is admissible and maximal wrt. to set inclusion (i.e. $\nexists S' \subseteq \mathcal{A}_{rgs}$ such that S' is admissible and $S \subset S'$.

To illustrate these notions we will use an example in which two sports fans discuss who is the greatest tennis player of all time.

Example 1 **Alice**: Roger Federer is the greatest player of all time. (a_1)

Bob: Pete Sampras dominated the game in a time when there were a lot more contenders. (b_1)

---

[2] Available at https://www.developers.meethue.com/philips-hue-api.

Alice: Federer defeated Sampras in their only Grand Slam match. (a_2)

Bob: Other players such as Rafael Nadal or Novak Djokovic lead Federer in head-to-head matches.(b_2)

The exchange of information can be represented through the argumentation framework $\langle \mathcal{A}_{rgs}, \mathcal{A}_{tt} \rangle$ where $\mathcal{A}_{rgs}$=$\{a_1, b_1, a_2, b_2\}$ and $\mathcal{A}_{tt}$=$\{(b_1, a_1), (a_2, b_1), (b_2, a_2)\}$. We have also represented the framework in the form of a graph as depicted in Fig. 1.

The admissible sets for this framework are $\{b_2\}$ and $\{b_1, b_2\}$, because in the first case b_2 is unattacked, and in the latter b_2 is able to defend b_1 from a_2's attack. The only preferred extension is $\{b_1, b_2\}$.

2.2 Weighted Argumentation Frameworks

Weighted argumentation frameworks [3, 9] have emerged as an answer to the criticism regarding the inability of abstract argumentation to model the differences in strength of various arguments.

Definition 3 ([3, 9]) A Weighted Argumentation Framework (*WAF*) is a tuple $\langle \mathcal{A}_{rgs}, \mathcal{A}_{tt}, w \rangle$ where $\mathcal{A}_{rgs}$ is a set of arguments, $\mathcal{A}_{tt}$ is a set denoting the binary attack relation between arguments, and $w : \mathcal{A}_{tt} \rightarrow \mathbb{R}^+ \cup \infty$ is a function of weights over attacks.

Example 2 In Fig. 2 we have depicted an example of a Weighted Argumentation Framework. The difference between this representation and that of a regular Abstract Argumentation framework is represented by the edge weights.

Let us examine the main definitions from [3] and [9]. We shall use the sets A = $\{a_1, a_2, a_3, a_4\}$ and B = $\{b_1, b_2, b_3, b_4\}$ for exemplification.

Definition 4 ([3]) Let S and S' be two sets such that $S, S' \subseteq \mathcal{A}_{rgs}$ and let $a \in \mathcal{A}_{rgs}$. We say that:

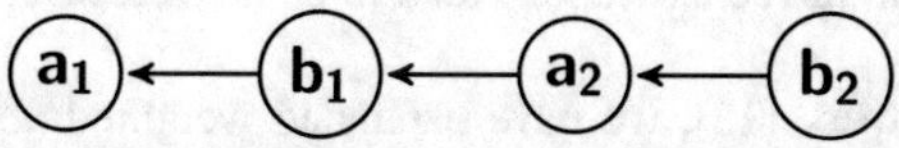

Fig. 1 Abstract argumentation framework for the "Inception" example

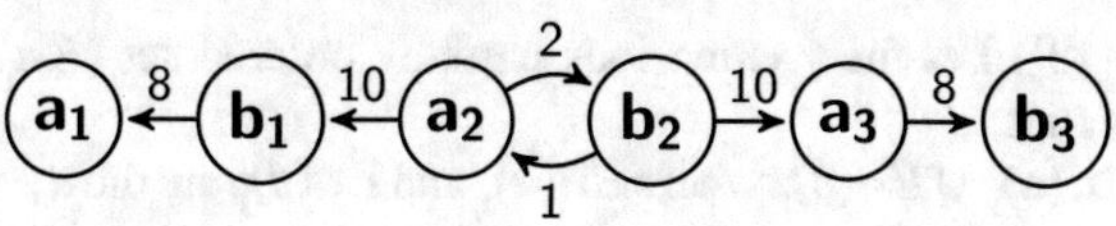

Fig. 2 A sample weighted argumentation framework

- S attacks a with strength k, iff $\sum_{b \in S} w(b, a) = k$;
- S attacks S' with strength k, iff $\sum_{a \in S', b \in S} w(b, a) = k$;

In our example in Fig. 2 set A attacks b_1 with strength 10. Set A attacks set B with strength $10 + 2 + 8 = 20$.

Definition 5 ([3]) An extension $S \subseteq \mathcal{A}_{rgs}$ is α-conflict-free if $w(S, S) \leq \alpha$.

Going back to our example, set A and B are 0-conflict-free. Set $\{a_2, b_2\}$ is 3-conflict-free and indeed z-conflict-free where $z \geq 3$.

Definition 6 ([3]) An extension $S \subseteq \mathcal{A}_{rgs}$ defends $a \in \mathcal{A}_{rgs}$ iff $\forall b \in \mathcal{A}_{rgs} \backslash S$ such that $(b, a) \in \mathcal{A}_{tt}$, $w(b, a) \leq w(S, b)$.

In Fig. 2, $\{a_2\}$ defends a_2 because $w(a_2, b_2) \geq w(b_2, a_2)$. The set $\{a_2\}$ does not defend a_3 because $w(a_2, b_2) < w(b_2, a_3)$.

Definition 7 ([3]) An extension $S \subseteq \mathcal{A}_{rgs}$ is α-admissible iff it is α-conflict-free and it defends each of its elements. An α-admissible extension $S \subseteq \mathcal{A}_{rgs}$ is α-preferred if it is maximal, with respect to set inclusion.

In our example, the 0-preferred extension is $\{a_1, a_2\}$, while the 3-preferred extension is $\{a_1, a_2, b_2, b_3\}$.

Definition 8 [9] Let **sub** be a function which takes an attack relation $\mathcal{A}_{tt}$, weight function $w : \mathcal{A}_{tt} \rightarrow \mathbb{R}^+$ and an inconsistency budget $\beta \in \mathbb{R}^+$ and returns the set of sub-graphs R of $\mathcal{A}_{tt}$ such that the edges in R sum to at most β.

$$sub(\mathcal{A}_{tt}, w, \beta) = \{R : R \subseteq \mathcal{A}_{tt} \ \& \sum_{e \in \mathcal{A}_{tt}} w(e) \leq \beta\}$$

Proposition 1 [9] *The β-extensions (admissible, preferred, grounded, etc.) are computed according to the original semantics, disregarding weights, on the abstract argumentation frameworks $\langle \mathcal{A}_{rgs}, \mathcal{A}_{tt} \backslash R \rangle$ where R is a set of edges returned by the* ***sub*** *function.*

In our example, for $\beta = 0$ we have two 0-preferred extensions, the sets we have initially named A and B. For $\beta = 3$ there is one 3-preferred set $\{a_1, a_2, b_2, b_3\}$. We can observe that α-preferred extensions tend to be more restrictive than β-preferred extensions.

In our previous work [13], we have expanded weighted argumentation frameworks by additionally computing γ-extensions and introduced *probabilistic argument weighted argumentation frameworks* (PAWAF).

Definition 9 ([13]) Let the γ-extensions partially ordered set $\langle Ext_\gamma, \leq_\gamma \rangle$ have the following properties:

(1) $Ext_\gamma = Ext(\alpha)_\gamma \cup Ext(\beta)_\gamma$ where $Ext(\alpha)_\gamma$ and $Ext(\beta)_\gamma$ are the α and β extensions respectively for $\alpha = \beta = \gamma$;

(2) A partial ordering relation $\leq_\gamma$ on the elements $a, b \in Ext_\gamma$ for which $a \leq_\gamma b$ iff

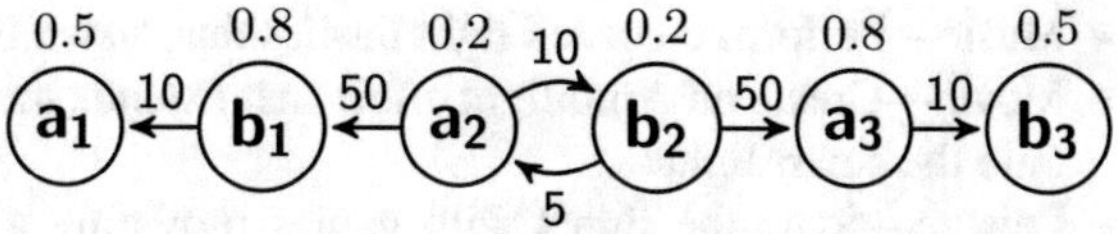

Fig. 3 A sample probabilistic argument weighted argumentation framework

- $a, b \in Ext_{\gamma'}, a \in Ext_{\gamma''}, b \notin Ext_{\gamma''}, \forall \gamma, \gamma', \gamma'' \in \mathbb{R}^+$ satisfying $\gamma'' \leq \gamma' \leq \gamma$;
- $a, b \in Ext_{\gamma'}, a, b \notin Ext_{\gamma''}, \forall \gamma, \gamma', \gamma'' \in \mathbb{R}^+$, $a \in Ext(\alpha)_{\gamma'}, b \notin Ext(\alpha)_{\gamma'}$ satisfying $\gamma'' \leq \gamma' \leq \gamma$.

In other words, we find the union of the α-solutions and β-solutions for a given $\alpha = \beta = \gamma$, and then we have a relation of preference among solutions for which we prefer solutions with lower γ score and in the case of equal γ scores we consider the α solutions first, which have the advantage of weight-defending each of their elements.

Definition 10 ([13]) A Probabilistic Argument Weighted Argumentation Framework (PAWAF) is a tuple $\langle \mathcal{A}_{rgs}, \mathcal{A}_{tt}, w, p \rangle$ where $\mathcal{A}_{rgs}$ is a set of arguments, $\mathcal{A}_{tt}$ is a set denoting the binary attack relation between arguments, $w : \mathcal{A}_{tt} \rightarrow \mathbb{R}^+ \cup \infty$ is a function of weights over attacks and $p : \mathcal{A}_{rgs} \rightarrow [0, 1]$ is a function of probability over arguments.

Proposition 2 ([13]) *A Probabilistic Argument Weighted Argumentation Framework can be reduced to a Weighted Argumentation Framework by composing each attack weight with the probability of the attacker:* $w_p(a, b) = w(a, b) * p(a)$ *where* $a, b \in \mathcal{A}_{rgs}$ *and* $(a, b) \in \mathcal{A}_{tt}$.

The Probabilistic Argument Weighted Argumentation Framework depicted in Fig. 3 can be reduced to the Weighted Argumentation Framework in Fig. 2 by multiplying each argument probability by the attack strength. For example, a_2 with probability 0.2 attacks argument b_2 with strength 10. Thus, the resulting attack strength is 2. The probability of a_1 and b_3 do not matter in this example because these arguments do not attack anything.

3 System Description

The purpose of the system we wish to design is to augment user activities while at his/her home office by connecting with a smart lighting solution such as the Philips Hue. We will consider four basic activities that each user can perform while in the home office: work, watch a movie, listen to music, engage in leisure activities (such as a casual game). Additionally, we must consider a fifth state, which we label AFK (away-from-keyboard) during which the system is on, but the user is not at the workstation.

Each of the five states will trigger a different behaviour for the lights:

- Work—Set the lights to the *Concentrate* preset to help the user focus

- Music—Perform coloured light flashes that match the rhythm of the music
- Movie—Create an Ambilight effect that extends the colours shown on the screen onto the smart lights
- Leisure—Sync the lights with games providing a more immersive experience through in-game effects (e.g. flash red when in danger)
- AFK—switch off bulbs

Currently, there are separate applications produced by third parties which provide the ability to sync the lights with the aforementioned activities, but our goal is to have a single application which is capable of invoking the desired behaviour depending on the activity that is performed. Moreover, the system needs to anticipate the next activity that will be performed in order to provide meaningful suggestions to the user.

3.1 Training Phase

Before the system can be utilized, it needs to be trained in order to categorize each application that is used on the user's machine. This can be done automatically, using an online database, and then can be manually fine-tuned by each user. For example, Fig. 4 represents a classification for a sample workstation using the Microsoft Windows operating system. While some applications have precise roles (e.g. Eclipse IDE is used for working, Spotify is used for listening to music, etc.), others can be used for multiple purposes. For example, Quicktime can be used for watching a movie,

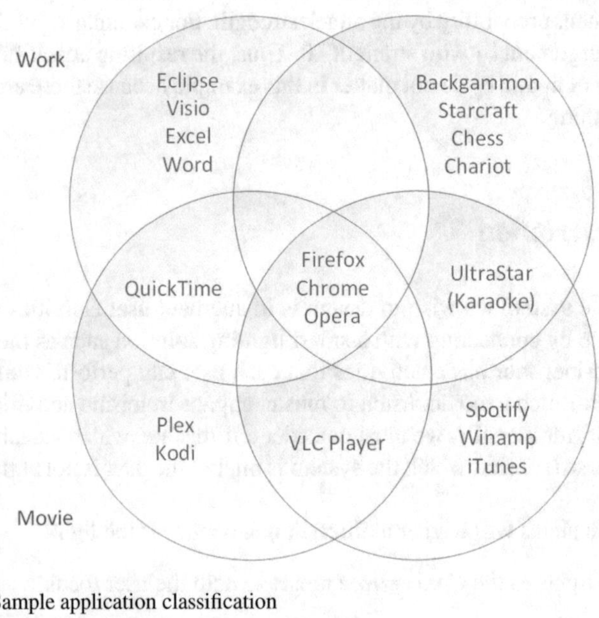

Fig. 4 Sample application classification

```
private static String getCurrentProcess()
{
  char[] bfr = new char[MAX_LEN];
  PointerByReference ptr = new PointerByReference();
 HWND frgndWnd = User32DLL.GetForegroundWindow();
  User32DLL.GetWindowThreadProcessId(frgndWnd, ptr);
  Pointer proc =
       Kernel32.OpenProcess(Kernel32.PROCESS_QUERY_INFORMATION |
       Kernel32.PROCESS_VM_READ, false, ptr.getValue());
  Psapi.GetModuleBaseNameW(proc, null, bfr, MAX_LEN);
  String procName = Native.toString(bfr);
  return procName;
}
```

Fig. 5 Java function for retrieving the active process

but can also be used for watching webinars for work. Furthermore, a web browser can be used for all forementioned activities.

In the same training phase, we must establish two measures that are central to our PAWAF-based model: how likely an activity is to follow another after a given amount of time (i.e. the argument probability), how unlikely it is for two activities to be simultaneous (i.e. the attack strength). The two measures have intuitive rationale behind them. For instance, if someone just started working it is unlikely that they will take a break right away for another activity. However, if someone has been working for a long time, it is quite probable that they will start doing something else soon. Similarly, working and listening to music are two activities that could go together, whereas working and watching a movie are more difficult to perform at the same time.

After several hunderd hours of training, a custom user model can be obtained that should reflect individual preferences. Over the course of time, the model will become more accurate, but the user is also given the choice to use the system right away with default parameters.

We also need a method for retrieving which process is active during a given moment. Since this is specific to each operating system, the solution involves using native system function calls. The Java Native Access[3] greatly simplifies the interaction by incorporating this functionality in natural Java method invocation. An example of such a method for the Microsoft Windows system is listed in Fig. 5, through which the foreground process name is retrieved.

3.2 *Computation Phase*

In Dispute Trees [8] the Proponent forwards a claim, while the Opponent focuses on attacking that claim. If the Proponent successfully counterattacks all Opponent arguments then the initial claim is deemed *acceptable*. We will modify their definition to incorporate argument probabilities and weighted attacks.

---

[3]https://github.com/java-native-access/jna.

Definition 11 A *Probabilistic Argument Weighted Dispute Tree* for an initial argument i is a possibly infinite tree $\mathcal{PAWDT}$ for which:

1. Every node of the $\mathcal{PAWDT}$ is labeled by an argument of probability p and can be an argument of the Proponent or Opponent.
2. The root is a Proponent node labeled by i (the initial claim).
3. For every Proponent node N_P labeled by an argument p_{arg}, and for every argument o_{arg} attacking p_{arg} with weight $w_{o,p}$, there exists a child of N_P, node N_O, which is an Opponent node labeled by o_{arg} with $weight(N_O, N_P) = w_{o,p}$.
4. For every Opponent node N_O labeled by an argument o_{arg}, and for every argument p_{arg} attacking o_{arg} with weight $w_{p,o}$, there exists a child of N_O, node N_P, which is an Proponent node labeled by p_{arg} with $weight(N_P, N_O) = w_{p,o}$.
5. There are no other nodes in $\mathcal{PAWDT}$ except those given by 1–4 above.

The root node in our case will always be the current detected activity. The Opponent nodes attacking the root are the other possible activities from the set *{work, music, movie, leisure, afk}*. Subsequent nodes are represented by Proponent or Opponent arguments attacking one or more nodes above them.

Definition 12 A **critical path** in a $\mathcal{PAWDT}$ is a complete path from root to leaves with maximal $\left| \sum_{a,b \in A_{rgs}} (-1)^{\mathbb{1}_{Proponent}(a)} * w(a,b) * p(a) \right|$.

Lemma 1 *The winning Proponent or Opponent arguments in the* **critical path** *belong to a 0-preferred γ-extension in the argumentation framework defined by the $\mathcal{PAWDT}$.*

The attacking strength is expressed on an increasing scale of 1–10. Depending on the user, a certain recommendation threshold must be passed in order for another activity to be suggested. In Fig. 6 a sample $\mathcal{PAWDT}$ is depicted after 30 min of working have elapsed. Suppose the user-defined threshold in this case is a medium range value of 5. The critical path in Fig. 6 is given by nodes {*Work, AFK*} with value 0.15 * 10 = 1.5, which is not enough to pass the threshold so no action will be taken.

In Fig. 7 2 h have passed since the user began working. The probability of going out jogging (AFK) has increased to 0.5, but the user needs to get some more work

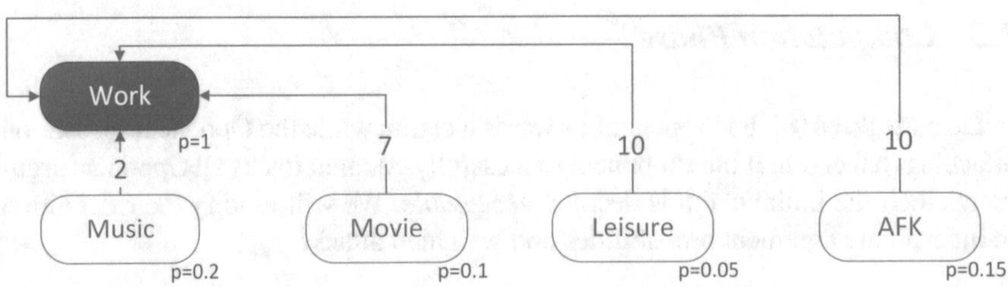

Fig. 6 Sample $\mathcal{PAWDT}$ after 30 min of work

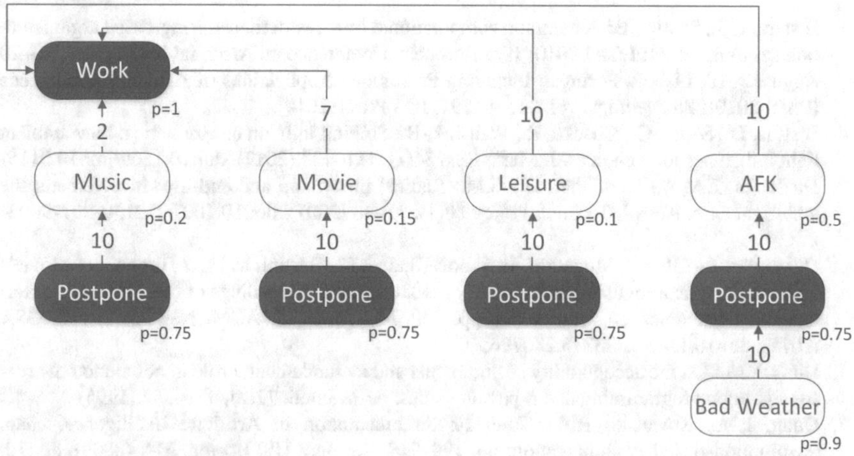

Fig. 7 Sample $\mathcal{PAWDT}$ after 2 h of work

done so he/she tries to postpone it. However, the system can detect that the probability of raining later is 90% so it persuades the user to go out now. The critical path is given by nodes {*Work, AFK, Postpone, Bad Weather*} with value $0.5 * 10 - 0.75 * 10 + 0.9 * 10 = 6.5$ which passes the user-defined threshold. Nodes *AFK, Bad Weather* represent a 0-preferred γ-extension, and therefore the *AFK* action is recommended.

4 Conclusions

This paper has presented the design of an intelligent system capable of synchronizing smart lighting solutions with the current activity performed at the home office. The system is able to detect the next likely activity and make justified recommendations through *probabilistic argument weighted dispute trees*, which the paper introduces. In the future, we intend to conduct a user evaluation which will indicate which improvements should be made.

References

1. Amores, J., Maes, P.: Influencing human behavior by means of subliminal stimuli using scent, light and brain computer interfaces. In: Proceedings of the 9th ACM International Conference on PErvasive Technologies Related to Assistive Environments, pp. 62:1–62:4. PETRA '16, ACM, New York, NY, USA (2016). doi:10.1145/2910674.2935853
2. Bench-Capon, T.J., Dunne, P.E.: Argumentation in artificial intelligence. Artif. Intell. **171**(10–15), 619–641 (2007)

3. Bistarelli, S., Santini, F.: A common computational framework for semiring-based argumentation systems1, 2. In: ECAI 2010: 19th European Conference on Artificial Intelligence, 16–20 August 2010, Lisbon, Portugal: Including Prestigious Applications of Artificial Intelligence (PAIS-2010): Proceedings. vol. 215, p. 131. IOS Press (2010)
4. Biswas, D., Szocs, C., Chacko, R., Wansink, B.: Shining light on atmospherics: how ambient light influences food choices. J. Mark. Res. **54**(1), 111–123 (2017). doi:10.1509/jmr.14.0115
5. Diederiks, E.M.A., Hoonhout, H.J.C.M.: Radical innovation and end-user involvement: the ambilight case. Knowl. Technol. Policy **20**(1), 31–38 (2007). doi:10.1007/s12130-007-9002-z
6. Dumoulin, J., Affi, D., Mugellini, E., Abou Khaled, O., Bertini, M., Del Bimbo, A.: Movie's affect communication using multisensory modalities. In: Proceedings of the 23rd ACM International Conference on Multimedia, pp. 739–740. MM '15, ACM, New York, NY, USA (2015). doi:10.1145/2733373.2807965
7. Dung, P.M.: On the acceptability of arguments and its fundamental role in nonmonotonic reasoning, logic programming and n-person games. Artif. Intell. **77**(2), 321–357 (1995)
8. Dung, P.M., Kowalski, R.A., Toni, F.: Argumentation in Artificial Intelligence, chap. Assumption-Based Argumentation, pp. 199–218. Springer US, Boston, MA (2009). doi:10.1007/978-0-387-98197-0_10
9. Dunne, P.E., Hunter, A., McBurney, P., Parsons, S., Wooldridge, M.: Inconsistency tolerance in weighted argument systems. In: Proceedings of The 8th International Conference on Autonomous Agents and Multiagent Systems-Volume 2, pp. 851–858. International Foundation for Autonomous Agents and Multiagent Systems (2009)
10. Fox, J., Glasspool, D., Patkar, V., Austin, M., Black, L., South, M., Robertson, D., Vincent, C.: Delivering clinical decision support services: there is nothing as practical as a good theory. J. Biomed. Inform. **43**(5), 831–843 (2010)
11. Holthaus, P., Leichsenring, C., Bernotat, J., Richter, V., Pohling, M., Carlmeyer, B., Köster, N., Meyer zu Borgsen, S., Zorn, R., Schiffhauer, B., et al.: How to adress smart homes with a social robot? a multi-modal corpus of user interactions with an intelligent environment. In: Proceedings of the 10th Language Resources and Evaluation Conference (2016)
12. Küller, R., Ballal, S., Laike, T., Mikellides, B., Tonello, G.: The impact of light and colour on psychological mood: a cross-cultural study of indoor work environments. Ergonomics **49**(14), 1496–1507 (2006). doi:10.1080/00140130600858142, pMID: 17050390
13. Mocanu, A.: Envisioning a collaborative smart home solution based on argumentative dialogues. In: Proceedings of the 7th Balkan Conference on Informatics Conference, p. 23. ACM (2015)
14. Morgner, P., Mattejat, S., Benenson, Z.: All your bulbs are belong to us: investigating the current state of security in connected lighting systems. CoRR abs/1608.03732 (2016). arXiv:1608.03732
15. Pfleging, B., Fekety, D.K., Schmidt, A., Kun, A.L.: A model relating pupil diameter to mental workload and lighting conditions. In: Proceedings of the 2016 CHI Conference on Human Factors in Computing Systems. pp. 5776–5788. CHI '16, ACM, New York, NY, USA (2016). doi:10.1145/2858036.2858117
16. Ronen, E., O'Flynn, C., Shamir, A., Weingarten, A.O.: IoT goes nuclear: creating a zigbee chain reaction. Cryptology ePrint Archive, Report 2016/1047 (2016). http://eprint.iacr.org/2016/1047
17. Weffers-Albu, A., de Waele, S., Hoogenstraaten, W., Kwisthout, C.: Immersive tv viewing with advanced ambilight. In: 2011 IEEE International Conference on Consumer Electronics (ICCE), pp. 753–754, Jan 2011

A Recommender System Based on Hierarchical Clustering for Cloud e-Learning

Krenare Pireva and Petros Kefalas

Abstract Cloud e-Learning (CeL) is a new paradigm for e-Learning, aiming towards using any possible learning object from the cloud in a smart way and generate a personalised learning path for individual learners. An issue that appears before the generation of the learning path through automated planning, is to filter a pool of resources that are relevant to the learners profile and desires in order to enhance their knowledge and skills at a higher cognitive level. In this paper, we present a Recommender System for Cloud e-Leaning (CeLRS) that uses hierarchical clustering to select the most appropriate resources and utilise a vector space model to rank these resources in order of relevance for any individual learner. We discuss the issues raised and we demonstrate how CeLRS works.

Keywords Intelligent e-learning · Recommender systems · Hierarchical clustering · Personalised learning

1 Introduction

Recommender Systems (RS) are software tools which provide smart suggestions for items to their users [15]. RS have been used mainly for e-commerce services, such as: Amazon, eBay, Netflix and others in order to propose items of interest to each one of them individually [17]. RS became popular and increased the research interest in the areas of human-computer interaction, machine learning, and information retrieval. In this paper, we will suggest how RS can be utilised in Cloud e-Learning.

Cloud e-Learning (CeL) is an advancement of e-Learning that aims to facilitate the learning process by personalising learning paths consisting of any available

K. Pireva (✉)
South-East European Research Center, 24 P. Koromila, 54622 Thessaloniki, Greece
e-mail: krpireva@seerc.org

P. Kefalas
The University of Sheffield International Faculty, City College, 3 L. Sofou, 54624 Thessaloniki, Greece
e-mail: kefalas@city.academic.gr

M. Ivanović et al. (eds.), *Intelligent Distributed Computing XI*,
Studies in Computational Intelligence 737, https://doi.org/10.1007/978-3-319-66379-1_21

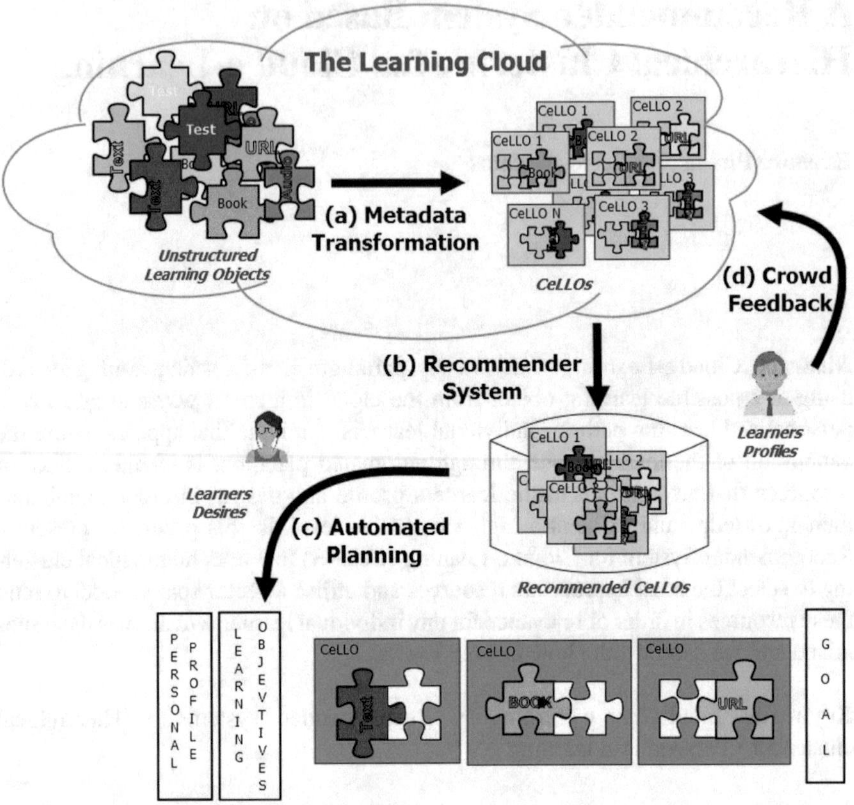

Fig. 1 The overall view of CeLand its processes: **a** metadata transformation. **b** recommender system, **c** automated planning, and **d** crowd feedback

source in the Cloud [12]. CeL uses *Learning Objects (CeLLOs)* which are structured electronic learning resources that can be found online and discussed thoroughly in [11]. Putting CeLLOs in some order that satisfy the learners profile and preferences, generates a learning path which suits the needs and desires of each individual. The construction of a such learning path is considered as an *automated planning process* which is discussed elsewhere [10]. Given the enormous amount of CeLLOs in the learning Cloud, the planning process can become a very complex and inefficient task. As a consequence, in order to address this issue, we propose a *CeL Recommender System (CeLRS)* aiming to facilitate planning by selecting only those CeLLOs which are relevant to a learner. The overall CeL process is depicted in Fig. 1. Clearly, CeL is an intelligent advancement of e-Leaning, that includes learning cloud, recommender system, planning and learning. In this paper, our aim is to present a part of CeL, namely CeLRS that is based on hierarchical clustering and is able to filter CeLLOs according to a user profile and desires, as a new state-of-the-art approach for personalising the content of learners.

The paper is structured as follows: Sect. 2 introduces the basic concepts of recommender systems. In Sect. 3, we present and describe the overall process of our idea, using CeLRS as a middle-layer in CeL. Section 4 presents an example that demonstrates how the ranking process takes place. Finally, we conclude and discuss further work.

2 Background and Related Work

Recommender systems use various techniques and algorithms, such as matrix factorization, apriori, k-nearest neighbors, Bayesian networks, neural networks [14] and so on, in order to make recommendations based on user's preferences, goals and desires. The algorithms used classify RS as Demographic-based, Case-based, Content-based, Collaborative filtering, Constrained-based, Community-based, Knowledge-based or Hybrid. For example, Amazon uses item-to-item collaborative filtering which produces recommendations in real time, scales to massive data sets, and generates high quality recommendations [8]. A model is proposed for improving the item-based recommendation for Amazon by considering the total number of feedback comments beside the rating data per item [6]. Internet radio services (e.g. Pandora) and movie providers (e.g. Netflix) use content filtering model [5]. Other services use collaborative filtering by considering similarity of user profiles for recommending audio CDs.

The same ideas applied in e-commerce are also applied in e-Learning platforms in order to guide the learners through the learning process by suggesting learning materials that fit to their desires [9]. These RS use fuzzy matching combined with a multi-attribute evaluation method. In another approach, a collaborative filtering algorithm combines multiple feedback measures for personalising the e-Learning environment based on users' preferences [7]. A system for monitoring users activities and tracks the navigation continuously is also proposed [3]. The system captures and processes data through data mining techniques and optimise similarities through association rules and collaborative filtering in order to recommend web pages. Finally, a large-scale Bayesian RS for more intelligent predictions is proposed in order to observe users preferences by monitoring the rated items [19].

The above RS rely on a single technique and they suffer from scalability issues. Some of them do not address effectively the sparsity challenge and the cold start situations (i.e. the state at the beginning of the usage of an RS, when the system does not have any information at all). Our contribution is to propose a CeLRS which is inspired from the above approaches and tries to combine the concept of matching with the idea of monitoring users progress. This is achieved through matching CeLLOs as part of a particular cluster considering as well the learning style, learners preferences expressed through ratings and learners desires. Also, it monitors the learners progress by updating the change observed in the learners profiles.

3 The Cloud e-Learning Recommender System

In CeL, a learner L_i expresses her desire to learn something new on a subject through an unstructured natural language query. The query together with any rating information for the subject as well as the overall learner profile (knowledge cognitive level of some subject expressed as in Bloom taxonomy and learning style [2]) constitute a learner's desire:

$desire_{Li} = \{query, subject, knowledge\_level, learning\_style, ratings\}$

The query text is processed through word segmentation, stopword removal, and stemming process. The *Term Frequency and Inverse Document Frequency (TF-IDF)* [13] technique is used to find the weighting factors in CeLLOs and decide how relevant a cluster of CeLLOs is for learner L_i and her desires $desire_{Li}$.

The CeLRS deploys a hybrid approach using content and collaborative filtering as combined techniques for providing smart prediction and ranking. The content-based filtering recommends the CeLLOs as part of a particular cluster based on the learners desires. Collaborative filtering is used to weight higher the most popular rated *CeLLOs* and recommend CeLLOs $\{c_1, c_2, \ldots, c_N\}$ based on the k-nearest neighbours user and item approach.

Before automated planning takes over to produce this personalised learning path (Fig. 1), the main problem to address is to recommend a reasonable pool size of appropriate CeLLOs from which the planning process will select as most relevant CeLLOs for a learner L_i and her desires $desire_{Li}$. Therefore, the CeLRS can be viewed as the operation:

$CeLRS(desire_{Li}, CeLLOs) = \{c_1, c_2, \ldots, c_N\}$.

Each of the CeLLOs $c_1, c_2, \ldots, c_N$ provide the smallest granularity of a particular subject and must satisfy at least a single learning objective within the desired *knowledge_level* of the *subject*.

3.1 The CeLRS Process

CeLRS is defined through a number of steps (Fig. 2). During the *information retrieval phase* the existing learners data and the *learners desires* are specified. During the *text*

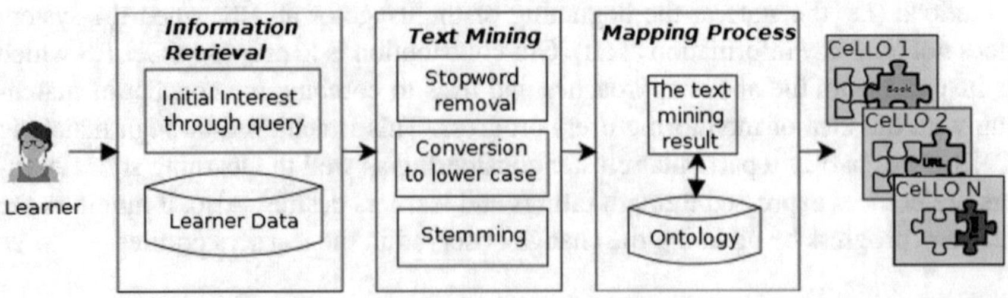

Fig. 2 The CeLRS process

mining process, the query is segmented to single words. The remaining word list is processed further through the stemming process using Porter algorithm [20]. Finally, through the *mapping process*, is generated a ranked list of all relevant CeLLO as part of the most similar cluster, by mapping the resulting list of the word generated from the text mining process from one side, and a vocabulary of the clusters using relevant ontology from the other side. Further below, we are going to explain in detail the last step of this process.

Data retrieval and ranking of information have impact to processing time and classification of most relevant from least or non-relevant information. In this context, clustering helps to partition the input space of CeLLOs into subsets on the basis of similarity metrics, such a learners desires. In CeL, we wish to use any learning object available in the cloud and thus engage with an enormous number of CeLLOs. The combination of clustering algorithms and ranking techniques are applied so that CeLLOs are listed in order of relevance per cluster. Clustering algorithms are categorised in hierarchical, partial, density-based, grid-based, graph-based etc., whereas the ranking algorithms could be content or linked-based [16]. In CeL, we propose a hierarchical approach for clustering the CeLLOs as part of topics and sub-topics of computer domain based on a specific ontology and a vector space model, a kind of content-based approach, for ranking most relevant CeLLOs within the selected clusters.

3.2 Hierarchical Clustering in CeLRS

Hierarchical clustering techniques produce a sequence of clusters within a domain knowledge [4]. There are two categories of algorithms used for hierarchical clustering, agglomerative (bottom-up, starts with singleton clusters and continues by merging clusters that are the most similar) and divisive (top-down, starts with a macro cluster and splits it furthers at it progresses). In the CeL context, the hierarchical clustering of CeLLOs produces a tree, representing parent-child relationship among the entities, which could be expressed as topic-subtopic relationship in a subject hierarchy. This is done through the use of an ontology that defines a knowledge domain.

For example, assume that a learner wishes to acquire new knowledge or skills through CeL in the Computing domain, e.g. "Software and its engineering". The ACM Computing Classification Taxonomy[1] used for the hierarchical clustering process defines the vocabulary used for various sub-domains within computing. The coverage, the user-friendliness of the interface, the use of hierarchical approach of controlled vocabulary and the well-planned classification system, are few sound arguments why we decided to use the ACM CCS Ontology among other existing ones [1]. The taxonomy is partly depicted in Fig. 3.

[1] ACM CCS, https://www.acm.org/publications/class-2012.

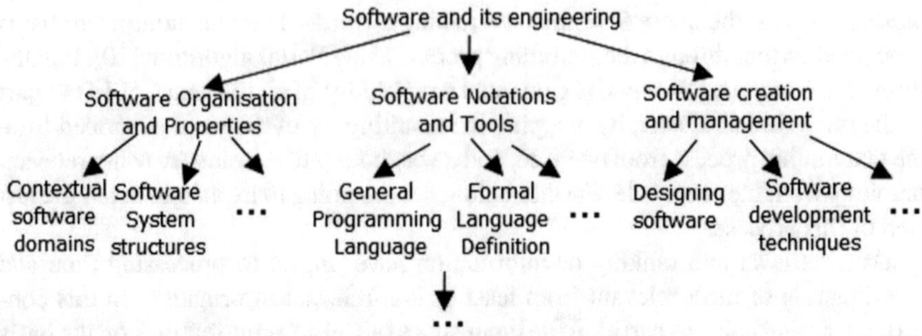

Fig. 3 A sample of the hierarchical classification of ACM ontology

3.3 Vector Space Modeling in CeLRS

The *Vector Space Model (VSM)* is an algebraic model commonly used for information retrieval [18]. The idea behind the model is to represent the clusters and the data through vectors in a multi-dimensional space and to compute their similarity through cosine similarity measure. For simplicity reason, the concepts of the clusters shown in Fig. 3 are denoted as follows: assume C1 represents "Software and its engineering" as the main topic of interest (Fig. 3), then C1 will be partitioned further into C2 which represents "Software Organisation and Properties", C3 "software notations and tools" and C4 "software creation and management". In turn, C2 continues to be partitioned further to subtopics, C5 "Contextual Software Domains" and so on. Applying the devisive clustering approach to ACM taxonomy (partly shown in Fig. 3) results in a structure clustering as shown in Fig. 4.

Then, the cosine similarity between the $desires_{Li}$ and the ACM cluster terms as resulted in Fig. 4 are weighted as product between the *Term Frequency (TF)* and the *Inverse Document Frequency (IDF)* [13]. The TF-IDF weight is a statistical measure used in retrieval process for evaluating how important a specific word is to a particular CeLLO that is part of a cluster. *TF* [13] provides the information how often the required term is found in a CeLLO:

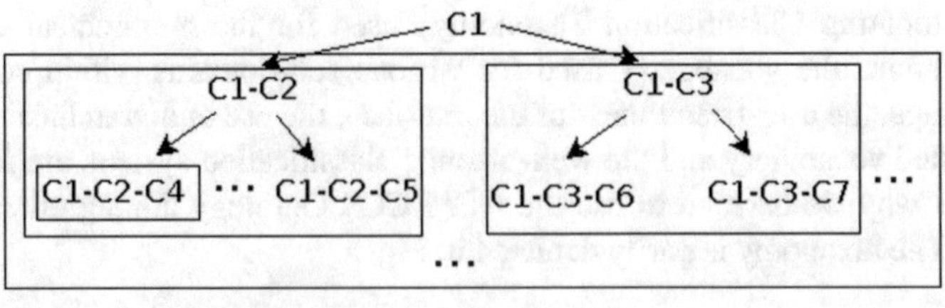

Fig. 4 The divisive clustering used for CeL based on ACM computing classification taxonomy

$$TF(t) = \frac{\sum t}{\sum atd} \tag{1}$$

where Σt represents the number of times the term t appears in a CeLLO. Whereas Σatd represents the total number of all terms in the CeLLO. In contrast, *IDF* [13] expresses through log, the total number of CeLLOs in CeL divided by the total number of the CeLLOs in which the term occurs. $IDF(t) = log(\sum c / \sum c(t))$, where Σc represents the total number of CeLLOs, and $\Sigma c(t)$ represents the number of CeLLOs containing the term t. Finally, the weighting of CeLLOs is calculated through:

$$weight(t, c) = TF(t) \times IDF(t) \tag{2}$$

The whole process is repeated until all the terms within a query are covered, and the final weighting is accumulated and represented as the final result:

$$Res(query, cellos) = \sum_{(t \in query)} TF(t) - IDF(t) \tag{3}$$

The recommended CeLLOs are listed from most to least relevant CeLLOs by computing the function of cosine similarity of the angle between respective vectors in the VSM using $\cos\theta = [v(q) \times v(c)]/[|v(q)| \times |v(c)|]$, where θ is the angle between the $desire_{Li}$ represented as $v(q)$ and the respective CeLLOs $v(c)$ as part of a particular cluster.

4 CeLRS: Example and Implementation

Assume the list of available CeLLOs in Table 1, with different format type such as: videos, audios, podcast and texts. In addition to those, there are also self-evaluation tests CeLLOs in order to assess the progress of the learners. Results from such tests are updated in learners profile on a continuous basis. Also, the pre-requisites and cognitive level are defined according to Bloom Taxonomy. For example, in order for a learner to be able to work with numbers in "Java", she should be able to understand the basic concepts of "General Math" defined as *math(1)* prerequisite for CeLLO c_6.

Let's present an example, in which a learner L_1 is interested to learn how to create classes in object-oriented programming (cognitive level 5, i.e. Synthesis or Creating). Her current knowledge is only "programing language features" at cognitive level 1, which means she is able to understand the basic concepts of the overall "programing features" such as: data types, control structures, constraints and so on. In contrast, learner L_2 currently has "General Programming Language" knowledge (cognitive level 4, i.e. analysis), and would like to learn about "interfaces" (cognitive level 5). Both learner profiles are listed in Table 2.

Table 1 Sample CeLLOs in some abstract format

Type of learner	Available format	Cello ID	Bloom level	Topic	Prerequ-isites
Audio	Podcast	c_1	1	Data abstraction	None
Audio	Podcast	c_2	3	Instance variables	algorithm(1)
Audio	Podcast	c_3	3	Objects and classes	None
Audio	Podcast	c_4	2	Control statements	None
Visual	Text	c_5	1	Data types and variables	None
Visual	Video	c_6	1	Arithmetic in java	math(1)
Audio	Podcast	c_7	6	Objects and classes	None
Audio	Podcast	c_8	1	Classes	None

Table 2 Sample learner profiles

Learner	Knows	Type	Desires to learn
L_1	Programing language features at level(1)	Audio	Classes at level(5)
L_2	General programming language at level(4)	Visual	Interfaces at level(5)

Therefore, the CeLRS filters the number of relevant CeLLOs to the $desires_{L1}$ and $desires_{L2}$ according to the process that is explained in the previous sections. In order to determine the relevance of CeLLOs as part of particular topic, the ACM Computing Classification System[2] is used. For example, $desires_{L1}$, is mapped with the cluster of "Language Feature" which contains the "Classes and Objects" as part of it (found as follows: "General → Sofware and its Engineering → Software notation and Tools → General Programming Languge → Language Features → Classes and Object"), and will filter a list of CeLLOs that are part of 'Language Feature'. Hence, all CeLLOs with cognitive level 2 to 5, that are part of the "Language Feature" cluster, containing audio materials from "Abstract Data Type", "Control Structures", "Constraints", "Classes and Objects" and so on, which are under the "Language Feature" umbrella in ACM CCS will be listed, predicted and ranked. The final result is listed in Table 3.

From the above example, c_7 and c_8 listed in Table 1, even though they contain materials related to "objects and classes" which are part of "Language Feature" clus-

[2]ACM CCS, https://www.acm.org/publications/class-2012.

Table 3 List of recommended CeLLOs by CeLRS

Cello ID	Topic
c_2	Instance variables
c_3	Objects and classes
c_4	Control statements

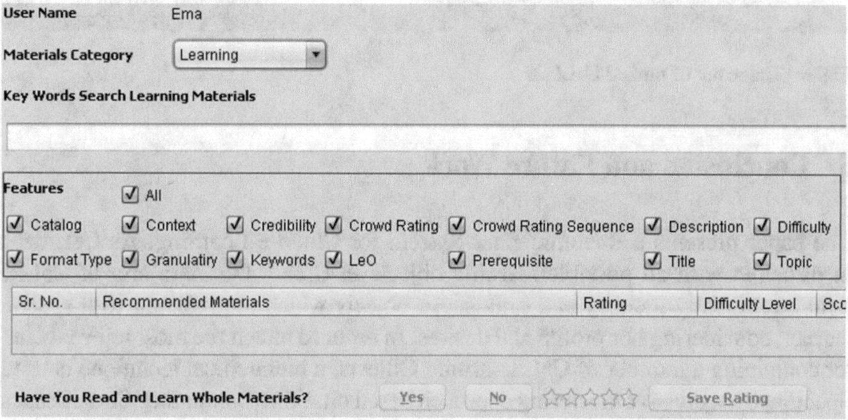

Fig. 5 A screenshot of CeLRS

ter, and both of them are "audio format types", they are omitted because the cognitive level 1 and 6 are out of the desired range 2–5. Knowledge at level 1 is already acquired (Table 2), whereas the cognitive level 6 overcome the cognitive level of learners desire. Similar to this, if a learner searches a particular topic that might be of cognitive level 6, but in the meantime the learner has the knowledge expertise of level 3, the CeLRS will propose the CeLLOs, from level 4 to 6. Proposing the CeLLOs from cognitive level 4 and 5 as well, tends to avoid the gaps that could be generated inadvertently (currently having knowledge at cognitive level 3, and immediately jumping to knowledge level 6), and make possible the progress of the learner more grounded.

So far, we have implemented the CeLRS and created initial tests which successfully recommend CeLLOs that form the pool out of which automated planning will be based to create the personalised path. CeLLOs are represented with XML metadata which contain special features to facilitate synthesis at the automated planning stage [11]. In CeLRS implementation (Fig. 5), the learner determines which feature to use for the text mining process while querying the particular topic. The list of ranked CeLLOs is returned as part of the most similar cluster (Fig. 6). Finally, these CeLLOs are passed to the automated planner which generate a personalised sequence of learning path, as discussed in [10].

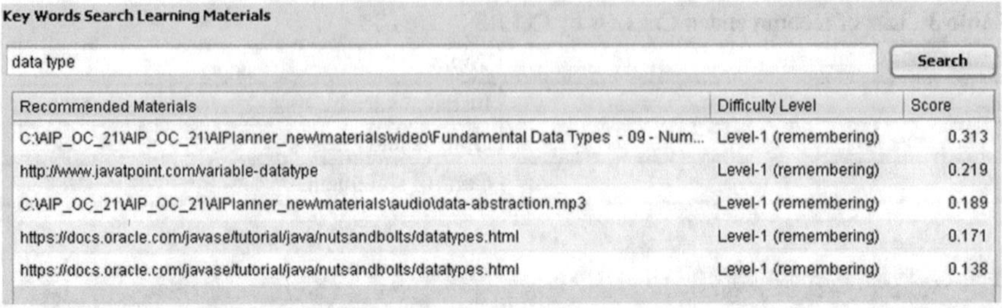

Key Words Search Learning Materials
data type
Search
Recommended Materials
Difficulty Level
Score
C:\AIP_OC_21\AIP_OC_21\AIPlanner_new\materials\video\Fundamental Data Types - 09 - Num...
Level-1 (remembering)
0.313
http://www.javatpoint.com/variable-datatype
Level-1 (remembering)
0.219
C:\AIP_OC_21\AIP_OC_21\AIPlanner_new\materials\audio\data-abstraction.mp3
Level-1 (remembering)
0.189
https://docs.oracle.com/javase/tutorial/java/nutsandbolts/datatypes.html
Level-1 (remembering)
0.171
https://docs.oracle.com/javase/tutorial/java/nutsandbolts/datatypes.html
Level-1 (remembering)
0.138

Fig. 6 The result of ranked CeLLOs

5 Conclusion and Future Work

The paper presents a Recommender System for Cloud e-Learning. As CeL deals in principle with all possible learning objects in Cloud, the main aim of CeLRS is to intelligently identify and rank those objects which are relevant to a specific learner, considering her profile and desires. In order to match the most relevant cluster containing a number of CeL Learning Objects, a hierarchical technique is used. Specifically, a divisive clustering approach based on ACM Computing Classification Taxonomy which defines various sub-domains within computing domain. In addition, in order to rank the appropriate CeLLOs within the clusters, the vector space modeling is used, particularly the cosine similarity algorithm. The resulted CeLLOs of CeLRS, will serve as an input of CeL automated planner in order to generate a sequence of CeL Learning Objects as a personalised learning path. The paper presented an example which demonstrated the applicability of the proposed approach. Future work includes the expansion of user data as well as the improvement of current system through learning from users feedback in order to improve the ranking and prediction of CeL Learning Objects that have been rated positively from similar group of learners. As a result of this, the collection of more data directly from the users activities will improve the evaluation process, which will be feasible through confusion matrix, thus defining the recall and precision ratio.

References

1. Buzydlowski, J.W., Lin, X., Zhang, M., Cassel, L.N.: A comparison of a hierarchical tree to an associative map interface for the selection of classification terms. Proc. Am. Soc. IST **50**(1), 1–4 (2013)
2. Felder, R.M., Brent, R.: Understanding student differences. J. Eng. Educ. **94**(1), 57–72 (2005)
3. Fu, X., Budzik, J., Hammond, K.J.: Mining navigation history for recommendation. In: Proceedings of IC on Intelligent UI, pp. 106–112. ACM (2000)
4. Fung, B.C., Wang, K., Ester, M.: Hierarchical document clustering using frequent itemsets. In: Proceedings of IC on Data Mining, pp. 59–70. SIAM (2003)

5. Gomez-Uribe, C.A., Hunt, N.: The netflix recommender system: algorithms, business value, and innovation. ACM TMIS **6**(4), 13 (2016)
6. Jabakji, A., Dag, H.: Improving item-based recommendation accuracy with user's preferences on apache mahout. In: 2016 IEEE International Conference on Big Data (Big Data), pp. 1742–1749. IEEE (2016)
7. Li, X., Chang, S.K.: A personalized e-learning system based on user profile constructed using information fusion. In: DMS, pp. 109–114 (2005)
8. Linden, G., Smith, B., York, J.: Amazon.com recommendations: item-to-item collaborative filtering. IEEE Internet Comput. **7**(1), 76–80 (2003)
9. Lu, J.: Personalized e-learning material recommender system. In: IC on information technology for application, pp. 374–379 (2004)
10. Pireva, K., Kefalas, P., Cowling, A.: A Review of Automated Planning and its Application to Cloud e-Learning. Work in progress, Paper Submitted (2017)
11. Pireva, K., Kefalas, P., Stamatopoulou, I.: Representation of Learning Objects in Cloud e-Learning. Work in progress, Paper Submitted (2017)
12. Pireva, K., Kefalas, P.: The use of multi agent systems in cloud e-learning. In: Doctoral Student Conference on ICT, pp. 324–336 (2015)
13. Polettini, N.: The vector space model in information retrieval-term weighting problem. Entropy, 1–9 (2004)
14. Portugal, I., Alencar, P., Cowan, D.: The use of machine learning algorithms in recommender systems: a systematic review (2015). arXiv:1511.05263
15. Ricci, F., Rokach, L., Shapira, B.: Introduction to Recommender Systems Handbook. Springer (2011)
16. Rokach, L., Maimon, O.: Clustering Methods. In: Data Mining and Knowledge Discovery Handbook, pp. 321–352. Springer (2005)
17. Schafer, J.B., Konstan, J., Riedl, J.: Recommender systems in e-commerce. In: Proceedings of Conference on Electronic Commerce, pp. 158–166. ACM (1999)
18. Singh, V.K., Singh, V.K.: Vector space model: an information retrieval system. Int. J. Adv. Eng. Res. Stud. **141**, 143 (2015)
19. Stern, D.H., Herbrich, R., Graepel, T.: Matchbox: large scale online bayesian recommendations. In: Proceedings of IC on WWW, pp. 111–120. ACM (2009)
20. Willett, P.: The porter stemming algorithm: then and now. Program **40**(3), 219–223 (2006)

A Taxonomy of Anomalies in Distributed Cloud Systems: The CRI-Model

Kim Reichert, Alexander Pokahr, Till Hohenberger, Christopher Haubeck and Winfried Lamersdorf

Abstract Anomaly Detection (AD) in distributed cloud systems is the process of identifying unexpected (i.e. anomalous) behaviour. Many approaches from machine learning to statistical methods exist to detect anomalous data instances. However, no generic solutions exist for identifying appropriate metrics for monitoring and choosing adequate detection approaches. In this paper, we present the CRI-Model (Change, Rupture, Impact), which is a taxonomy based on a study of anomaly types in the literatureand an analysis of system outages in major cloud and web-portal companies. The taxonomy can be used as an anlaysis-tool on identified anomalies to discover gaps in the AD state of a system or determine components most often affected by a particular anomaly type. While the dimensions of the taxonomy are fixed, the categories can be adapted to different domains. We show the applicability of the taxonomy to distributed cloud systems using a large dataset of anomaly reports from a software company. The adaptability is further shown for the production automation domain, as a first attempt to generalize the taxonomy to other distributed systems.

Keywords Anomaly detection · Distributed cloud systems · Mitigation approaches · System failures · Feature selection · Taxonomy of anomalies

K. Reichert (✉) · T. Hohenberger
Adobe Systems Engineering GmbH, 22767 Hamburg, Germany
e-mail: reichert@adobe.com

T. Hohenberger
e-mail: hohenber@adobe.com

A. Pokahr · C. Haubeck · W. Lamersdorf
University of Hamburg, 22527 Hamburg, Germany
e-mail: pokahr@informatik.uni-hamburg.de

C. Haubeck
e-mail: haubeck@informatik.uni-hamburg.de

W. Lamersdorf
e-mail: lamersdorf@informatik.uni-hamburg.de

M. Ivanović et al. (eds.), *Intelligent Distributed Computing XI*,
Studies in Computational Intelligence 737, https://doi.org/10.1007/978-3-319-66379-1_22

1 Introduction

Anomalous or unwanted behaviour in *Distributed Cloud Systems*(DCS) can lead to user dissatisfaction (e.g., decreased performance) [12], loss of data or incursion of penalties due to a broken *Service Level Agreement*(SLA).

Until the reason for the anomaly is found and can be remedied, the system might be vulnerable to attacks or require additional resources. According to [13], e.g., Amazon lost roughly $25.000 per minute during a Thanksgiving Weekend in 2001 due to a series of outages.

Thus, early detection of system outages and automated recovery are very valuable for software companies. *Anomaly Detection*(AD) refers to identifying patterns in data, which do not conform to expected behavior [4]. Detection happens across many different domains such as finance [5], healthcare [8] or network traffic [3].

The definition of anomaly often takes on a numerical nature. [4] speak of patterns in data, [6] distinguish global and local anomalies. Contrary to this statistical perspective, we adopt the more holistic view of [2], considering anomalies as *"random, nonconforming, or unexpected behaviour within a system"* [2, p. 29].

Following this broad definition, any system might exhibit a potentially large number of anomalies. e.g., [7, p. 7] find that the a priori definition of all possible behaviours of an application is unrealistic. However, every anomaly, regardless of its type, will happen within a known system (the DCS in question), which puts constraints on what and how behaviour within that system can occur. While it may not be possible to predict specific anomalous behaviours of a system, it might be feasible to provide an a priori classification for them.

In their survey paper, [7] determine that *"a taxonomy of performance issues [...] will be highly essential for industry and academia."* [7, 4: 24] A taxonomy for anomalies in DCS could enable developers to define anomalies they want to detect, verify which types are already covered by detection or prevention policies and which are yet untreated. [2, p. 30] find, that *"[...], when the types of anomalies are not known a priori [...] selecting an anomaly detection technique is not trivial."* With an encompassing taxonomy, anomaly detection techniques could be mapped to different anomaly classes for improving the selection process of AD approaches.

Given the variety of anomaly types and manifold of metrics available, it is hard to choose one singular machine learning algorithm or statistical approach for all. Rather, multiple approaches using different metrics might be more successful. Our taxonomy of anomalies was built to support this process as an analysis-tool by classifying types of already identified anomalies in a system and laying the groundwork for a mapping of approaches to detect future ones. It allows a structured discussion of anomalies within a given system, by being adaptable at its lower level.

This paper is structured as follows: in Sect. 2 we explain our approach to designing the taxonomy. In Sect. 3 we present the resulting CRI-Model, and examine it's applicability in Sect. 4. Related work is discussed in Sect. 5. Finally, in Sect. 6 we discuss our takeaways, shortcomings of the model and future prospects.

2 Approach

To develop a taxonomy of anomalies in DCS, we considered anomalies from both perspectives of research and practice. Thus, we studied the literature on AD to investigate which types of anomalies the different studies mentioned and detected. For the practice perspective, we explored reports on anomalies (namely system outages) from major online software companies. Altogether we

1. scoped *20* papers from the literature related to AD
2. and studied *16* reports of system outages in major software companies.

2.1 *Scoping the Literature*

Papers were selected based on their focus on anomaly detection in distributed systems, cloud systems or micro-services (Appendix B). They had to contain some description of the anomalies, the data used for detection or any classification of anomalous system behaviour. We focused on how anomalies were described, what root causes were mentioned, what broke down or failed to function, what kind of effect was determined and which names were chosen to describe anomaly types. Many of these names differentiate anomalies according to dimensions such as root cause (*Intrusion Anomaly, Bottleneck-anomaly, Contention Anomaly, Flood-based Anomaly, Execution Anomaly*), the effect they have (*Performance Anomaly, Busy Loop Fault*), or the location where they happen (*Network Anomaly, System Anomaly, Application-Anomaly, Physical Layer Anomaly, Utility Cloud Anomaly*).

2.2 *Analysis of Real World Cases*

We analysed sixteen reported anomalies that happened in companies such as Google, Amazon or Yahoo (Table 3). In almost all cases, these anomalies were full system outages (the impact needed to be intense enough to merit public reporting). Nonetheless, these reports added valuable input for the design of our taxonomy, particularly with regard to the different types of root causes (if they were reported), their manifoldness, and the possible failures they induced.

3 The CRI-Model

We condensed our findings from research and practice into a taxonomy, called CRI-Model (see Fig. 1). It captures *what* happened, *where* it happened and *when* it happened for an individual anomaly. Regarding *when*, we identified three phases:

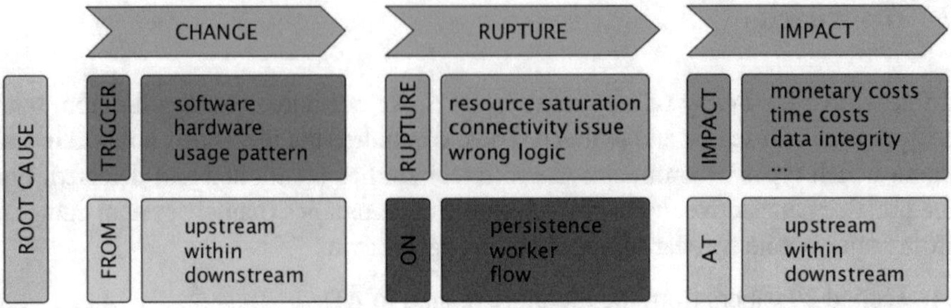

Fig. 1 The CRI-Model populated with our sample categories

Change, *Rupture*, and *Impact*, accounting for the name (CRI). The *what* describes the types of change, rupture or impact, whereas the *where* identifies the components or stakeholders, where the change, rupture or impact occurred. Following this model, an anomaly is described as one or multiple *changes* in or outside the system, which trigger a chain of *ruptures* at one (or multiple) components in the system, followed by one or multiple *impacts* on the system and other stakeholders.

The resulting CRI-Model provides six *dimensions* for characterising an anomaly: change type, rupture type and impact type, as well as change location, rupture location and impact location. The dimensions are domain independent, whereas the *categories* for each dimension are not (e.g., which types of changes can occur).

Each dimension can be filled repeatedly (e.g., if multiple triggers are known, which cumulatively provoked a rupture, they should all be listed under trigger). However, in many cases, only parts of the reaction chain are known in retrospect. They can nonetheless be used to describe an anomaly in the CRI-Model. For instance, an increased request count which triggered a rupture on the load balancer is known, but the original root cause is a change in the public API, which provoked the users to request a service more often. Even if the original trigger is never detected, filing the increased request count as a trigger can still be useful to determine metrics for future detection or discuss resilience strategies.

Categories for the six dimensions should be adapted to each system. As an example for validation, we modelled a generic DC (see Fig. 2) based on a software system which provided the validation dataset of reports (Sect. 4). It consists of Workers (parts executing the application logic), Flow (components for transport between workers) and Persistence (any kind of data storage). The individual categories were derived from the literature and the practical reports on distributed cloud systems. In the following, each of the dimensions is described in more detail, and the categories which we derived for our sample system are presented.

Trigger (Change Type) is determined by the acute change(s) that trigger the anomaly. The change itself does not have to be malicious, but must be acute, because the anomaly itself only appears after the anomalous behaviour is triggered. This trigger *can* but does not have to be equal to the root cause. Since the root cause could be, for instance, a faulty code change, the anomaly may not be triggered until a client

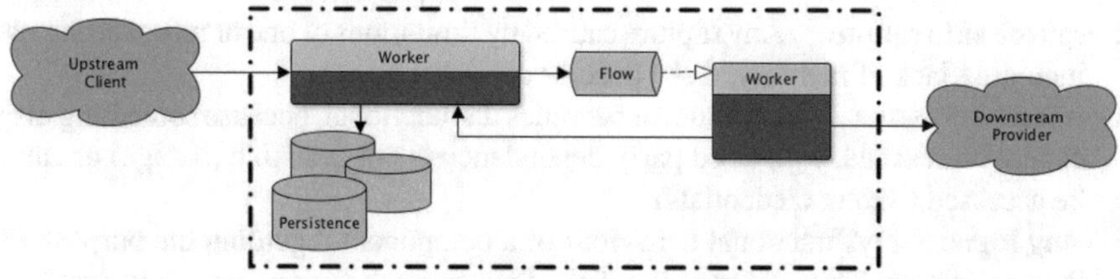

Fig. 2 An abstract view on the different components of a distributed cloud system

makes a specific call to the DCS. Identifying and classifying the trigger of an anomaly is often a necessary step in solving the underlying problems, because anomalies in each of the categories might be monitored using different metrics like system load, error-messages or text-logs. Categories for the trigger dimension are:

Software change: Any change which affects the logic of the application(s). For instance, a developer deploying new code to the system or a downstream dependency changing its interface.

Hardware change: Changes regarding the underlying hardware of the system, including unplanned changes like corrupt disk write, but also planned changes like scaling down a worker component.

Usage pattern change: Changes in the amount or type of data flowing through the system, including, e.g., unusually large files, or a flash crowd event.

Triggered From (Change Location) captures where the change comes from (or who triggered it). It gives additional indications to the selection of metrics. For instance, if most anomalies are triggered from upstream, it makes sense to apply AD on client-related metrics combined with metrics from components were the rupture usually happens. Also, it can be important to remedy the anomaly or find a short-term fix as other parties might be necessary to solve the outage. Categories of the triggered-from dimension are:

Upstream: Any change from a client system or user of the DCS functionality, e.g., intentional DoS-attacks, wrong usage of an interface (increased load, increased trigger of error-responses) or unnecessary retries from an impatient user.

Within: A change in the DCS itself, including a changed worker configuration (e.g., scaling, expired keys), a crash of a persistence component, or an update to routing tables for one of the load-balancers controlling the flow in the system.

Downstream: Changes which are triggered by dependencies of the DCS, e.g., an outage of a cloud provider's queue system or a change to a public HTTP-API.

Rupture (Rupture Type) also called failure or error in the literature, describes how the system reacts to the changed behaviour in a problematic way. A trigger does not have to happen on the same component as the rupture. Often, the trigger(s) set off a chain of events that eventually lead to the rupture(s) somewhere else in the system. The rupture dimension has the following sample categories:

Resource saturation: Any rupture caused by limitations of one or more resources, including lack of memory, disk space or cpu-bottlenecks.

Connectivity issue: A component becomes disfunctional, because something else cannot be reached, e.g., a 3rd party dependency is not available (outage) or can't be accessed (wrong credentials).

Wrong logic: Dysfunctional behaviour of a component regarding the purpose of the application. This category fits if the flow through the system works just fine (connectivity and resource saturation), but there is a flaw in the handling of the data content, for example, due to a bad commit that was deployed.

Ruptured On (Rupture Location) distinguishes the types of components which are hit by the rupture. This dimension is useful to identify bottlenecks, problems, and respective metrics in a specific DCS. Categories follow the components of Fig. 2:

Worker: Any affected component with application logic, which is hit by the anomaly, i.e. basically all components that transform data.

Flow: Components that deal with the flow of the system, forwarding, collecting requests or messages, such as load balancers, queues or event buses. Typical anomalies would be loss of events, bad throttling, overloading of queues.

Persistence: Any persistence component, like databases, system files or caches, typically concerning wrong storage of data, or bottleneck problems.

Impact Type is determined by the effect that an anomaly has on stakeholders. This dimension allows distinguishing between, e.g., financial impact or reputation. Some exemplary categories for this dimension are:

Monetary costs: This includes anomalies which lead to broken SLA-agreements, or require an upscaling of resources, which costs more money.

Image costs: Often an anomaly leads to decreased usability of a service (e.g., slow response time), which can lead to unsatisfied users.

Vulnerability: This category covers anomalies which trigger system states that make the system vulnerable to attacks.

Impact At (Impact Location) determines who or what is impacted by the anomaly. Similarly to impact, an anomaly can hit multiple places at once. We follow our abstract architecture from Fig. 2 for the categories:

Upstream: An anomaly can have negative impacts for upstream clients. Increased error-rate or loss of data that impact the client directly.

Distributed System: An anomaly can also negatively impact the DCS itself like any decrease regarding non-functional properties.

Downstream: Any downstream provider of the DCS could be impacted, e.g., if an anomaly leads to an increase in requests sent to the downstream client.

4 Applicability to Different Environments

Mapping anomalies of software applications into the CRI-Model brings the benefit of identifying (1) common causes of anomalies and outages, (2) areas where the monitoring of the DCS does not catch issues before a human does, and (3) a better understanding of the reaction chain of common anomalies. Based on these identified fields, actions can be derived to improve quality as well as monitoring.

4.1 Mapping of an Anomaly Dataset to the Taxonomy

To explore the applicability of the taxonomy for DCS, we decided to map a dataset of 51 anomaly reports from a large software company against the different categories of the CRI-Model. Our goal was to gain insight into which types of anomalies exist in the system, and which would merit more thorough attention. The reports consist of incident summaries and their impacts.

Mapping Example Due to space limitations, we present one exemplary mapping, based on a singular report: the root cause was a bug which was introduced by a third-party vendor of a persistence component. This bug influenced how the queries were handled when a cluster of persistence nodes is not able to communicate with the primary node. A network issue was the trigger (hardware change from within), that forced all nodes to determine a new primary node. This procedure in turn triggered the bug and led to a rupture in the system (connectivity issue on persistence components), since the individual nodes could no longer contact each other. The resulting impact was a delayed response time combined with an increase in error messages for the users (upstream) and required a temporary fix of upscaling the node cluster (costs for within).

Applicability of Categories To further verify the soundness of our categories, Table 1 presents an overview on how many reports could be mapped to which types of categories within our six dimensions. While most categories seem to be sensible, monetary costs and Downstream Impact are lacking mappings. However, the monetary impact of anomalies was usually not noted in the reports, and the impact on downstream services was also not communicated. In a different company this information might be available and very interesting to analyze.

Taxonomy Usage In order to determine which types of anomalies might be particularly relevant to the system in question, and where anomaly detection should be implemented first, we picked two dimensions from the taxonomy (trigger and rupture), and sorted the reported anomalies into the cross-matrix (see Table 2). We further distinguished which of these anomalies were detected manually (this means by an engineer or client) and which had been detected by monitoring. The bold numbers indicate how many cases were detected by a human (which ideally should never happen, if the human is a client or user).

Table 1 Count of reports mapped to our categories of the CRI-Model

Trigger		Triggered from		Rupture	
Hardware change	22	Downstream	7	Connectivity issue	24
Software change	21	Upstream	22	Wrong logic	23
Usage pattern change	22	Within	36	Resource saturation	18
Rupture at		Impact		Impact on	
Flow	4	Data integrity	38	Downstream	0
Persistence	18	Time costs	27	Upstream	54
Worker	43	Money costs	1	Within	11

Table 2 Mapping of reported anomalies to the dimensions trigger and rupture. The bold numbers indicate how many of the anomalies were detected by a human, while the other numbers indicate the total number of incidents

Trigger x rupture	Software	Hardware	Usage pattern	Sum
Connectivity issue	**1**/2	**11**/17	**3**/3	22
Wrong logic	**7**/7	**2**/2	**2**/3	12
Resource saturation	**2**/2	**0**/1	**12**/14	17
Sum	11	20	20	51

Looking at Table 2, it becomes clear that this system has most ruptures concerning connectivity issues and resource saturation. Based on system operator knowledge, we know that most connectivity issues are triggered by sudden network problems, making it difficult to detect and mitigate. The resource saturation anomalies, on the other hand, could be investigated further.

This could indicate to the engineer, that anomaly detection should be implemented with a focus on hardware-level metrics, (rather than error-logs, application-level metrics).

4.2 Applicability to Other Domains

The CRI-Model follows an abstracted view on DCS as shown in Fig. 2, but the core model can be generalized.

To show this we briefly introduce *Cyber-Physical Production System* (CPPS)s which are characterized by interconnected physical and cyber components in the domain of production plants. CPPSs work in realtime and are contrary to pure software systems structured along a physical equipment hierarchy and operate in real-time cycles of Programmable Logic Controllers. Due to its inherent interdisciplinarity, anomaly detection of CPPSs is mostly model-based, which means that anomalies are identified by comparing the actual behaviour with an intended behaviour or environment model [9].

But even under different detection approaches, the core aspects of the CRI-Model remain. In a CPPS, changes are much more hardware dependent and the usage is often exclusively based on local communication of machineries operating together. The taxonomy regarding rupture remains, even if the resources, their logic implementation and connection significantly differ due to domain-specific conditions, because the general failures are comparable. Possible impacts of anomalies are broad and can impact cost, the product quality or other (non-)functional requirements of a plant [9].

The CRI-Model could be extended to better represent failure scenarios in CPPSs. The rupture locations, for example, could be further divided into mechanical hardware, electrical or pneumatic connections as well as in the automation control software. But even under more complex location dimensions, the CRI-Model provides benefits by allowing a categorization of anomalies, identification of a lack of monitoring signals in specific parts of a CPPS or better localization of common anomalies.

5 Related Work

Looking at the literature, attempts have been made to create taxonomies of anomalies in (distributed) software systems. In [10] Mazel et al. define a taxonomy for anomalies in backbone network traffic, Plonka and colleagues introduce a taxonomy which focuses on network anomalies in general ([14]) and Mirkovic et al. present a classification of dDos-Attack anomalies [11].

Contrary to our point of view, Mazel's anomaly also contains normal events (since an important part of their taxonomy is the mapping of signatures, it makes sense to consider normal signatures for reasons of exclusion as well.) [10]. Furthermore, their result contains classes and subclasses of anomalies, while our model allows arbitrary combination of dimensional categories. Where they give a name to each distinct anomaly, we use the CRI-model to describe it. In our opinion, this makes it far more likely to cover any type of anomaly that can occur.

In their Attack-Taxonomy, Mirkovic et al. describe anomalies by intent and trigger, only. Their dimensions can have sub-classes, e.g., a classification by degree of automation has a subclass scanning strategy. Contrary to Mazel's taxonomy, their categories are not exclusive. Some anomalies can be sorted under different combinations of classes: e.g., Hitlist Scanning anomalies can be semi-automatic or automatic. The CRI-Model, on the other hand, does not consider intention, but therefore considers the trigger. While it can be helpful to understand an intentionally caused anomaly (e.g., for defence mechanisms), as a more generic attempt to describe anomalies, this dimension of intent seems too detailed for a generic classification.

Tobergte and Curtis. [15] discuss two dimensions: the cause of a failure and the location of the component where it happens. They determine eight categories of cause (e.g., operator error, node hardware, node software…) und four for location (front-end node, back-end node, network, unknown) where the anomaly first appears. They also differentiate between impact that leads to service failure (in our words impact at the upstream client) and impact that makes only singular components fail. Finally, like us, they distinguish an underlying flaw (the root causes in our case), which can become active in a certain system state, from the cause (our trigger).

Just like us, they also create two-dimensional mappings of dimensions (cause-location) but mention their trouble to correctly map network-problems (mapping to network and 'unknown' as cause). In the CRI-Model, the rupture would be 'Connectivity', and the trigger could be labelled as hardware change (since the network could be seen as part of the underlying hardware of a distributed software system).

Contrary to the CRI-Model, their taxonomy might be most suited for mitigation strategies, not necessarily for anomaly detection, since its focus is on the impact.

In their survey paper on anomaly detection techniques, [4] do define different types of anomaly types (point anomalies, contextual anomalies and Collective anomalies). However, these only take on the numeric aspect of an anomaly, and leave out the system and its components completely.

Avizienis et al. [1], finally, present the most detailed and simultaneously widespread taxonomy we could find. They distinguish three main components that are part of an anomaly, a *fault* which causes an *error* which can lead to a *failure*. Accordingly, they present a classification of faults and one of failures, the error itself, however, is not further distinguished. From the AD perspective, the monitoring data which the error (rupture) creates can be most essential. This lack of further distinction makes the taxonomy from Avizienis et al. unsuitable for a classification of anomalies with the purpose of automated detection.

The CRI-Model, on the other hand, provides the rupture and component dimensions, which can help narrow down what kind and where from to collect data for detection.

6 Conclusion

Based on research as well as publicly available outage reports of system failures, we introduce the CRI-Model, a taxonomy for mapping anomalies in DCS. Anomalies are classified based on their causing change, the rupture indicated by the anomaly and their impact. This general mapping gives a good overview of anomaly types in a field of DCS.

The model offers extensibility, e.g., by adding new dimensions or categories as well as by further specification of component types. This allows users of the model to adjust it to their needs, while keeping a common understanding and language of anomalies.

Further research could augment the model by adding a common set of metrics or anomaly detection techniques to each category of the taxonomy, easing the detection of anomalies of each kind, also studying which types of anomalies are notoriously hard to detect or have big impact to the company running the system.

Besides showing the applicability of the model to DCS, we examined its usage on a production system. It would be interesting to see its application in other fields of (distributed) systems, or possible extensions for such.

Similarly, a mapping of different metric types (system-level vs. application-level metrics for instance) and their contribution to effective detection with different anomaly types could create new insights and be a valuable tool.

Finally, [15] mention the need for an industry-wide repository of anomaly descriptions and [7] request a publicly available datastore for performance datasets. Creating such databases with standardized formats (and dimensions, such as those from the CRI-Model) could foster a better understanding of anomaly types, confirm choices of anomaly detection approaches and allow a wider evaluation and comparison of those approaches.

To conclude, the taxonomy can be used for building and improving monitoring of anomalies as well as improving processes to remedy anomalies in case of system outages. While a general purpose monitoring is desired, it may be hard to catch all kinds of anomalies. Retroactively mapping anomalies to the CRI-Model helps in identifying areas that demand improvement, either because of a multitude of occurred anomalies or because of low detection rates, with the goal to improve quality of the distributed system.

The CRI-Model lays the ground-work for a future mapping of different anomaly classes against existing detection approaches and creates a vocabulary and research tool for further work in the field of AD.

Appendix A: Literature Sources Used for the Taxonomy

Barford, P. et al.: A signal analysis of network traffic anomalies. In: the second ACM SIGCOMM Workshop, 71–82 (2002)

Cheng, H. et al.: Detection and Characterization of Anomalies in Multivariate Time Series. In: Proc. SIAM, 413–424 (2009)

Cohen, I. et al.: Correlating Instrumentation Data to System States: A Building Block for Automated Diagnosis and Control. In: Proc. of the 7th symp. on Operating systems design and implementation. 4. (2004)

Düllmann, T.: Performance Anomaly Detection in Microservice Architectures Under Continuous Change. Masterthesis. University of Stuttgart, 2017

Dunning, T. and Friedman, E.: Practical Machine Learning: A New Look At Anomaly Detection. (2014), p. 65

Fu, Q. et al.: Execution anomaly detection in distributed systems through unstructured log analysis. In: Proce.—IEEE International Conference on Data Mining December, 149–158 (2009)

Goldstein, M. and Uchida, S.: A Comparative Evaluation of Unsupervised Anomaly Detection Algorithms for Multivariate Data. In: PloS one 11.4 (2016): e0152173 11.April, 1–31 (2016)
Gu, X. and Wang, H.: Online anomaly prediction for robust cluster systems. In: Proc.—International Conference on Data Engineering, 1000–1011 (2009)
Guan, Q. et al.: Efficient and accurate anomaly identification using reduced metric space in utility clouds. In: Proc.—IEEE 7th Int. Conf. on Networking, Architecture and Storage, 207–216 (2012)
Gupta, M. et al.: Context-Aware Time Series Anomaly Detection for Complex Systems. In: Proc. of the SDM Workshop on Data Mining for Service and Maintenance, 14–22 (2013)
Hole, K.: Anomaly detection with htm. In: Anti-fragile ICT Systems. Springer International Publishing, 2016. Chap. Anomaly de, pp. 125–132
Ibidunmoye, O., Hernández-Rodriguez, F., and Elmroth, E.: Performance Anomaly Detection and Bottleneck Identification. In: ACM Computing Surveys 48.1, 1–35 (2015)
Munawar, M. et al.: Filtering system metrics for minimal correlation-based self-monitoring. In: IEEE Int. Conf. on Self-Adaptive and Self-Organizing Systems, 233–242 (2009)
Sharma, A. et al.: Fault detection and localization in distributed systems using invariant relationships. In: IEEE/IFIP Int. Conf. on Dependable Systems and Networks 1, 1–8 (2013)
Sheth, A. et al.: Mojo: A Distributed Physical Layer Anomaly Detection System for 802.11 WLANs. In: Proc. of the 4th int. conf. on Mobile systems, applications and services, 191 (2006)
Smith, D., Guan, Q., and Fu, S.: An anomaly detection framework for autonomic management of compute cloud systems. In: Proc.—Int. Computer Software and Applications Conf. 376–381 (2010)
Takeishi, N. and Yairi, T.: Anomaly detection from multivariate time-series with sparse representation. In: Proc.—IEEE Int. Conf. on Systems, Man and Cybernetics 2014-Janua.January, 2651–2656 (2014)
Tan, Y. et al.: PREPARE : Predictive Performance Anomaly Prevention for Virtualized Cloud Systems. In: Distributed Computing Systems (ICDCS). Vcl, pp. 285–294 (2012)
Thottan, M. and Ji, C.: Proactive anomaly detection using distributed intelligent agents. In: Network, IEEE October, 21–27 (1998)
Wang, T. et al.: Fault Detection for Cloud Computing Systems with Correlation Analysis. In: 652–658 (2015)

Appendix B: Reports of System Outages

Table 3 The List of system outages reports from (software) companies

Company	Date	Source
HealthCare.gov	01.10.2013	https://www.theguardian.com/world/2013/oct/30/health-secretary-sebelius-healthcare-website-hearing-live
Motorola	02.12.2013	https://www.cnet.com/news/motorola-delays-moto-x-cyber-monday-deal-after-site-crash
Bank of America	01.02.2013	http://www.cbsnews.com/news/bank-of-america-web-portal-back-online-after-outage
Yahoo	09.12.2013	http://www.theverge.com/2013/12/11/5201146/yahoo-apologizes-for-email-outage
AWS	13.09.2013	http://www.usatoday.com/story/tech/2013/09/13/amazon-cloud-outage/2810257/
Google Amazon Microsoft	19.08.2013	http://blog.smartbear.com/performance-testing/any-given-monday-google-microsoft-and-amazon-all-experience-outages/
Nasdaq	22.08.2013	http://www.networkcomputing.com/government/nasdaq-outage-explored-7-facts/276050510
LinkedIn	23.10.2013	http://www.onlinesocialmedia.net/20131023/linkedin-outage-today-follows-facebook-going-down/
LinkedIn	18.06.2013	https://techcrunch.com/2013/06/19/linkedin-outage-due-to-possible-dns-hijacking/

(continued)

Table 3 (continued)

Company	Date	Source
Verizon	18.06.2013	http://www.datacenterdynamics.com/content-tracks/security-risk/95538.fullarticle
Twitter	19.01.2016	http://www.techradar.com/news/world-of-tech/twitter-is-down-for-most-people-right-now-1313338
Microsoft	18.01.2016	https://www.theregister.co.uk/2016/01/25/office
Salesforce	03.03.2016	http://www.v3.co.uk/v3-uk/news/2449449/salesforce-suffers-cloud-outage-and-service-disruption-in-europe
Salesforce	09.05.2016	http://www.zdnet.com/article/circuit-breaker-failure-was-initial-cause-of-salesforce-service-outage/

References

1. Avizienis, A., Laprie, J.-C., Randell, B.: Dependability and its threats: a taxonomy. In: Proceedings of IFIP 18th World Computer Congress, pp. 91–120 (2004)
2. Baddar, S., Merlo, A., Migliardi, M.: Anomaly detection in computer networks: a state-of-the-art review. J. Wirel. Mobile Netw. Ubiquitous Comput. Dependable Appl. (JoWUA) **5**(4), 29–64 (2014)
3. Barford, P. et al.: A signal analysis of network traffic anomalies. In: The Second ACM SIGCOMM Workshop, pp. 71–82 (2002)
4. Chandola, V., Banerjee, A., Kumar, V.: Anomaly detection: a survey. ACM Comput. Surv. **41**(3), 1–15 (2009)
5. Ghosh, S., Reilly, D.L.: Credit card fraud detection with a neural-network. In: 1994 Proceedings of the Twenty-Seventh Hawaii International Conference on System Sciences, vol. 3, pp. 621–630 (1994)
6. Goldstein, M., Uchida, S.: A comparative evaluation of unsupervised anomaly detection algorithms for multivariate data. PloS one **11**(4), 1–31 (2016). e0152173
7. Ibidunmoye, O., Hernández-Rodriguez, F., Elmroth, E.: Performance anomaly detection and bottleneck identification. ACM Comput. Surv. **48**(1), 1–35 (2015)
8. Kumar, M., Ghani, R., Mei, Z.-S.: Data mining to predict and prevent errors in health insurance claims processing. In: The 16th ACM SIGKDD International Conference, pp. 65–74 (2010)
9. Ladiges, J., et al.: Evolution management of production facilities by semiautomated requirement verification. at-Automatisierungstechnik **62**(11), 781–793 (2014)
10. Mazel, J., Fontugne, R., Fukuda, K.: A taxonomy of anomalies in backbone network traffic. In: IWCMC 2014—10th Int. Wireless Communications and Mobile Computing Conference, pp. 30–36 (2014)
11. Mirkovic, J., Reiher, P.: A taxonomy of DDoS attack and DDoS defense mechanisms. SIGCOMM Comp. Comm. Rev. **34**(2), 39–53 (2004)

12. Nielsen, J.: Usability Engineering. Elsevier (1994)
13. Pertet, S., et al.: Causes of failure in web applications, Parallel Data Laboratory December, pp. 1–19 (2005)
14. Plonka, D., Barford, P.: Network anomaly confirmation, diagnosis and remediation. In: 47th Annual Allerton Conference on Communication, Control, and Computing, pp. 128–135 (2009)
15. Tobergte, D., Curtis, S.: Why Internet services fail and what can be done about these. J. Chem. Inf. Model. **53**(9), 1689–1699 (2013)

Modeling and Analysis of IoT Energy Resource Exhaustion Attacks

Vasily Desnitsky and Igor Kotenko

Abstract Subjection of wireless Internet of Things (IoT) devices to energy resource exhaustion attacks gets increasing importance. Being stealthy enough for an attack target and systems of its monitoring such attacks are capable to exhaust energy of the device in a relatively short period and thereby impair the function and availability of the device. The paper analyzes possible types of ERE attacks, proposes an intruder model regarding this kind of attacks and provides experimental studies on the basis of a developed use case.

Keywords Security · Energy resource exhaustion · IoT · Modeling and analysis

1 Introduction

Various IoT systems with mobile devices, sensors, actuators as well as normal communications and computing equipment are getting widespread. Beside target attacks using the specificity of a particular devices, universal attacks aimed at disrupting the availability are gaining a crucial importance. The mobility of devices, their autonomous working and limitations on energy resources make them vulnerable to energy resources exhaustion (ERE) attacks. Depending on a wide range of IoT applications, availability violation leads to serious man-made disasters and high financial damage.

V. Desnitsky · I. Kotenko (✉)
St. Petersburg Institute for Informatics and Automation of the Russian Academy of Sciences, 14-th Liniya, 39, 199178 St. Petersburg, Russia
e-mail: ivkote@comsec.spb.ru

V. Desnitsky
e-mail: desnitsky@comsec.spb.ru

V. Desnitsky · I. Kotenko
St. Petersburg National Research University of Information Technologies, Mechanics and Optics, (ITMO University), 49, Kronverkskiy Prospekt, St. Petersburg, Russia

M. Ivanović et al. (eds.), *Intelligent Distributed Computing XI*,
Studies in Computational Intelligence 737, https://doi.org/10.1007/978-3-319-66379-1_23

The core ERE attack features are *difficulty* of their detection, their *effectiveness* and *variability*. The difficulty stems from the facts that (1) the impact on a victim is often indirect—it comprises sending a series of false requests to the device, which are not easy identified as an attack, (2) it is needed not only to capture the battery discharging process, but also to analyze changes in the discharging speed and (3) ERE attacks may be accompanied by casual factors of battery discharge due to legitimate user actions and applications. Besides, an attacker is able to choose variably the most convenient and effective ways to carry out the rapid energy exhaustion.

In published sources ERE attacks are represented at a fairly low degree, mostly in the analysis of certain types of ERE attacks on particular devices [1, 2]. Together with a trend to poor IoT security and a lack of an effective ERE attack monitoring it demands further advance in the field. The contribution of this paper is an analysis of ERE attacks, a ERE intruder model and experimental studies. The novelty lies in an unified ERE attack classification within the framework of an approach to modeling and evaluating ERE attacks. The rest of the paper is structured as follows. Section 2 overviews the related works. Section 3 covers the proposed intruder model. Section 4 presents a developed use case. Sections 5 and 6 expose experiments and discussion.

2 Related Works

This paper considers detecting faulty process and reducing energy consumption by stopping or delaying suspicious activity. Moyers et al. describe ERE attacks on Wi-Fi and Bluetooth interfaces as well as some mixed attacks to deduce their impact on the battery lifetime, having simulated particular attacks accelerating battery drain vastly [3]. Battery-Sensing Intrusion Protection System (B-SIPS) warns about the change of power on small wireless devices by algorithm for dynamic calculation Threshold [4].

Boubiche et al. analyze the following attack types at the intersection of the physical and data link layers—Sleep Deprivation, Barrage, Replay, Broadcast, Collision and Synchronization attacks [5]. Racic et al. describe the implementation of an ERE attack by exploiting a vulnerability of MMS messages [2]. At that, an attacker collects specific data on the network and phones by using MMS notifications. After that the attacker drains the battery of the device by sending periodic UDP packets by using stored PDP contexts and search call channel. Karpagam et al. investigate Selective Jamming Attacks in a wireless sensor network (WSN) to physically impact on the most crucial packets transmitted wirelessly [6]. The key element of such attack is a real-time packet classification at the physical layer. An intellectual swarm based technique is proposed to identify these attacks [7]. It allows analyzing data from sensors against jamming, dynamically changing the used communication channel.

Goudar et al. analyze Denial-of-Sleep attacks in WSN, influencing through manipulating network packets [1]. Specifically *collision attacks* represent a simultaneous delivery of multiple packets to a node, causing their discard, retransmission and extra energy costs. *Control Packet Overhead attack* involves sending specific service commands in a broadcast, such as RTS (i.e. a notification on the preparedness of a node to pass data) and CTS (i.e. a notification on a feasibility to receive data) hindering a recipient node to switch to the sleep mode. *Over Emitting Attacks* represent sending data to a node that is not ready to receive them. Capossele et al. [8] outline an approach to building a WSN with nodes operating in an energy efficient *wake-up-receive mode* (WuR), using of event-based management. *Vampire attacks* affecting transmitted packets at a layer of a routing protocol are regarded in [9]. These attacks are targeted on the whole WSN to drain energy of all its nodes. At that an attacker exploits formally correct traffic with some small changes in routing headers, thereby extremely complicating attack detection. A *carousel attack* intentionally introducing loops in packet routes allows messages passing via the same WSN nodes for several times. A *stretch attack* runs by forcing the use of the longest routes in WSN.

3 Intruder Model

The proposed analytical intruder model is represented by the following formal tuple (*G, O, A, R, F, E*). Attack goal *G* defines breaching availability of an autonomously operating device by draining the energy resource in a fast gradual or spasmodic manner, depending on the specificity of the target device. *O* specifies objects of direct or indirect effects of ERE attacks, including sensors of the physical environment, communication channels, software components, operating system processes, etc.

A represents actions an attacker fulfills to achieve the goals *G*. *R* describes resources, tools and starting possibilities of the intruder. In the analysis we differentiate five intruder types according to a classification of the intruder access to an embedded device, namely $Type_0$—no access (social engineering), $Type_1$—no direct access (TCP/IP based attacks from Internet), $Type_2$—remote access from a small distance (Wi-Fi, IR, Bluetooth, etc.), $Type_3$—direct access to wired interfaces (e.g. RS-232, I_2C, etc.) and $Type_4$—full access (tamper with microchip) [10]. Besides we rely on a classification of intruder capabilities that comprises $Level_1$—public accessed tools, well-known vulnerabilities, $Level_2$—specialized tools, previously unknown vulnerabilities, $Level_3$—group of intruders level 2 (unlimited resources) [11]. *F* describes specific ERE features (e.g. a policy for transition between a sleep and normal modes of the device. *E* determines efficiency of an ERE attack. E is computed experimentally and represents an average increase of the battery discharging speed. Analyzing state-of-the-art the most important ERE attacks are

Table 1 Denial-of-Sleep attacks

Goal of attack	Reducing the time of the device being in a sleep mode to increase energy costs
Attack features	1. Presence of device idle states of a small energy consumption 2. Using energy consuming wireless interfaces such as Bluetooth, Wi-Fi, etc 3. Attack category <*Types$_1$ & Type$_2$, Level$_1$ & above*> [10, 11]
Actions of intruder	Increase of the power consumption of the device by changing its modes (*idle mode → active mode*)
Required resources, tools and starting possibilities of intruder	1. Superficial knowledge of Linux. Downloading and installing typical software means into the operating system. Capabilities of reproducing manuals 2. Minimal time for deployment of a new attacking device is required after the precursive software and hardware preparation 3. Usually it is sufficient to use a typical laptop/single-board computer. A rooted Android device may be required to run some specific soft
Conclusion	1. An attack is carried out without direct impact on the device (i.e. wirelessly) 2. The attack distance depends on wave frequencies of the wireless protocol and the power of the antenna of the intruder 3. There is no need for an intruder to be authorized on the target device, which complicates an effective protection from such attacks

Table 2 Attacks of wireless traffic increasing

Goal of attack	Increasing amounts of income/outcome data and decreasing their speed
Attack features	1. Typically devices transmit data not permanently. An attacker is to enlarge amounts of data transmitted and time of the transmission 2. Attack category <*Types$_1$ & Type$_2$, Level$_1$ & above*>
Actions of intruder	Intruder logs into the device and starts messaging. To defeat authorization the one breaks the key or makes a replay attack by some past legitimate traffic
Intruder abilities	Basic knowledge on the target system. A typical laptop/single-board computer is enough. A rooted Android device may be required to run some specific soft
Conclusion	An attack can be carried out without direct influence on the device. Attack distance depends on wave frequencies and the antenna of the intruder

singled out: (1) *forced waking of a sleeping device* (Denial-of-Sleep) [3, 4] (see Table 1); (2) *growth of wireless traffic* (Table 2); (3) *electromagnetic jamming* [6] (Table 3); (4) *misuse of the device* (Table 4).

Table 3 Creation of electromagnetic interference on wireless channels (jamming)

Goal of attack	To force the device to increase the signal power during the communication
Attack features	Normally wireless modules try to transmit at the minimum power to reduce energy costs. Attack category <$Type_2$, $Level_2$ *& above*>
Actions	Electromagnetic noising on wireless data transmission channels
Abilities	Tools to affect wireless channels. Location in a short distance from the device
Conclusion	An attack can be carried out without direct influence on the device Attack distance depends on wave frequencies and the antenna of the intruder

Table 4 Untypical usage of a device and its software

Goal of attack	Waste energy by forcing the device to run some unnecessary functions
Features	Attack category <$Type_0$–$Type_4$, $Level_1$ *& above*>
Actions of intruder	Extra CPU load, access to energy consuming memory, packet transmission via communication channels, multiple launch of applications, breaking/bypass optimizations and non-typical use of the software, remote desktop session, etc
Abilities	Penetration skills for direct/remote access to the device and run malware on it
Conclusion	The attack assumes the deepest affection of the intruder to the device

4 Experimental Results

For modeling ERE attacks and analysis we have developed a use case—ZigBee mesh network on wireless XBee s2 modules able to work without any external microcircuit in a power saving mode. XBee is configured for automatic reading analog/digital input data and their periodic sending out. In the sleep mode the core XBee elements are disabled, keeping the current consumption up to a few *mA*. The misuse attack has fairly limited application due to an ability to change some XBee configuration settings only by intruders of $Type_1$, $Type_2$ and $Type_3$. An increase of noises/shielding is a quite universal attack and could be applied.

We simulated a combined attack on a target XBee of an End Device role operating in the sleep mode by sending queries from a false XBee. On the target XBee the following energy saving settings were assigned, namely SM = 4 (cyclic sleep mode), ST = 1000 ms (time before sleep), and SP = 10000 ms (Cyclic Sleep Period). The intruder performs regular sending messages to the target node (2/s). Such message resets the ST timer, hindering the node to return to the sleep mode (Denial-of-Sleep). The combined character of the attack is that besides preventing the XBee transition to the sleep mode some additional power consumption is performed also as a result of data reception instead of the XBee being in the idle mode (i.e. attack of wireless traffic growth). Figure 1 exposes the scheme of the test bench modeling this attack.

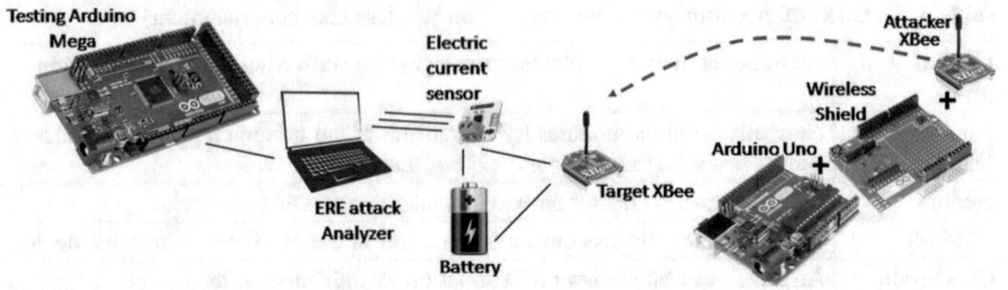

Fig. 1 Test bed for modeling ERE attack on XBee

Table 5 The results of the XBee energy consumption in normal mode and under attack

Time gap, msec	0–1000	1000–11000
I_{IDLE}, mA	45	8
I_{ATTACK}, mA	51	51

Readings from MAX471 sensor are taken by a testing Arduino microcircuit and passed to a PC via the UART port in shape of logs for subsequent processing and analysis. The attacking sketch on Arduino Uno regularly sends data to a specified ZigBee destination address. The current consumption measurements in the idle mode and ones under the attack are presented in Table 5. This is the average data for one wakefulness cycle of 1 s activity and 10 s sleep. The consuming current during the simulated attack of a constant intensity is measured as a constant value $I_{ATTACK} = 51$ mA, while in the sleep mode it is 8 mA. The attack efficiency $E = I_{ATTACK} \cdot (t_2 - t_1) \; / \; \int_{t_1}^{t_2} I_{IDLE}(t)dt$, where the consumption in the idle mode I_{IDLE} is computed as an average value by Lagrange interpolation. The experiment was conducted repeatedly on time interval $t_2 - t_1 = 600$ s. The efficiency $E = 4.488$ means the attack exhausted the battery of the target XBee more than 4 times faster.

5 Discussion

ERE attacks are applicable to devices using autonomous exhaustible energy sources. The experimental study showed the strength of such attacks in reducing the lifetime significantly. ERE attacks are relevant especially for critical application domains, which business processes cannot be suspended to recharge or replace batteries. Exploiting an ERE attack on an unmanned aerial vehicle allows draining hiddenly its battery used by the motor and raising a risk of crashing the drone. Autonomous modules of a Digital City, such as mobile interactive road signs/traffic lights, pollution sensors, etc. defeated by a dishonest competitor can lead to frequent maintenance and extra costs in the system management.

Detection should be specific to an expected types of ERE attacks. For an attack of a wireless incoming traffic increase, the traffic should be evaluated for its legitimacy, starting from its source, application protocol, other header and payload, regarding the history of interaction. In contrast to the pure software measurements of the current the hardware sensors are considered as less energy consuming and hence more precise as well as more customizable. However hardware sensors may require its physically soldering to the battery pins not always achievable for chip-based batteries.

Being an application-specific process ERE attack detection includes detailed tracking data on energy consumption of the device, its software and hardware modules transitions between specific modes, logs on starting and pausing applications, local storage calls, etc. All these data should be used to analyze events and verify feasibility of correlation rules to identify symptoms of ERE attacks. Besides the monitoring, specific security means against ERE attacks are assumed. Particularly we regard intermediate hardware-based energy-efficient units to filter input data, recognize and discard ERE data flows effectively before the flows hem rich the nodes.

6 Conclusion

The paper comprises modeling and analysis of ERE attacks on two IoT use cases. The results obtained express possibility of ERE attack modeling both analytically and on physical IoT equipment. Such modeling could be introduced into composite techniques of construction and configuration of secure IoT devices [12, 13]. Further analysis as well as experiments and comparisons on use cases are planned in future.

Acknowledgements The work is supported by RSF #15-11-30029 in SPIIRAS.

References

1. Goudar, C.P., Kulkarni, S.S.: Mechanisms for detecting and preventing denial of sleep attacks and strengthening signals in wireless sensor networks. Int. J. Emerg. Res. Manag. Technol. **4** (6) (2015)
2. Racic, R., Chen, D.M., Chen H.: Exploiting MMS vulnerabilities to stealthily exhaust mobile phone's battery. In: 2006 Securecomm and Workshops, pp. 1–10 (2006)
3. Moyers, B.R., Dunning, J.P., Marchany, R.C., Tront J.G.: Effects of Wi-Fi and bluetooth battery exhaustion attacks on mobile devices. In: 43rd Hawaii International Conference on System Sciences, pp. 1–9 (2010)
4. Buennemeyer, T.K., Gora, M., Marchany R.C., Tront, J.G.: Battery exhaustion attack detection with small handheld mobile computers. In: IEEE International Conference on Portable Information Devices, pp. 1–5 (2007)
5. Boubiche, D.E., Bilami, A.: A defense strategy against energy exhausting attacks in wireless sensor networks. J. Emerg. Technol. Web Intel. **5**(1) (2013)

6. Karpagam, R., Archana, P.: Prevention of selective jamming attacks using swarm intelligence packet-hiding methods. Int. J. Eng. Comput. Sci. **2**, 2774–2778 (2013)
7. Periyanayagi, S., Sumathy, V., Kulandaivel, R.: A defense technique for jamming attacks in wireless sensor networks based on sensor networks. In: International Conference on Process Automation, Control and Computing, pp. 1–5 (2011)
8. Capossele, A.T., Cervo, V., Petrioli, C., Spenza, D.: Counteracting denial-of-sleep attacks in wake-up-radio-based sensing systems. In: 13th Annual IEEE International Conference on Sensing, Communication, and Networking (SECON), pp. 1–9 (2016)
9. Farzana, T., Babu, A.: A light weight PLGP based method for mitigating vampire attacks in wireless sensor networks. Int. J. Eng. Comput. Sci. **3**(7) (2014)
10. Rae, A.J., Wildman, L.P.: A taxonomy of attacks on secure devices. In: Australian Information Warfare and IT Security, pp. 251–264 (2003)
11. Abraham, D.G., Dolan, G.M., Double, G.P., Stevens, J.V.: Transaction security system. IBM Syst. J. **30**(2), 206–228 (1991)
12. Desnitsky, V., Kotenko, I., Chechulin, A.: Configuration-based approach to embedded device security. In: Lecture Notes in Computer Science, vol. 7531, pp. 270–285. Springer (2012)
13. Desnitsky, V., Levshun, D., Chechulin, A., Kotenko, I.: Design technique for secure embedded devices: application for creation of integrated cyber-physical security system. J. Wireless Mob. Netw. Ubiquitous Comput. Dependable Appl. (JoWUA) **7**(2), 60–80 (2016)

Part VII
Service-Based Distributed Systems

Service Discovery in Megascale Distributed Systems

Kai Jander, Alexander Pokahr, Lars Braubach and Julian Kalinowski

Abstract Service discovery is a well-known but important aspect of dynamic service-based systems, which is rather unsolved for megascale systems with a huge number of dynamically appearing and vanishing service providers. In this paper we deduce requirements for service discovery in megascale systems and analyze existing approaches with these in mind. Shortcomings of existing solutions are explained and a novel solution architecture is presented. It is based on the idea that service description data can be subdivided into static and dynamic properties. The first group remains constant over time while the second is valid only for shorter durations and has to be updated. Expressive service queries rely on both, e.g. service location as example for the first and response time for the latter category. In order to deal with this problem, our main idea is to also subdivide the architecture into two interconnected processing levels that work independently on static and dynamic query parts. Both processing levels consist of interconnected peers allowing to auto-scale the registry dynamically according to the current workload. Finally, some parts of the ongoing implementation based on Jadex agent technology will be explained.

Keywords Service discovery · Services · SOA · Cloud · Megascale

K. Jander (✉) · A. Pokahr · J. Kalinowski
Distributed Systems Group, University of Hamburg, Hamburg, Germany
e-mail: jander@informatik.uni-hamburg.de

A. Pokahr
e-mail: pokahr@informatik.uni-hamburg.de

J. Kalinowski
e-mail: kalinowski@informatik.uni-hamburg.de

L. Braubach
Complex Software Systems Group, City University of Applied Sciences,
Bremen, Germany
e-mail: lars.braubach@hs-bremen.de

M. Ivanović et al. (eds.), *Intelligent Distributed Computing XI*,
Studies in Computational Intelligence 737, https://doi.org/10.1007/978-3-319-66379-1_24

1 Introduction

In recent years, many types of distributed applications have become increasingly large-scale. This trend is primarily driven by the number of users accessing such applications but complexity and intelligence of the applications themselves are a driving factor as well. While basic applications with thousands of users can be scaled easily, applications such as Facebook, Twitter and eBay have millions to hundreds of millions of simultaneous users and the functionality of the applications also allow a high degree of interaction between them.

This means that such distributed systems require a high level of scalability. While scalability of algorithms has always been a focus for research, even carefully designed system are often limited in practice when scaling to such large scale applications, often requiring the developing company to optimize the existing software for higher scalability, despite careful previous design.[1]

Current systems primarily deal with large numbers of users using relatively simple services. In this type of application, scalability can be managed using simple architectures (client-server models, massively replicated server instances) that reduce the risk of introducing poorly scaling components. For example, while a large number of users interact with systems like Facebook, the number of services offered to the user is actually below 10^3 or lower. In such systems, service providers are usually directly included, e.g., as REST-URLs in the client-side software package. This approach works well on centralized systems where users interact with a limited set of server-based service and do not directly interact with each other. However, such an approach is not suitable for every type of application. Systems like smart homes and smart cities demand considerable autonomy between components. Enabling components and users to directly interact without a central mediating component means that each part in the system has to offer services for use by other parts. In such a system, finding the right service for interaction can be a challenge to the components of the system. Of course, also other aspects like e.g. the trust of service providers are important for large open systems but are out of the scope of this paper, see. e.g. [4].

This paper will specifically look at the practical implications of service discovery in systems with a large number of services, in particular in the range of 10^6 up to 10^9 simultaneous services providers. It can also be expected that the services appear and disappear dynamically at any time. Example application areas for these type of systems are as follows:

- Increased automation in homes ("smart home") and cities ("smart cities") result in large sets of smart components dealing with aspects of the automation. While a centralized approach may be feasible for homes, increased scales in case of cities as well as privacy concerns would suggest give advantages to a peer-to-peer-based approach.

---

[1] E.g., http://highscalability.com/blog/2016/1/11/a-beginners-guide-to-scaling-to-11-million-users-on-amazons.html.

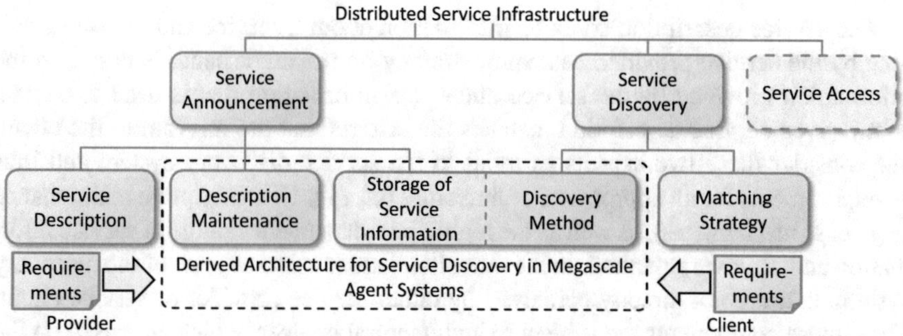

Fig. 1 Deriving requirements for a service discovery (upper part adapted from [11])

- In some cases, decisions have to be made locally and communication being used opportunistically to enhance capabilities. For example, autonomous vehicles need to be able to act safely even if communication is interrupted but should be capable to leverage communication to enhance its capabilities by maintaining distance to other vehicles, avoiding heavy traffic and adjusting speed to match upcoming traffic lights.
- Even in the case of existing centralized megascale applications, increasing application complexity encourages the use of an internal peer-to-peer model like microservices [5]. In contrast to more traditional architectures such as the three-tier architecture with presentation layer, business layer and data layer, the application consists of small, integrated services calling each other.

In the following, we first identify requirements for service discovery in megascale distributed systems (Sect. 2). In Sect. 3, existing approaches to service discovery are analyzed and a new solution architecture is presented in Sect. 4. The paper concludes with a summary and an outlook in Sect. 5.

2 Requirements

Any kind of distributed service infrastructure requires mechanisms for the three fundamental steps: *announce*, *discover*, and *access* (cf. e.g. [11]). Whereas the last step—service access—is concerned with the technical details of service invocation, the focus of this paper is on the first two steps: announcement and discovery. As shown in Fig. 1, [11][2] consider *service description*, *description maintenance* and *storage of service information* as relevant components of the service announcement part. Service discovery is subdivided into *discovery method* and *matching strategy*. In the following, the requirements for service discovery in megascale agent systems are derived by considering these components from a use-case-driven view.

[2]We use different terms compared to [11] in an attempt for more concise terminology.

The service description contains information about a service and is made available by the service provider. Matching strategy on the other hand, is based on the information provided by the service client, i.e. information that is used to decide, if any given service description matches the criteria that are relevant to the client. We consider these two aspects as input to the service discovery system and thus as requirements for an appropriate architecture (cf. Fig. 1). Description maintenance (e.g. lease mechanisms), as well as the highly interdependent storage of service information and discovery method on the other hand are internal aspects of the discovery system, that are not directly perceived by either service provider or service client. Thus, these components are subject to architectural choices, which are based on the requirements derived from the service provider and service client side.

2.1 Provider-Side Requirements

Service providers want to be found by potential clients and delegate this responsibility to the discovery infrastructure. To support providers in a megascale system, the following requirements apply to the discovery infrastructure:

Many Providers In contrast to the state of the art, a megascale distributed service infrastructure should not only support millions of service clients, but also *millions of service providers* on (potentially) millions of nodes.

Auto Maintenance The service infrastructure should provide mechanisms for maintaining the consistency of descriptions of available services, even in case of failures (e.g. when service providers are unable to deregister their description due to network or node failures).

2.2 Client-Side Requirements

Service clients access the discovery infrastructure for obtaining information about service providers that match specific needs of the client. Thus, the client-side requirements are derived from use cases with regard to service search specifications. To cover a broad range of use cases, the content of a search request is classified into different dimensions in the following. The first three dimensions are concerned with the expressivity of the search specification. Two other dimensions consider the life time of a search request, as well as quality expectations by the client:

Service Matching The search specification should allow referring to both *functional and non-functional properties* for expressing the needs of the client. Functional properties describe the general applicability of a service to a specific need, whereas non-functional properties allow capturing conditions describing how the service should be provided (e.g. costs, security constraints, or response times). *Exact matching* of properties can be based on known service type names or, e.g., tags. *Semantic matching* uses, e.g., logic-based reasoning techniques or statistical methods. As most practical use cases prefer exact matching for reliable system behavior, we do not explicitly consider semantic matching in the remainder of this paper.[3]

Result Size Common use cases can be distinguished with regard to requiring only *one* service (e.g. for buying a product) versus looking for *all* services of a type (e.g. in a chat application). Given that a mega scale system can potentially have a huge number of services, clients may also want to *limit* the result size.

Result Ordering Service matching can be based on *boolean criteria* as well as *gradual measures*. Boolean criteria allow quickly selecting services based on, e.g., exact matching of functional properties like a service type name, whereas gradual measures, often applied to non-functional properties, allow ordering of search results based on user-specified fitness-functions (e.g. service provider with lowest utilization).

Query Persistence A service search can be *one-shot* or *continuous*. A one-shot search will only result in currently available services, whereas a continuous search will be stored as a persistent query in the infrastructure, such that the client will receive timely updates, whenever services matching the search specification appear or disappear.

Quality Levels Some clients may require to discover *all* services (or the globally *best* with regard to a given fitness function), whereas other clients just need *some* services. Specifying a quality level allows for optimizations inside the search system. Moreover, clients are usually interested in *result freshness*, which means that the search result should only contain services, for which a given max age has not been reached.

3 Analysis of Service Discovery Approaches

Service discovery mechanisms have originated in different research areas. Broadly one can distinguish hardware oriented service discovery, which is mainly motivated by the wish to find typical services in a local network, e.g. a print or scan service. This area has inspired discovery protocols with a focus on local area network technologies like IP/TCP multicast. Examples of this category are Jini [9]and UPnP [10]. These are not considered further, because they do not fit to the challenges of mega-scale and internetwork service environments. The second interesting area of research is centered around WSDL based web services. In this context service discovery played a central role and a lot of standards and proposals for service management emerged. Finally, practice paved the way towards the much simpler to use and build RESTful web services. In this field service discovery did not gain that much importance and even today many service providers are simply hardcoded in clients using the respective service URLs [1]. The novel trend of mircoservices changed this to some degree.

---

[3]Especially, in large-scale open environments semantic metaching becomes very important.

3.1 WSDL Web Service Solutions

From the beginnings UDDI (Universal Description, Discovery and Integration) [3] has been proposed as standard for service discovery as part of the so called SOA (Service Oriented Architecture) triangle [6]. The SOA triangle assumes provider register service at a central registry and user inquire the registry to get service descriptions including contact data to directly talk to the service. UDDI allows for storing services descriptions in WSDL together with meta information about these services including business as well as technical details. Registration and search is possible using white, yellow and green page service information via a dedicated UDDI API. In order to cope with large service amounts, publicly available enterprise UDDI registries had been set up by major players including Microsoft, IBM and SAP. Even though UDDI had been designed as open standard and major companies tries to push its usage both efforts failed to a large extent. The enterprise registries were silently taken down at end of 2005 and users searched for alternatives to UDDI.[4] UDDI failure had several reasons, but one important aspect is the huge complexity of the standard that required much effort to setup and use a registry.

The problems with UDDI led to the development of alternatives like WS-Inspection [2] and WS-Discovery [7]. WS-Inspection defines an XML format for listing references to existing service descriptions typically in WSDL. Hence, WS-Inspection documents can be read and services can be contacted based on the deposited addresses. WS-Inspection can be seen in contrast to UDDI as a handy and simple but also quite limited solution because it does not emcompass service search functionalities. Both solutions reviewed so far concentrate on rather static service networks, i.e. it is assumed that the rate of leaving and arriving services is rather low. WS-Discovery has been devised to fill this gap. It is a hybrid mode protocol that can work with multicasts as well as registries (called discovery proxies).

3.2 Cloud-Based Microservice Solutions

Microservices represent a specific interpretation of SOA in the sense that an application is seen as composition of rather small services [5]. In a microservice architecture each service may consist of the full vertical software stack from user interface to its own database making them independently developable. Business functionalities are realized by service interactions so that service discovery becomes a vital part also in these kinds of applications. Microservices are practice driven by major IT companies. Thus, in the following, architectures of registry software are evaluated.

Consul[5] is a cloud enabled service registry from Hachi Corp. Cloud-enabled means that functionalities facilitate the operation in cloud datacenters, e.g. support for different regions in queries. Its architecture is based on a strongly synchronized

[4]http://www.computerwoche.de/a/570059.

[5]https://www.consul.io/.

server cluster. Requests will be answered by a leader following the Raft consensus protocol. Primary goal of the system is high fault tolerance, because server failures can be compensated to some degree by the used consensus protocol. As long as a quorum of servers answers to a request the system keeps functioning. The size of the cluster can be configured, but higher numbers of servers in the cluster negatively impact the response time due to increased overhead. The system offers a RESTful API for (de)registering and searching services. Search is based on arbitrary key value pairs that the system checks for presence in registered entries.

Eureka 2[6] is a service registry for microservices developed by Netflix that is also destined for use in cloud environments. Like Consul it provides first-level support for cloud properties within service specifications and queries. Eureka 2 has a more advanced architecture than its predecessor as well as most other registries available. The Eureka 2 architecture comprises two disjunctive server clusters: a write and a read cluster. As the names suggest, these clusters are responsible for handling service (de)registrations and query processing separately. The advantage of this separation of responsibilities is that the read cluster can be auto-scaled independently from the write cluster in accordance to the current query workload. The write cluster must be set up manually and is not auto-scaled. Eureka 2 is also the first system supporting persistent queries, i.e. has a push model for service discovery. The queries itself are based on service properties described as key value pairs. In addition to checking for the presence of value simple operators are also supported.

Both, Consul and Eureka contain problematic design decisions for megascale systems. First, both registries use health checks to periodically ping services, which is costly with many services, because many open connections are held. Second, query processing currently is not based on indexed data structures, which renders it necessary to check each registered service in each query. For Consul, also the data storage model is problematic, because strong consistency incurs unnecessary overhead. As potential failure is network immanent, services can appear, disappear or become invalid at any point in time. Thus, an *eventually consistent* data storage model is preferable (as also used by Eureka).

3.3 Requirements Evaluation

The approaches presented beforehand have been also systematically evaluated with respect to the requirements from Sect. 2. Figure 2 shows the results of the evaluation and it becomes clear that none of the representatives is currently well suited for megascale systems. Regarding the challenge of handling many providers, UDDI, Consul and Eureka can at least deal with hundreds and even thousands of providers, but the latter two use persistent connections to all registered services limiting their overall capacities. Auto maintenance is only present in novel systems (Consul and Eureka 2) and realized using heart beat mechanisms based on the connection with

[6]https://github.com/Netflix/eureka/wiki/Eureka-2.0-Architecture-Overview.

	Many Providers	Auto Maintenance	Service Matching	Result Size	Result Ordering	Query Persistence	Quality Levels
UDDI	=	-	+ (operators on properties)	-	-	-	-
WS-Inspection	-	-	- (no queries)	-	-	-	-
WS-Discovery	-	= (in ad-hoc mode)	= type name	-	-	-	-
Consul	=	+	= (existence of properties)	-	-	=	-
Eureka 2	=	+	+ (operators on properties)	-	-	+	-

Fig. 2 Existing service discovery approaches (support: + full, = some, – none)

a service. Service matching is kept simple in most approaches and based on service type names and properties in many cases. Eureka 2 has the most flexible solution, because it offers operators that can be used to build queries on service properties. None of the approaches has flexible means for controlling the result size of a query. In addition also none considers result ordering and quality levels. Support for persistent queries has been integrated in Consul and Eureka 2. In Consul, the concept has been realized as so called blocking queues, i.e. a call to service endpoint can be made to persist and changes on that endpoint will be automatically published back to the caller.

4 Solution Architecture

The analysis from the last section has revealed, that current approaches lack support for the aspects *result size* and *result ordering*. The combination of both is a very useful feature because it allows to query the best n services with regard to specific properties. Such behavior can be emulated in existing registries only with much effort by retrieving all matching services and ranking them on client side. In megascale systems this can be impractical given large result sets that would have to be transferred, processed and afterwards pruned nearly completely on client side.

4.1 Distributed Query Processing Model

Query processing is the key functionality of the registry and is handled in a specific way to ensure scalability. A query itself is defined by *service type name*, *service tags*, *a filter function*, *a ranking function*, *result size* and *effort* (all optional). This structure and example query is depicted in Fig. 3. The filter function defines a boolean function which must evaluate to true in order to include the current entry. In contrast, the ranking function maps an entry to a [0,1] interval with 1 being the best value. The

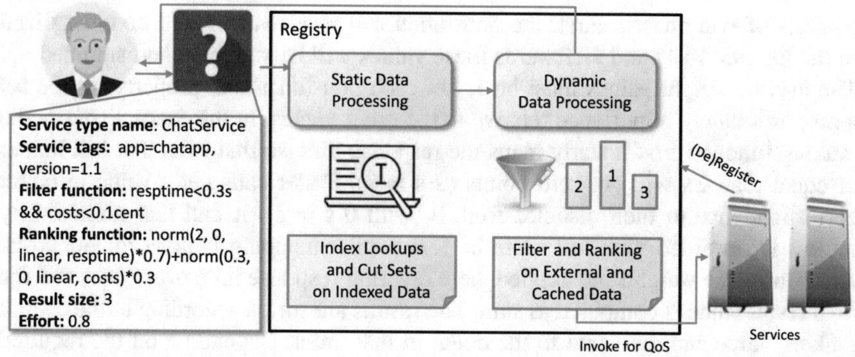

Fig. 3 Query processing model

example query requests the best three chat services of application 'chatapp' in version 1.1 with a response time lower than 0.3 s and costs lower than 0.1 cent ranked by both factors.

The query processing is done in a very specific way that differs from other systems. It is based on the observation that service data can be subdivided into two different groups: static data that remains constant over time and dynamic data which is subject to frequent changes. Hence, the query processing system consists of two subsystems each responsible for one of those categories. The static data (type name and tags) is used by the static query processors to build up full indexes for all elements. During processing the engine looks up all static properties and compares the results set sizes. It then starts with the minimal set and creates the cut-set with all other result sets one by one. Starting with the smallest result set ensures that the biggest reduction of result elements is performed first.

After the static query data has been evaluated the dynamic processors are headed over the result set of services. The non-functional properties of the query can be both, static and dynamic but are always treated using the dynamic processors. This is due to the fact that different operators can be used and a simple indexing is not possible any more.[7] If a property is static the query can be directly evaluated against the registered value of the service description. If not, it needs to be retrieved first. For this purpose the service description has to contain a REST-URL that can be queried for non-functional properties. The registry will evaluate the query against the fetched value and additionally store the value in a cache. This allows subsequent requests to reuse the fetched value during a freshness interval which can be defined in the service description. Requesting always fresh values can be enforced by explicitly setting the required maximum age to zero. The resulting services will then be ranked by the ranking function using static as well as dynamic properties. The ranking function

[7]One could use B* trees in future versions to further enhance indexing for comparative operators.

consists of two phases. First, the non-functional result values will be normalized to the interval [0,1] and afterwards these values will be weighted and summed up. The overall weight values must be 1, i.e. each non-functional property can be set a proportionately importance relative to the other values. In the example query the ranking function first linearly maps the response time so that calls that last longer or equal than 2 s will get zero points (2 s = 0). Faster calls get a value in direct correspondence to their distance from 0 (with 0 s = 1). A call that immediately returns is rewarded with full score of 1. A similar mapping is done for the costs. Afterwards the weights are applied, here favoring response time over costs, and the final result value is computed as sum. The results are sorted according to the overall ranking value and delivered to the client in that order. Depending on the required result size of the client, only a small fraction of the result set is finally transferred.

The registry also supports persistent queries via the same query definition, i.e. the query once submitted will be continuously be evaluated against all newly available services. In this case the ranking function is interpreted differently as the ordering of newly appearing services cannot be enforced when results are sent to the client right after discovery. Instead, the value of the ranking function is headed over to the client as mere quality value that it can use in comparison with formerly received services.

4.2 Implementation and Discussion

The implementation of the registry is performed in two distinct phases. In the first phase the interaction protocols for inter-superpeer synchronization as well as client-superpeer interactions are realized. This part has already been finished and allows for preparing larger experimental testbeds on the infrastructure layer. In the second phase, the two-staged query processing is implemented. The functionalities for indexing, performing queries and ranking are in place but are currently performed on the same nodes, i.e. the separation of query stages including a suitable protcol has to be realized. The registry is implemented using the Jadex active components framework [8] because it facilitates the automatic superpeer discovery and synchronization considerably.

In megascale service-based systems systems, search requests often match hundreds of thousands of services. Thus, result size and result ordering are important criteria to fetch the right services for each client. Although we are targeting highly dynamic systems, we expect the load to be dominated by search requests and not service (de-)registrations. i.e., we assume that the number of search requests is usually much larger than the number of service (de-)registrations in the same amount of time. As a result, the following discussion focuses on matching complexity only.

A naive matching approach would lead to $f_{matchnaive} \in \mathcal{O}\left(n_p * n_c\right)$, i.e. a matching complexity proportional to the number of all service providers in the system n_p for each request, times the number of search requests, which is proportional to the number of clients n_c. Moreover, checking dynamic properties incurs a lot of communication overhead for each service, such that not only a huge number of checks would need to be performed, but also each check of a single service against a single request would consume a considerable amount of resources.

The proposed architecture contributes to handling this complexity in two important ways. First, scaling processing nodes can reduce n_c: While increasing processing nodes will increase service registration cost, the number of clients per processing node becomes constant if scaled linearly with the clients. As a result, query processing cost is reduced and depends only on the providers, i.e. $f_{matchscaled} \in \mathcal{O}\left(n_p\right)$.

Second, the processing model uses indexing mechanisms reducing complexity with regard to n_p. Index lookups can be performed using, e.g., hashing mechanisms in constant time. As a result, only the smallest result set from all possible index lookups for a single search request needs to be searched linearly. We expect asymptotic behavior of the smallest-size result set being sublinear with service descriptions becoming more diverse with increasing numbers. Since the target scale is large but not infinite, it gives even slight sublinear complexities large advantage in pessimistic cases.

In addition, practical issues have to be considered as well: A large part of the matching effort is offloaded to the processing nodes. This avoids the need for transferring of large data sets as well as reducing client workload, which can be a considerable advantage in e.g. mobile clients. Preselecting a result subset (e.g. 10 “best”) also helps in this goal. Finally, persistent queries reduce bandwidth and offer an easier programming model.

5 Conclusion and Outook

This paper argues for the relevance of megascale systems, i.e. systems, in which 10^6 or more decentralized service providers exist and are used by potentially even larger numbers of clients. We expect such systems to appear in the near future due to ongoing trends such as smart cities or autonomous vehicles, where large numbers of existing devices are transformed into decentralized networks of service providers.

This paper tackles the problem of service discovery in such a system. Requirements with respect to the service provider and service client side are identified and existing approaches are analyzed with respect to their contribution to these requirements. Following this analysis, a solution architecture based on a scalable node structure and a request processing model is presented, that addresses open problems for a megascale service infrastructure. The request processing model allows distributing the matching load of service queries across different nodes in the network and thus allows handling static as well as dynamic service properties.The distributed service infrastructure is currently implemented using the *Jadex* framework. Future work will initially focus on testing the scalability of the infrastructure in increasingly larger scenarios.

References

1. Algermissen, J.: Using DNS for rest web service discovery (2010). https://www.infoq.com/articles/rest-discovery-dns
2. Ballinger, K., Brittenham, P., Malhotra, A., Nagy, W.A., Pharies, S.: Web services inspection language (2002). https://svn.apache.org/repos/asf/webservices/archive/wsil4j/trunk/java/docs/wsinspection.html
3. Bellwood, T., Capell, S., Clement, L., Colgrave, J., Dovey, M.J., Feygin, D., Hately, A., Kochman, R., Macias, P., Novotny, M., Paolucci, M., von Riegen, C., Rogers, T., Sycara, K., Wenzel, P., Wu, Z.: Uddi version 3.0.2 (Oct 2004). http://www.oasis-open.org/committees/uddi-spec/doc/spec/v3/uddi-v3.0.2-20041019.htm
4. Messina, F., Pappalardo, G., Rosaci, D., Santoro, C., Sarné, G.M.: A trust-aware, self-organizing system for large-scale federations of utility computing infrastructures. Future Gener. Comput. Syst. **56**, 77–94 (2016)
5. Newman, S.: Building Microservices—Designing Fine-Grained Systems. O'Reilly Media (2015)
6. OASIS: Reference Model for Service Oriented Architecture. Organization for the Advancement of Structured Information Standards (OASIS), version 1.0 edn. (2006)
7. OASIS: Web Services Dynamic Discovery (WS-Discovery). Organization for the Advancement of Structured Information Standards, version 1.1 edn (2009)
8. Pokahr, A., Braubach, L.: The active components approach for distributed systems development. Int. J. Parallel Emerg. Distrib. Syst. **28**(4), 321–369 (2013)
9. Sun Microsystems: Jini Architecture Specification version 2 (2003)
10. UPnP Forum: UPnP device architecture version 1 (2000)
11. Zhu, J., Oliya, M., Pung, H.: Service discovery for mobile computing—classifications, considerations, and Challenges. In: Handbook of Mobile Systems Applications and Services, Chap. 2, pp. 45–90. Auerbach Publications (2012)

Context-Aware Access Control Model for Services Provided from Cloud Computing

Ichiro Satoh

Abstract Since computing devices in IoT tend to have only limited computational resources, to provide enrich context-aware services, e.g., location-aware user assistant services, from IoT environments, such services should be offloaded to be executed on server-sides, including cloud computing platforms. However, there are differences between access control models in context-aware services and cloud computing platforms, where the former needs context-aware access models and the latter widely uses role/subject-based access control models. This paper aims to bridging the models. We present a model for spatially specifying containment relationships of persons, physical entities, spaces, and computers to specify contextual information about the real world. Our approach connects between the world model and services offloaded to cloud computing as an access control mechanism. This paper presents the basic notion of the model and its prototype implementation.

Keywords Access control · Context-awareness · Cloud computing

1 Introduction

The Internet of Things (IoT) environments enable things to connect to devices with awareness/sensing ability, so that IoTs can know contextual changes in the real world. The notion of context-awareness was a core feature of ubiquitous computing systems since the early 1990s. The focus on context-aware computing evolved from ubiquitous computing to IoT over the last decade. Computing devices in IoT, in addition to ubiquitous computing, tend to have a variety of constraints: non-powerful processors, small memory and storage, and battery-powered portable devices. On the other hand, modern applications running on IoT computing devices with such constraints are required to provide more enriched and sophisticated functions than ever. Therefore, such applications cannot be longer constructed as standalone ones in IoT. Instead, applications running on computing devices in IoT often access infor-

I. Satoh (✉)
National Institute of Informatics, 2-1-2 Hitotsubashi, Chiyoda-ku, Tokyo, Japan
e-mail: ichiro@nii.ac.jp

M. Ivanović et al. (eds.), *Intelligent Distributed Computing XI*,
Studies in Computational Intelligence 737, https://doi.org/10.1007/978-3-319-66379-1_25

mation from servers available in their current networks and delegate heavy tasks to the servers. The concept of *offloading* data and computation in server-sides, including cloud computing, is used to address the inherent problems by using resource providers other than the computing device itself to host the execution of applications.

IoT and cloud computing need to be operated based on multi-tenant models IoT and cloud computing are used by multiple users. Nevertheless, context-aware services should be accessed according to the current contexts detected by IoT and be provided for users from IoT. Access control models for context-aware services in IoT tend to be context-dependent, but access control models designed for cloud computing are independent of any context. As mentioned previously, computational resources available from context-aware services in IoT tend to be limited. To better cope with the limited resources in IoT intensive tasks can be delegated to cloud computing. Since these tasks are executed on servers in cloud computing on behalf of computing devices in IoT, we need to make sure the tasks that access computational resources from the servers are under appropriate access control.

This paper presents an approach to bridge a gap between access control models in cloud computing and context-aware computing. Most existing access control models tend to be subject-centric in the sense that permissions are provided according to the subjects, e.g., the users. However, we need context-centric ones. In such models, *context* is the first-class principle that explicitly guides both policy specifications and the enforcement process, and it is not possible to define a policy without the explicit specifications of the context that makes the policy valid.

When software for defining context-aware services is offloaded to cloud computing, it should utilize the knowledge created by context providers that is accessible through a communication interface between programs running on computing devices in IoT and servers in cloud computing platforms. The approach enables context-aware services executed at cloud computing to access computational resources and information under an access control model for IoT environments.

This paper consists of the following sections. Section 2 presents example scenarios and Sect. 3 describes the basic ideas of the approach presented in this paper. Section 4 describes the design and implementation of the system. We show the systems' evaluation in Sect. 5 and survey related work in Sect. 6. We give some concluding remarks Sect. 7.

2 Example Scenarios

Before describing the approach, we present example scenarios. There are several electric lights in a room. When a user has his/her smart phone and enters the room, he/she wants to turn the lights on from his/her smart phone. (1) While he/she in the room, his/her smart phones should have a capability to control electric appliances in the room. Other people who are not in the room should have no capability to control them. (2) If there is an administrator responsible for managing the house contains the room, the administrator should have the capability. The first scenario needs a context-centric model for access controls and the second is a typical example of subject-based models, which are widely used in cloud computing platforms. Furthermore, we need to support IoT computing devices whose resources tend to be limited. These scenarios should be provided with a large number of devices in a huge area.

3 Approach

This section outlines the approach proposed in this paper. Like the second scenario in the previous section, conventional approaches for access controlling are often based on subject-based access control systems in the sense that they exploit user identity or role information to determine the set of user permissions. Permissions for access controls are tightly coupled to the identity or role of the subject requesting a resource access, whereas contextual information can only further limit the applicability of the available permissions. Therefore, they require administrators to know all contexts and users. However, in IoT environments, context-aware services, e.g., location-aware services, should be selected and configured according to contextual changes, e.g., users, their goals and locations. If conventional approaches for access controlling are used in such environments, they may lead a combinatorial explosion of the number of policies. Furthermore, when offloading task intensive services to cloud computing, such services may be managed in an access control model for cloud computing independently of any context.

To support context-centric access control model with subject-based one, we introduce an interface, called *spatial connector*. It is used to bind front-services executed at computing devices in IoT and back-end-services executed at servers in cloud computing platforms. It also bridges between context-centric access control and subject-based access control models. We also introduce a symbolic world model to represent context like other existing world models for ambient intelligent/pervasive/ubiquitous computing [1, 8, 10]. Each connector is maintained in such a model according to its contextual existence. The model introduces a containment relationship between spaces, because physical spaces are often organized in a containment relationship. For example, each floor is contained within at most one building and each room is contained within at most one floor (Fig. 1). The model reflects on changes in the real world by using the external sensing systems. The model spatially binds the positions of entities and spaces with the locations of their virtual counterparts and, when they move in the physical world, it deploys their counterparts at proper locations within it. The model supports a context-centric access control model as a containment relationship between devices and spaces (Fig. 2). The model can activate/deactivate context-aware services in accordance with contextural changes in the real world. The model can bridge between a context-centric access control model and subject-based access control models in cloud computing platforms.

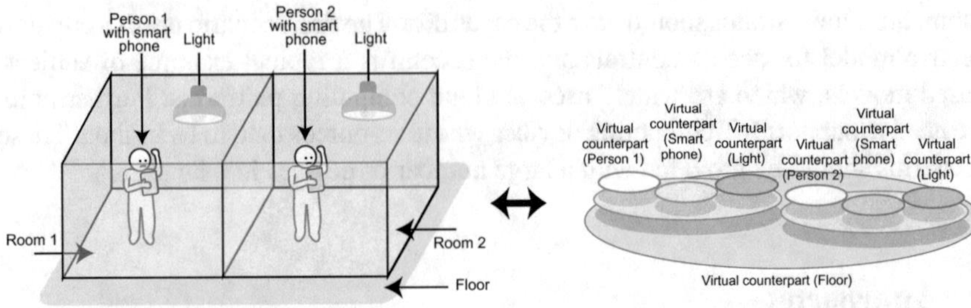

Fig. 1 World model for contexts and access controls

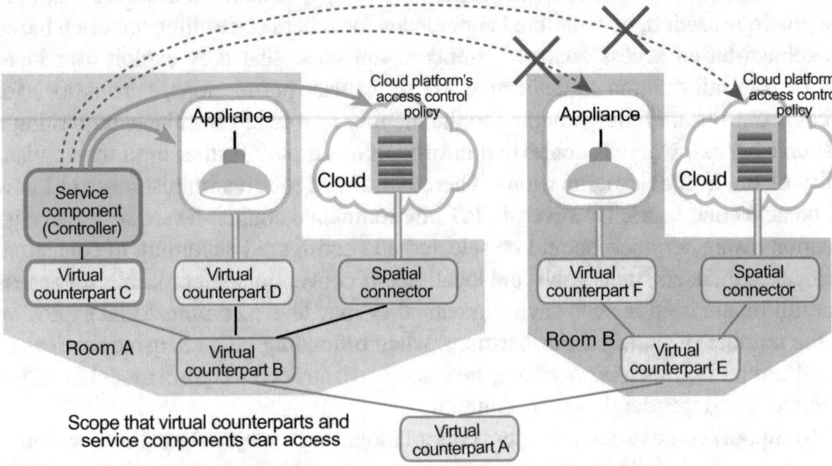

Fig. 2 Context-centric access control model

3.1 Design Principles

By using location-sensing systems, our approach introduces the following three elements:

- *Virtual counterpart* is a representation of its target, e.g., a user, physical entity, or computing device and contains profiles about the target, e.g., the name of the user, the attributes of the entity and the network address of the device. It always deploys at IoT computing devices close to the current location of its target.
- *World model* is constructed as an acyclic-tree structure of virtual counterparts corresponding to spaces, entities, and computing devices according to their containment relationships. The model is maintained by using location-sensing systems.
- *Spatial connector* is used to connect between at most one virtual counterpart and one or more services, which are provided from either computing devices in IoT or servers in cloud computing platforms. It bridges between access models in context-aware services and cloud computing. It also enables services, which may

be executed at IoT or cloud to connect other services provided in the virtual counterparts that contain its virtual counterpart.

For example, when a user enters a room with his/her smart phone, his/her virtual counterpart is migrated in the counterpart corresponding to the room. The spatial connector that is connected to the virtual counterpart corresponding to his/her smart phone enables services, which may be executed at computing devices in IoT and servers in cloud computing platforms to access services provided in the virtual counterparts corresponding to the room. Therefore, the smart phone can use the services via its connector, but other devices, which are not in the room, cannot use the services in the room. Therefore, our model can be used as an access control model in addition to a world model. Our connectors do not distinguish between services executed at computing devices in IoT and servers in cloud computing platforms. However, access controls for services are context-aware because virtual counterparts corresponding to a target, e.g., users, physical entities, and computing devices, enable their services to access other services within the spaces that spatially contain their targets.

4 Design and Implementation

To prove the effectiveness of the approach described in the previous section, we implemented a prototype system. The approach itself is independent of any programming languages but the current implementation uses Java as an implementation language for components. Our approach maintains a world model for context-aware services and access controls. In addition to the model, the approach has five elements: *Context manager* is responsible for reflecting changes in the real world and the locations of users, entities, and IoT computing devices. *Virtual counterpart* is a digital representation of a user, physical entity, or computing device, where each counterpart can contain other counterparts or service components inside it, but service components cannot. *Service component* is software components for defining services and is exited on a computing device in IoT, or is offloaded to and executed at a server in cloud computing platforms. *Spatial connector* is responsible for connecting between one or more service components uploaded to cloud computing platforms and at most one virtual counterpart on the model. *Service runtime systems* is responsible for executing service components and runs on computing devices in IoT or servers in cloud computing platforms and The model is organized as an acyclic-tree structure of virtual counterparts and service components, like Unix's file-directory.

4.1 Connecting Between IoT and Cloud

When physical entities, spaces, and computing devices move from location to location in the physical world, the model detects their movements through location-sensing systems and changes the containment relationships between counterparts

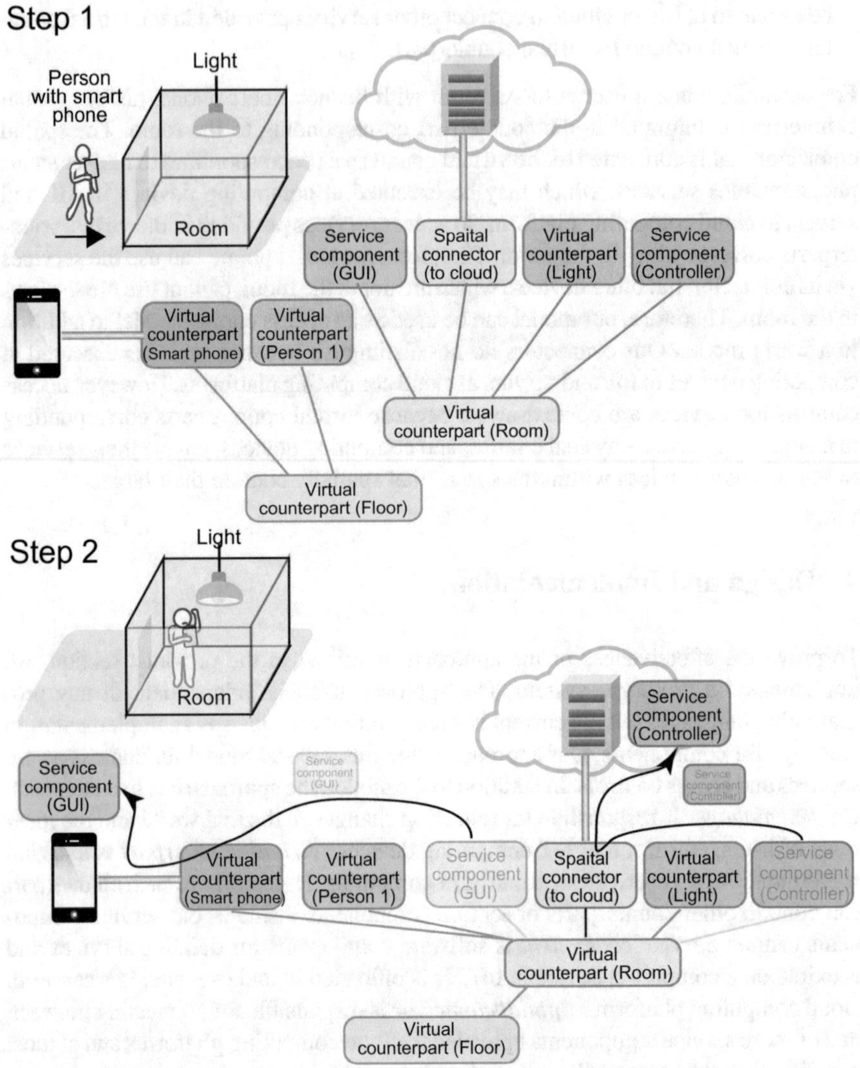

Fig. 3 Containment and deployment of virtual counterparts and service components when person enter in room

corresponding to moving entities, their sources, and destinations by using the external sensing systems. When a counterpart is moved to another counterpart, a subtree consisting of it and its descendent counterparts and components is moved to a subtree representing the destination (Fig. 3). When a component is transferred over a network, the runtime system stores the state and the code of the component, including the components embedded within it, into a bit-stream formed in Java's JAR file format that can support digital signatures for authentication.

Each counterpart can be attached to at most one target, e.g., a user, physical entity, or space. In the current prototype implementation of the approach, each counterpart keeps the identifier of its target or RFID tag attached to the target. The implementation uses to monitor location-sensing systems, e.g., active RFID tag systems, and spatially to bind more than one counterpart to each user or physical entity. Virtual counterparts for computers can automatically forwards their service components to their target computers by using the component migration mechanism, when it receives the service components.

Each *spatial connector* binds between a virtual counterpart and service components offloaded to the cloud through a network between the computer that maintains the model and a server on the cloud. When the counterpart contains a service component in the model, it automatically forwards the component to a server that can satisfy the requirement of the component on the cloud.

4.2 Binding Access Control Models

Access control models of many commercial cloud computing platforms, their access control models are subject-based, e.g., RBAC and RBAC's extensions, in the sense that permissions are provided according to subjects, e.g., users. Our model are based contexts in the sense that contexts are used as the first-class principle for access controls to services and resources. There is a gap between access control models in IoT environments and cloud computing platforms. Spatial connectors have protocols for authentication used in cloud platforms. For example, the current implementation supported interfaces for AWS Multi-Factor Authentication (MFA) to use Amazon's cloud computing services, including EC2. When a service in our model signs in to an AWS web-site, they will be prompted for their user name and password, as well as for an authentication code from their AWS MFA device. Taken together, these multiple factors provide increased security for your AWS account settings and resources. Our world model allows each spatial connector to bind between services according to their locations (Fig. 2).

- When a service component is executed at a computing device in IoT, the component is contained in the counterpart corresponding to the computer. They can access the resources provided from the computer under the computer's access control policy.
- Each counterpart is contained in another counterpart, where the former corresponds to a target, e.g., person, physical entity, space, or computer as a spatial containment relationship and the latter corresponds to the space that contains the target. It can access service components contained in the latter.

We here explain the model with an example scenario. Suppose that a television in a room delegates its services to cloud computing platforms. When a user a smart phone enters the room, his/her counterpart is contained in the counterpart corresponding to the room. Our model enables counterparts to bind spatial connectors to the services contained in the room. Therefore, his/her smart phone can connect the services provided from cloud computing platforms, while the phone and television are in the same room.

4.3 Service Component and Runtime System

Each service component is constructed as a collection of Java objects. When the life-cycle state of a service component is changed, the runtime system issues certain events to the service component and the service component's descendent components. Service components are deployable software that can travel from computing device and computing device archived by using mobile agent technology [11]. Each service component can be dynamically deployed at computing devices and consists of service methods and is defined as a subclass of a built-in abstract class. Most serializable JavaBeans can be used as service components. When a service component migrates to another computer, not only the program code but also its state is transferred to the destination.

Each service runtime system runs on a computing device in IoT or cloud. It is responsible for executing and migrating service components with containers to other service runtime systems running on different computers through a TCP channel. It governs all the components inside it and maintains the life-cycle state of each service component via its component. The system has a mechanism for transmitting the bit-stream over the network through an extension of the HTTP protocol. The current system basically uses the Java object serialization package for marshaling components. The package does not support the capturing of stack frames of threads. Instead, when a component is serialized, the system propagates certain events within its embedded components to instruct the component to stop its active threads. Since each runtime system is connected to other runtime systems in a peer-to-peer manner, our system can be scalable in the sense that it has no bottleneck.

5 Current Status

The current prototype implementation was built on the Java virtual machine (Java VM version 1.8 or later), which concealed differences between the platform architectures of computers. Although the current implementation was not constructed for performance, we evaluated that of only basic operations in a distributed system where eight computers (Intel Core i5 2.6 GHz with MacOS X 10.9 and J2SE version 8) were connected through a giga-ethernet. The cost of component migration was 38 ms through a giga-ethernet and The cost of component migration to Amazon EC2 was 240 ms. The prototype implementation also supported two commercial tracking systems to locate persons, physical entities, and computers. The first is the Spider active RFID tag system, which is a typical example of proximity-based tracking. It provides active RF-tags to users. Each tag has a unique identifier that periodically emits an RF-beacon (every second) that conveys an identifier within a range of 1–20 m. The current implementation is a prototype system. Nevertheless, it has several security mechanisms. For example, it can encrypt components before migrating them over the network and it can then decrypt them after they arrive at their destinations. Moreover, since each component is simply a programmable entity, it can explicitly encrypt its individual fields and migrate itself with these and its own cryptographic procedure. The JVM could explicitly restrict components so that they could only access specified resources to protect computers from malicious components. Although the current implementation cannot protect components from malicious computers, the runtime system supports authentication mechanisms to migrate components so that all runtime systems can only send components to, and only receive them from, trusted runtime systems.

We implemented an example of the proposed model, to proof the utility and validation of the model. By using this model, we could implement a *disaggregated computing*, which is an approach to dynamically composing devices, e.g., displays, keyboard, and mice that are not attached to the same computer, into a virtual computer among distributed computers in ubiquitous computing environments [2, 12]. Our system introduced a context-centric access control and offloading services to cloud computing into a disaggregated computing system. The system consists of three kinds software components, which support *model*, *view*, and *control* behaviors based on a model-view-control (MVC) pattern. The *model* component manages and stores drawing data and should be executed on a server equipped with a powerful processor and a lot of memory. The *view* component displays drawing data on the screen of its current host and should be deployed at computers equipped with large screens. The *control* component forwards drawing data from the pointing device, e.g., mouse, of its current computer to the first behavior.

We executed these components at three computing entities: smart TV and tablet in a room and and a server in cloud computing environment. In the model, the counterpart corresponding to the smart TV and the counterpart corresponding to the tablet were contained in a counterpart corresponding to the room. There was also a spatial connector between the their counterpart and the server. We initially deployed the *view* component at the counterpart corresponding to the smart TV and the *control* component at the counterpart corresponding to the tablet. The two counterparts forwarded the *view* component to the smart TV and *control* component to the tablet. The *model* component was deployed at the counterpart corresponding to the room and then was forwarded to the server through the connector. Since the smart TV's counterpart, the tablet's counterpart, the connector to the server were contained in the same counterpart, they could coordinate with one another. We also added another computer outside the room. The counterpart of the computer was not contained in the counterpart corresponding to the room. Therefore, the computer could not access the three components, i.e., the *model*, *view*, *control* components and the three computers, i.e., smart TV, tablet, and servers.

6 Related Work

This section briefly highlights several existing access models that have influenced our work with access control models for cloud computing and context-aware access control models.

Conventional access control models can be classified into three types: Mandatory Access Control (MAC) Discretionary Access Control (DAC), and RBAC (Role-Based Access Control) [4, 6, 9]. RBAC is an alternative to traditional approaches, i.e., DAC, and MAC. In RBAC, users are assigned roles and roles are assigned permissions. A principle motivation behind RBAC is the ability to specify and enforce enterprise specific security policies in a way that maps naturally to an organization structure.

In cloud computing access control is one of the most important issues. Cloud computing is often characterized by its multi-tenancy and virtualization features. These features need unique security and access privilege challenges due to sharing of resources among potential untrusted tenants. RBAC is a key technology for cloud cloud platforms, well suited for multi-domain architecture, and applicable in cloud systems relevant to health records, stock trading and pairing, and social networking. The RBAC model is employed by many cloud computing platforms, e.g., Microsoft Azure are OpenStack. To multi-tenancy at a single data storage in cloud computing, several researcher proposed approaches to encrypting data before uploading to the cloud by using some cryptographic algorithms so that the data were protected from other tenants. Since access control models in cloud computing platforms are often forced by the platforms, it is difficult to introduce other models into the platforms.

Many researchers have been several attempts to extend RBAC with the notion of context-awareness, e.g., [7]. By using the uniform notion of a role to capture both user and environmental attributes, our model allows for the definition of context-aware security policies. Roles can also make it easy to define and understand complex security policies; adding environment roles to the model was necessary to support the advanced access control requirements that we are faced with in ambient intelligent/pervasive computing. However, RBAC approaches assume that permissions are first associated with roles, and subsequently subjects are assigned to roles. In context-aware services, permissions should be first associated to contexts and subsequently subjects are associated to the contexts they are currently operating in. To solve these problems, Covington et al. [3] allows administrators to specify environmental context through a new type of role called environmental role to generalize traditional RBAC. Their approach aimed at overcoming the inherent subject-centric nature of RBAC. Georgiadis et al. [5] proposed a context-based term based control by integrating of RBAC and Team based Access Control (TMAC) [13].

7 Conclusion

We constructed an approach for enabling software for defining context-aware services to be provided from cloud computing platforms according to a context-aware access control model. The approach assumed such software to be offloaded to servers on cloud computing planforms in addition to computing devices in IoT, because such devices have too limited computational resources to execute enrich services. Context-aware services in IoT should be managed based on a context-aware access model, whereas cloud computing platforms are managed based on subject-based access control models. This approach was useful to bridge between access control models in IoT and cloud, when providing location-aware services.

References

1. Beigl, M., Zimmer, T., Decker, C.: A location model for communicating and processing of context. Pers. Ubiquit. Comput. (Springer) **6**(5–6), 341–357 (2002)
2. Brumitt, B.L., Meyers, B., Krumm, J., Kern, A., Shafer, S.: Easy living: technologies for intelligent environments. In: International Symposium on Handheld and Ubiquitous Computing, pp. 12–27. Springer (2000)
3. Covington, M.J., Long, W., Srinivasan, S., Dev, A.K., Ahamad, M., Abowd, G.D.: Securing context-aware applications using environment roles. In: Proceedings of 6th ACM Symposium on Access Control Models and Technologies (SACMAT'2001), pp. 10–20 (2001)
4. Ferraiolo, D.F., Barkley, J.F., Kuhn, D.: A role based access control model and reference implementation within a corporate intranet. ACM Trans. Inf. Syst. Secur. **2**(1), 34–64 (1999)
5. Georgiadis, C.K., Mavridis, I., Pangalos, G., Thomas, R.K.: Flexible team-based access control using contexts. In: 6th ACM Symposium on Access Control Models and Technologies (SACMAT'01), pp. 21–27 (2001)
6. Giuri, L., Iglio, P.: Role templates for content-based access control. In: 2nd ACM Workshop on Role Based Access Control (RBAC'97), pp. 153–159 (1997)
7. Hulsebosch, R.J., Salden, A.H., Bargh, M.S., Ebben, P.W.G., Reitsma. J.: Context sensitive access control. In: 10th ACM Symposium on Access Control Models and Technologies (SACMAT '05), pp. 111–119 (2005)
8. Leonhardt, U., Magee, J.: Towards a general location service for mobile environments. In: IEEE Workshop on Services in Distributed and Networked Environments, pp. 43–50, IEEE Computer Society (1996)
9. Sandhu, R.S., Coyne, E.J., Feinstein, H.L., Youman, C.E.: Role-based access control models. IEEE Comput. **29** (1996)
10. Satoh, I.: A location model for pervasive computing environments. In: Proceedings of IEEE 3rd International Conference on Pervasive Computing and Communications (PerCom'05), pp, 215–224. IEEE Computer Society (2005)
11. Satoh, I.: Mobile agents. In: Handbook of Ambient Intelligence and Smart Environments, pp. 771–791. Springer (2010)
12. Tandler, P.: The BEACH application model and software framework for synchronous collaboration in ubiquitous computing environments. J. Syst. Softw. **69**(3), 267–296 (2004)
13. Thomas, R.K.: Team-based access control (TMAC): a primitive for applying role-based access controls in collaborative environments. In: 3nd ACM workshop on Role-based Access Control, pp. 13–19 (1997)

Reference Architecture for Self-adaptive Microservice Systems

Krasimir Baylov and Aleksandar Dimov

Abstract Microservice architectural style emerged as a way of building highly scalable and flexible systems as opposed to the standard monolith approach. Despite the multiple benefits, as the number of services increase, the cost of service management and support also raises. In this paper we propose reference architecture for microservice systems in order to find a solution to the problem. The architecture approach is based on the notion of autonomic computing. It allows services to register or search for self-adaptation mechanisms when they need to respond to external environment changes.

Keywords Self-adaptation · Microservices · Reference architecture · SOA

1 Introduction

The notion of microservices represents an architectural style that aims at solving many challenges (like effective support and evolution), implied by large monolith applications. Microservices are defined as small, autonomous services that work together [1]. They are often compared to Service Oriented Architectures (SOA), although, microservices have some characteristics that are specific only for this style of architectures. For instance, they are organized around business capabilities and mostly used in product development [2] and require higher level of infrastructure automation. Additionally, microservice architectures are highly decentralized in terms of governance and data management.

Those specific characteristics imply that microservices have a completely new set of problems that prevent their straightforward application in industry. At first,

K. Baylov (✉) · A. Dimov
Faculty of Mathematics and Informatics, Department of Software Engineering,
Sofia University "St. KlimentOhridski", Sofia, Bulgaria
e-mail: krasimirb@uni-sofia.bg

A. Dimov
e-mail: aldi@fmi.uni-sofia.bg

M. Ivanović et al. (eds.), *Intelligent Distributed Computing XI*,
Studies in Computational Intelligence 737, https://doi.org/10.1007/978-3-319-66379-1_26

they are very complex in terms of infrastructure and administration. The large number of services requires significant operational and administrative efforts. The distributed nature of microservices makes this even harder to support. Secondly, microservice applications are constantly evolving. This requires utilization of new tools and techniques that support rapid automated deployment to productive environments.

In this context, this paper presents an ongoing work to address the aforesaid challenges, by development of a reference architecture that supports microservice self-adaptation. It is based on existing SOA reference models and the idea that self-adaptive systems can manage themselves with minimal or no human intervention and according to some high level objectives [3], this way aiming to handle system complexity.

The rest of the paper is organized as follows: Sect. 2 makes an overview of the related work in the field; Sect. 3 introduces the reference architecture for self-adaptation; Sect. 4 shows an example for application of our approach and discuss some generic implementation details and Sect. 5 concludes the paper.

2 Related Work

There are many interesting works in the field that provide guidelines or solve only specific problems related to microservices self-adaptation. Some of them are briefly described in this section.

Architecture for self-managing microservices is presented at [4]. It proposes a distributed architecture that can take advantage of the consensus algorithm in order to select a leader. This leader can assign management functionalities to the nodes that build the systems.

An approach for dealing with the multiple service updates in service-oriented systems is presented in [5]. The authors propose a design that can handle customizations by defining new test cases for determining mismatches between services that comply with identical contracts.

Authors of [6] provide the results of a case study on the anatomy of cloud monitoring and metering. Automated monitoring is one of the key aspects of microservice architectures and the proposed solution reveals that the monitored data can be reduced by up to 80%. This can have a strong positive effect on the overhead of self-adaptation in microservice architectures.

Another reference architecture for configuration and behavior self-adaptation is MORPH [7]. It is based on the MAPE-K model [3] and proposes three layers for self-adaptation—goal management, strategy management and strategy enactment. The three layers use a common knowledge repository.

Authors of [9] propose a multi-agent approach for building distributed systems. Each agent is a separate component that can solve problems in an autonomous ways. Agents can interact in order to solve problems that are beyond their own capabilities. Authors argue that multi-agent systems have features that are key to building self-adaptive systems—loose coupling, robustness, context sensitivity, etc.

Only few of the studied approaches directly address microservices.

3 Reference Architecture for Self-adaptive Microservices

In this section we present the reference architecture for self-adaptive microservice systems. Reference architecture [9] is a kind of pattern for creation of specific architectures in a particular domain, i.e. it provides guidelines on how to design, develop and evolve systems in that domain.

Our architecture is based on the classical for self-adaptive systems, autonomic control loop [8] and the MAPE-K (Monitor-Analyze-Plan-Execute plus Knowledge) model [3]. It considers specifics of microservices like technology heterogeneity, the large number of productive services, highly distributed nature, service componentization, etc. Therefore, it can be applied regardless of the technology that is used for developing the services. The proposed architecture (Fig. 1) targets microservices specifics, like incremental evolution of services, takes advantage of container monitoring, and allows easy integration for large number of fine-grained services. It has 5 main components, described below.

Service Consumer. Service consumers are the actors that consume the functionality exposed by the services. They make service invocations in order to use the desired functionality.

Service Registry. Service registry is a catalog that provides information about the services, like service provider and the service itself. Consumers search the

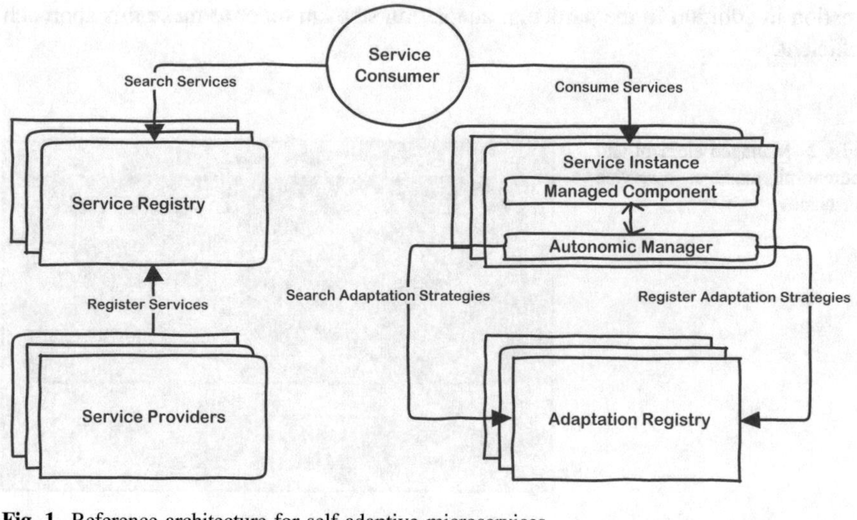

Fig. 1 Reference architecture for self-adaptive microservices

service registry in order to find the required services and determine what is the best way to invoke them is.

Service Provider. Service providers are responsible for service development and making them accessible over the network (internally or externally). Service providers should register their services in the service registry to make them reachable for service consumers.

Service Instance. Service instances, together with the adaptation registry are the core components in the reference architecture. It provides business functionality needed by service consumers. Due to the small code base of microservices, an autonomic control loop could easily be applied within the scope of each developed service.

Each service instance consists of two main components—*managed component* and *autonomic manager*. Managed component represents the whole or part of the service functionality. Autonomic manager is a service itself and is responsible for monitoring the managed component and application of any changes when adaptation needs to be triggered. The adaptation process is achieved using the mechanism of feedback loop [8]. It consists of four phases—measurement of managed element parameters, and then according to the observation—eventual analysis, planning and execution of an adaptive change into the managed element (Fig. 2).

Adaptation Registry. Adaptation registry serves as a repository of adaptation mechanisms. It is very likely that different microservices may have faced identical problems and have some optimized solution. Thus, it is useful if service autonomic manager can search the adaptation registry for successful adaptations mechanisms in identical operational context. This helps by saving time and avoiding intermediate steps in the adaptation process.

Service autonomic managers could also register such adaptation approaches. This way other autonomic managers will have knowledge about optimization of managed components. Autonomic managers also should provide contextual information in addition to the particular adaptation steps in order to make this approach efficient.

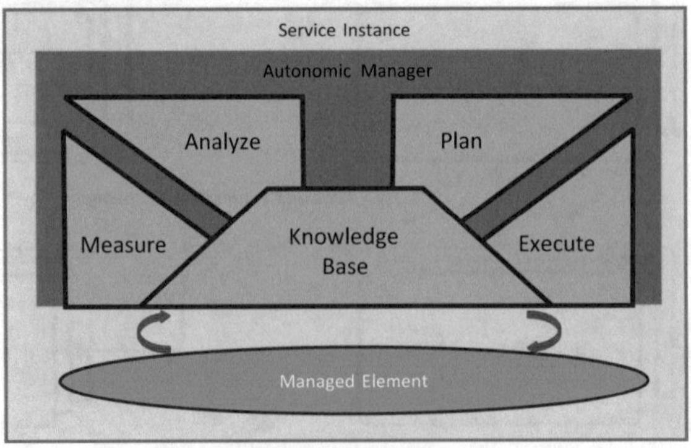

Fig. 2 Managed element and autonomic manager in service instances

4 Application of the Approach

4.1 Sample Architecture Use Case

This section describes basic adaptation use case that our architecture supports. Consider a business process, based on microservices as shown on Fig. 3. It represents a basic e-trade case (select, order and deliver goods), and initially, does not apply any self-adaptive mechanisms.

Possible application of our approach to the eTrade system would have the following steps:

(1) *"Shipping" service is the first to be transformed to self-adaptive service*—an autonomic manager is implemented and deployed as part of the service. Now the autonomic manager can monitor the managed component of the "Shipping" service and adapt it when required.
(2) *The "Shipping" autonomic manager detects a trigger and starts self-adaptation process*—operational environment of the service has changed and the number of read requests has increased by 200%, which overloads the server. The autonomic managers tries to increase the service cache size to 2, 5, 10 MB, etc. When the size is set to 20 MB, the server returns to normal average load.
(3) *Developers enhance other services with self-adaptive mechanisms*
(4) *Adaptation Registry is integrated into the system*—initial set of common adaptation practices may be deployed into the registry.
(5) *"Shipping" service uploads its adaptation approach to the adaptation registry*—a specific pattern for the information may be used.
(6) *"Orders" service is loaded with 1100 requests per minute*—the autonomic manager searches the adaptation registry for similar situations, finds the adaptation approach applied by the "Shipping" service and applies it.

4.2 Service Instance Reference Implementation

Here we describe a possible implementation of the service instance, following the proposed self-adaptive architecture. It is based on Jersey (https://jersey.java.net/)—A RESTful web services Java library. For implementation of the MAPE-K (Monitor-Analyze-Plan-Execute plus Knowledge) loop, we have implemented several service interceptors and filters (Fig. 4). They are used to collect service

Fig. 3 eTrade customer orders business process

Fig. 4 Service instance interceptors and knowledgebase database

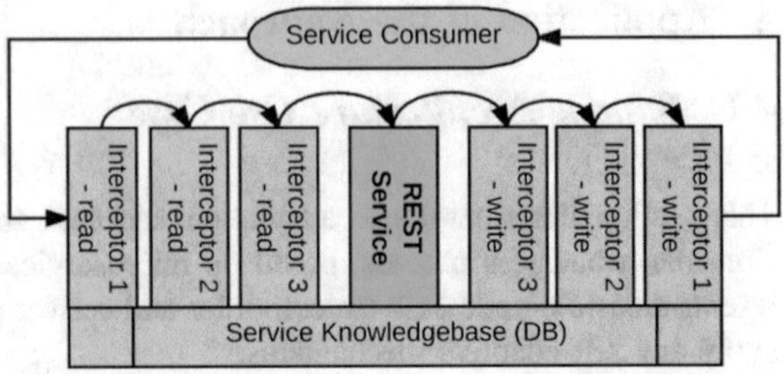

execution data (the first step of the MAPE-K loop) like number of calls, size of messages, response time, throughput, and others.

The interfaces implemented by interceptor classes are called javax.ws.rs.ext. WriterInterceptor and javax.ws.rs.ext.ReaderInterceptor. Similarly for filter classes, we use the interfaces javax.ws.rs.container.ContainerRequestFilter and javax.ws. rs.container.ContainerResponseFilter. Application of changes to services (the Execute step of the MAPE-K loop) is done through the service interceptors. They have access to the context of the service and can manipulate it at runtime. The service context is a java object that provides access to specific properties of the service—request body, request headers, operation type, etc.

For storage of collected measurements we have used Apache Derby (http://db. apache.org/derby/), although implementation is not dependent on database management system.

5 Conclusion

This paper describes reference architecture for self-adaptive microservice systems that should handle complexity of management and evolution of large-scale microservice based software systems. The architecture is based on SOA, but introduces additional components to support self-adaptation like autonomic manager and adaptation registry. The proposed architecture doesn't require all services to be self-adaptive. This allows transforming existing productive systems to self-adaptive ones in multiple iterations and low risk.

Future research includes analysis of self-adaptation within compositions of microservices. The proposed architecture should also be deployed to real systems in order to analyze its benefits and determine further points for improvement and optimization.

Acknowledgements Research, presented in this paper was partially supported by the DFNI I02-2/2014 (ДФНИ И02-2/2014) project, funded by the National Science Fund, Ministry of Education and Science in Bulgaria.

References

1. Newman, S.: Building Microservices. O'Reilly Media, Inc. (2015)
2. James, L., Fowler, M.: Microservices: a definition of this new architectural term. http://martinfowler.com/articles/microservices.html (2014)
3. Kephart, J., Chess, D.: The vision of autonomic computing. Computer **36**(1), 41–50 (2003)
4. Toffetti, G., et al.: An architecture for self-managing microservices. In: Proceedings of 1st International Workshop on Automated Incident Management in Cloud. ACM (2015)
5. Denaro, G., Pezze, M. Tosi, D. Designing self-adaptive service-oriented applications. In: Fourth International Conference on Autonomic Computing, ICAC'07. IEEE (2007)
6. Anwar, A., et al.: Anatomy of cloud monitoring and metering: a case study and open problems. In: Proceeding of the 6th Asia-Pacific Workshop on Systems. ACM (2015)
7. Braberman, V., et al.: Morph: a reference architecture for configuration and behaviour self-adaptation. In: Proceedings of the 1st International Workshop on CTSE, pp. 9–16. ACM (2015)
8. Brun, Y., et al.: Engineering self-adaptive systems through feedback loops. In: Software Engineering for Self-adaptive Systems, pp. 48–70. Springer, Berlin, Heidelberg (2009)
9. Gomaa, H., Hashimoto, K.: Dynamic self-adaptation for distributed service-oriented transactions. In: Proceedings of the 7th International Symposium SEAMS. IEEE Press (2012)

Part VIII
WASA 2017 (7th Workshop on Applications of Software Agents)

Agent-Based Computing in the Internet of Things: A Survey

Claudio Savaglio, Giancarlo Fortino, Maria Ganzha, Marcin Paprzycki, Costin Bădică and Mirjana Ivanović

Abstract The Internet of Things is a revolutionary concept, within cyberphysical systems, rich in potential as well as in multifacet requirements and development issues. To properly address them and to fully support IoT systems development, Agent-Based Computing represents a suitable and effective modeling, programming, simulation paradigm. As matter of facts, agent metaphors, concepts, techniques, methods and tools have been widely exploited to develop IoT systems. Main contemporary contributions in this direction are surveyed and reported in this work.

Keywords Internet of Things · Agent-based Computing · Modeling · Architectures · Simulation · Methodology

C. Savaglio (✉) · G. Fortino
DIMES, Università Della Calabria, Rende (CS), Italy
e-mail: csavaglio@dimes.unical.it

G. Fortino
e-mail: g.fortino@unical.it

M. Ganzha
Warsaw University of Technology, Warsaw, Poland
e-mail: m.ganzha@mini.pw.edu.pl

M. Paprzycki
Systems Research Institute Polish Academy of Sciences, Warsaw, Poland
e-mail: marcin.paprzycki@ibspan.waw.pl

C. Bădică
CIT, University of Craiova, Craiova, Romania
e-mail: cbadica@software.ucv.ro

M. Ivanović
DMI, University of Novi Sad, Balkans, Serbia
e-mail: mira@dmi.uns.ac.rs

M. Ivanović et al. (eds.), *Intelligent Distributed Computing XI*,
Studies in Computational Intelligence 737, https://doi.org/10.1007/978-3-319-66379-1_27

1 Introduction

Since early 2000s, technological advances in wireless communication, embedded processing, sensing and actuation, are fueling rapid spread of novel cyberphysical artifacts. Ranging from simple movement detectors and temperature sensors, to more sophisticated smartphones and smart cars, they can sense the physical world, process data, and impact the surrounding environment in different ways, for example by triggering actions through actuators or engaging customized users interactions. In the context of the Internet of Things (IoT) [1], such devices have been massively networked and provided with (different degrees of) intelligence, being defined as "Smart Objects" (SOs) [2]. They communicate with each other, as well as with conventional computing systems, and cooperate in a synergic fashion, both on local and global scales, to implement cyberphysical applications in multitude of scenarios, e.g., industrial automation, logistic optimization, energy management, public security, entertainment, ambient assisted living and wellness, to name just a few.

To comprehensively support needs arising in complex development of heterogeneous IoT applications and systems, different mainstream paradigms and approaches (especially in closely related fields of wireless sensor networks, distributed systems, ubiquitous and pervasive computing), have been jointly exploited [3]. Among these, Agent-Based Computing (ABC) [4] has been widely recognized as full-fledged, effective support for development of decentralized, dynamic, cooperating and open IoT systems, particularly in conjunction with other complementary paradigms, e.g., cloud [5] and autonomic [6] computing.

In this paper, our intention is to show how ABC has been effectively exploited for modeling, programming and simulating IoT systems. Indeed, the ABC provides ideas, metaphors, techniques, methods and tools for systematically conceptualizing, realizing and simulating distributed systems composed of heterogeneous interacting entities [7]. Therefore, in Sect. 2, we first provide some insights specifically focused on the main IoT development issues and distinctive features of ABC. In Sect. 3, we survey several contributions exploiting ABC in the IoT context, for modeling, programming and simulation purposes. Ultimately, a brief analysis of surveyed state-of-the-art and some final remarks conclude the work.

2 Background

2.1 Internet of Things Development Challenges

The Internet of Things consists of great number and variety of components (RFiD, sensors, conventional laptops, micro and super computers, smartphones, robots, home appliances, vehicles, etc.), network types (Bluetooth-based personal area networks, ZigBee-based industrial networks, 5G and Wifi-based very dense metropolitan area networks, etc.), and stakeholders (citizens, private companies, public administrations, other digital systems, etc.), thus constituting an extremely multi-facet global ecosystem [1]. Because of heterogeneity of IoT building blocks, lack of standards, massive scale (the total number of "things" is forecasted to reach 20.4 billion in 2020) and rapid evolution, development of IoT applications and systems involves large number of requirements and issues [3]. In particular, IoT devices, also denoted as Smart Objects (SOs, in contrast with simple resources as sensors, actuators, databases, etc.) are expected not only to be intelligent, context-aware and autonomous, but also easy to use, reliable and secure. The same desiderata should apply to all IoT systems that are expected to be autonomic, scalable and open, thus avoiding the proliferation of poorly interoperable "intra-nets of things" [6]. Beside the fulfillment of such requirements, unexpected issues related to different development phases need to be considered and dynamically handled (e.g., an already deployed application needs to expose new functionality and interfaces for interacting with novel SOs). In this context, a multidisciplinary and systematic approach, involving different expertise for coping with the cyberphysical nature of IoT ecosystem [8], is necessary. Henceforth, full-fledged IoT methodologies are gaining traction [9], aiming at systematically supporting all development phases, addressing mentioned issues, and reducing time-to-market, efforts and probability of failure.

2.2 *Agent-Based Computing Paradigm*

Agent-based Computing is centered around the concept of an agent [4], a sophisticated software abstraction defining an autonomous, social, reactive and proactive entity. Agents are situated in some environment (namely, world of perceived resources) and act to achieve their design objectives, exhibiting flexible problem solving behaviors. Agents, interacting and cooperating to solve/realize problems/ services that are beyond the capabilities of a single agent, constitute a MAS (Multi Agent System) [4]. MASs are distributed and self-steering societies, featured by a strong situatedness and well-defined organizational relationships, covering variety of domains (e.g., sociology, economy, logistics). The above characterization, although not exhaustive, indicates that ABC provides a set of key abstractions and metaphors for straightforwardly *modeling* complex systems, their components, interactions and organizational relationships.

Beside modeling, ABC is also a well-established *programming* paradigm for concretely implementing agents' advanced features, and effectively addressing key requirements typical of modern (distributed) applications. Indeed, agent's, society's and environment's modeling abstractions have been exploited to devise a high-level, distributed programming paradigm, centered around two cornerstones [10]: (1) encapsulation of control (that consists in giving each agent its own thread of control and reasoning capabilities, thus designing context-aware entities with autonomous behaviors), and (2) interaction (including coordination and cooperation mechanisms, based on high-level asynchronous message passing). Here, adoption of shared communication standards and management specifications (e.g., the IEEE FIPA-based system platforms and communication languages [11, 12]) allows agents to act also as interoperability facilitators, by incorporating within the agent society a variety of resources and existing legacy systems. Such advantages enable agent-based programming paradigm to enhance system's performance (i.e., computational efficiency, reliability, responsiveness, etc.), interoperability and scalability, specially with respect to the centralized approaches.

Finally, computing systems, modeled and programmed following the agent-oriented approach, can be straightforwardly simulated, for effectively studying macro phenomena and patterns, as well as individual behaviors and environment evolution [13]. Indeed, agent-based *simulation* allows evaluating agent-based systems exposing discrete, not linear, adaptive behaviors even in highly interacting, distributed, scaling-up, virtual scenarios. To properly exploit the surveyed agent-oriented metaphors, techniques and tools, thus providing a systematical approach to the agent-based modeling, programming, and simulation, several agent-oriented development *methodologies* have been designed and successfully applied [14]. However, as highlighted in [15, 16], ABC is neither a universal nor necessarily effective development solution, since agent-level and society-level pitfalls can occur from different perspectives (management, conceptual, design, etc.), thus outweighing any agent-related benefits. Therefore, the adoption of ABC paradigm needs to be carefully assessed.

3 Agents' Contribution in Developing IoT Systems

The agent-oriented view of the world is perhaps the most natural way of approaching several types of (natural and artificial) systems, featured by a relevant complexity, dynamicity, situatedness and autonomy [7]. In particular, strong conceptual relation exists between agents and SOs, as well as between MAS and IoT systems [43]. Thus, considering the entire set of requirements and issues related to the development of IoT systems, ABC has been exploited for modeling, programming and simulating IoT applications and systems, and thus systematically driving and speeding-up their development. The most relevant contributions which exploit ABC for these purposes have been surveyed and compared, according to their provided agent-based features in Table 1, summarizing the outcomes of Sects. 3.1–3.4. In detail, for each contribution, Table 1 indicates if it performs a fine or coarse grained agent-based IoT entity modeling, if it implements mechanisms for (technological/syntactical/semantic) interoperability, autonomicity, cognitivity, virtualization or security, if (and how) it performs IoT system simulation, and finally, if it presents an agent-based IoT development methodology.

Table 1 Surveyed works and provided agent-based features - *T* = *technological, Sy* = *syntactical, Se* = *semantic interoperability; A* = *autonomicity; C* = *cognitivity; V* = *virtualization; S* = *security.*

Surveyed work <name, ref. >	Agent-based IoT model		Agent-based IoT implementation							Agent-based IoT simulation		Agent-based IoT Methodology
	Fine grained	*Coarse grained*	*T*	*Sy*	*Se*	*C*	*A*	*V*	*S*	*Pure*	*Hybrid*	
Cascadas, [17]	X		X	X			X					
iCore, [21]	X		X	X	X	X			X			
ACOSO [6], [9], [22], [42], [43]	X		X	X	X	X	X	X			X	X
UBIWARE, [19]; UBIROAD, [26]	X		X	X	X	X		X	X			
[29]		X	X	X				X				
[44]	X		X	X	X		X	X				X
AoT, [27]		X	X	X		X	X	X				
Smart Grids, [40]		X		X						X		
[41]		X	X	X				X			X	
TAEC, [39]		X	X	X	X		X	X	X			
CIoT, [32]	X		X	X	X	X		X	X			
iSapiens, [23]	X		X	X				X	X			
[31]		X	X	X				X				
Radigost, [38]		X	X	X				X				
ASSIST, [45]	X		X	X	X	X		X		X		
BEMOSS, [25]		X	X	X				X	X			
INTER-IoT, [50]	X		X	X	X	X	X	X	X		X	X
VICINITY, [33]	X		X	X	X	X		X	X			

(continued)

Table 1 (continued)

Surveyed work <name, ref. >	Agent-based IoT model		Agent-based IoT implementation							Agent-based IoT simulation		Agent-based IoT Methodology
	Fine grained	*Coarse grained*	*T*	*Sy*	*Se*	*C*	*A*	*V*	*S*	*Pure*	*Hybrid*	
SOL, [28]	X		X	X	X		X					
[20]	X		X	X	X							
[24]		X	X	X			X					
[30]		X	X	X		X	X					
Smart Santander, [35]		X	X	X	X			X	X			
[34]		X	X	X	X			X	X			
Prometheus, [46]	X		X	X	X			X	X			X
ASEME, [18]	X		X	X	X			X	X			X
SAMSON, [51]	X						X			X		

3.1 ABC as IoT Modeling Paradigm

Agent-based modeling allows capturing key characteristics of SOs and IoT systems, at different degrees of granularity and in a technology-agnostic way. Indeed, SO autonomicity, proactiveness and situatedness are implicitly embedded in the agent model, while other important SO features can be explicitly described through agent-related concepts. This is the case of [17, 24, 51], which express SO functionalities in terms of goals, SO working plan in terms of behaviors, and SO augmentation-related components (like knowledge bases, sensors and actuators) in terms of dynamically bindable agent resources. However, these works adopt different mechanisms for specifically characterizing SOs/agents. In particular, in [18, 20] each agent/SO has a role (taken from a scenario-dependent repository, e.g., smart car, smart driver-support or smart road for the transportation context) that determines, by default, its own behaviors, goals and communications paradigms; similarly, in [21], SO/agent plans and goals are encoded in templates reflecting their functionalities. Other contributions do not reference a-priori defined roles or templates. For example, in [17], each agent/SO has a self-model (an automaton) driving its actions according to stimuli (modeled as messages) from other agents or the environment. Similarly, in [22, 23], SOs' actions/reactions are encoded in behaviors, driven by incoming (internal/external) events and design goals (encapsulated in state-based tasks). Finally, in [24], SO/agent self-state is dynamically determined by combining its real-time sensor data, position and status of computational units.

Furthermore, surveyed agent-oriented SO models are particularly suitable for supporting preliminary development phase of analysis, abstracting main SOs features from low-level details or specific implementation constraints. Beside the satisfactory "per-se" agent-based SOs descriptions, however, further research efforts are necessary to thoughtfully model relationships among cyberphysical agents/SOs interacting within physical everyday environments. Nevertheless, agent-based IoT models represent a convenient starting point for subsequent phases of agent-oriented programming and simulations [9].

3.2 ABC as IoT Programming Paradigm

Because of deep heterogeneity of resources and communication protocols in the IoT context, many authors propose an agent-oriented approach for programming uniform interfaces and thus transparently interacting with resources and SOs. Authors of [19, 21, 22, 26, 29], exploit software adapters (developed for specific technologies and coordinated internally by a device manager [22]) for accessing agent/SO augmentation devices. This approach improves modularity and extendibility, since it leverages pluggable software components that can be defined when needed, and customized within the target resource. Instead, [20, 25], follow a different approach: each resource is directly coupled with one agent that interfaces the resource itself with the related SO, or with rest of the system. This solution completely hides the underlying technological heterogeneity, but it is not suitable for those constrained devices constrained devices that cannot support an agent-based architecture. Apart from resource handling, agent-oriented programming contributes to overcoming lack of communication/coordination standards within the IoT arena: (i) by implementing the IEEE FIPA 'de facto' standard specifications [11], and (ii) by supporting the SOs virtualization [21], thus paving the way towards integration of the agentified SOs within the Cloud [5, 23], (outsourcing computation/storage, and thus mitigating the SO hardware/software limitations) and with Web Services [27] (thus enhancing the SOs accessibility). Indeed, FIPA specifications standardize message format (specifically, the Agent Communication Language [12], ACL, is used for encoding message envelope) and message content (whose concepts, typically expressed through metadata-oriented languages, refer to ontology for facilitating data and the context management), and provide effective message transport service (leveraging on both semi/centralized and distributed services of agent discovery). Conversely, SOA and REST make SOs functionalities accessible under the form of Web Services over standard Internet protocols that are platforms and programming languages independent [31, 38].

Summarizing, agents are powerful mechanisms that realize the following functions:

- technical interoperability through shared resource/communication interfaces, as emphasized in [17, 22, 28, 30, 35, 38] (however, agents developed by different organizations are unable to interoperate well, as FIPA standard are not ready to support full interoperability in real-time/distributed control and diagnostic [47]);
- syntactical inteoperability through a shared message format, because ACL is adopted across FIPA standard obeying platforms for message envelope, while XML and JSON are used for message content in [19, 20, 29, 31, 33, 35] (but it is worth noting that ACL is a 'de facto' standard and other languages, like Knowledge Query and Manipulation Language, KQML, have some success [15]); and
- semantic interoperability through shared ontology and knowledge representation, as particularly done in [19, 21, 26, 32, 34, 36] (although this function is quite limited and underveloped due to the scarcity of grounded domain-specific ontology and semantics [16]).

At a higher level of conceptualization, agents allow to straightforwardly instill smartness and autonomy within a single SO, and realize cognitive and autonomic IoT systems [6, 37]. In fact, agent-based programming paradigm enables development of (i) self-configuring, self-healing, self-protecting and self-optimizing SOs/IoT systems, that are manageable with a minimum human intervention [17, 28, 30, 39]; and (ii) self-learning, context-aware and adaptive SOs/IoT systems [21, 32, 33], capable of solving problems without requiring human assistance. Autonomic and cognitive features are particularly important for ensuring self-management, distributed intelligence and scalability, but also in the perspective a secured and trusted IoT scenario, supporting the implementation of conventional, as well as unconventional, trust mechanisms. The first case refers to certificate-based reputation systems

[21, 39]; the second one refers to the Social IoT approach [23, 25, 33, 45] that leverages on the inter-SOs relationships (e.g., location, ownership, chronology of mutual interactions).

3.3 *ABC as IoT Simulation Paradigm*

Particularly in the IoT context, where interactions are subject to variety of contingent factors [51] (e.g., SOs density, physical network design, traffic congestion, wireless signal attenuation and coverage), and deployment is often error prone and time consuming, being able to simulate the system plays a crucial role [42, 43]. In fact, it allows understanding overall dynamics, estimating performance, and validating models, protocols and algorithms featuring under-development SOs and IoT systems. Although agent-based simulators allow effectively inspecting high-level aspects such as the raise of collective dynamics and behavioral patterns, they typically neglect, or coarsely handle, low-level communication issues, thus resulting in quasi aseptic simulation environments that are far from the actual cyberphysical IoT scenarios [40, 45]. Therefore, authors in [6, 41, 43], propose an hybrid approach, based on joint exploitation of agent-oriented modeling and network-based simulation. Such complementary approach allows mitigating limitations of pure agent-based simulation (but maintaining advantages derived from ABC) and effectively simulating IoT systems of different scales (from small ad-hoc networks, to large and dense Smart Cities) with variety of configurations (different communication patterns, protocols, parameters) [6, 42, 43].

Differently from well-established agent-based modeling and programming paradigm, research in agent-based IoT simulation is in its infancy. However, considering that currently IoT-specific simulators are not available, the hybrid agent-based approach represents one of the few state-of-the-art candidates for supporting the crucial activity of IoT system simulation. Overlooking such aspect could be a critical pitfall that compromises the agents acceptance in the IoT context, since MASs have no central control and thus unpredictable and emergent behaviors are likely [47].

3.4 *Agent-Based Methodology for IoT*

Some agent-oriented methodologies have been specifically extended for the IoT context [18, 46], or ex-novo designed [9, 44]. However, beside disciplining the exploitation of agent-based suite of models, programming techniques and simulation tools [46], requirements that are typically overlooked by agent-based methodologies need to be considered. In this direction, to support the IoT system development in all its phases, the aforementioned agent-based methodologies:

- thoroughly consider the cyberphysical nature of the involved entities and environments, foreseeing by design, solutions for interoperability, security and scalability [9, 44, 46];
- emphasize identification of IoT users and stakeholders, depicting significant use cases through textual descriptions [44] and technical notations, like UML (Unified Modeling Language) [18] or BPMN (Business Process Model Notation) [46], for meeting different expertise and perspectives;
- define the proper management, coordination and virtualization mechanisms [9], typically situating them at the middleware level [48, 49], for gluing hardware and software components;
- analyze infrastructural features and limitations according to the specific IoT system requirements [18, 44], since these factors cannot be considered independently;
- provide guidelines and best-practices for unbinding developers from a specific technology or protocol, driving and promoting integration of different computing paradigms and application contexts [9, 50].

Without extensively dealing with all these factors, even effective and well-known conventional agent-based software development methodologies like Tropos [52] are definitively inadequate, and unable to actually unfold the full IoT potential.

4 Analysis and Concluding Remarks

IoT full realization is not hindered by hardware constraints or computational/storage /communication limitations, but by some requirements that have not been totally or simultaneously addressed. Leveraging agents key features of autonomy, proactiveness, intelligence and sociability, and according to the numerous contributions surveyed in this work, we believe that ABC can be effectively exploited as modeling, programming, and simulation paradigm for developing IoT ecosystems. Indeed, better than other computing paradigms (object-oriented, service-oriented, component-oriented) and both at things and at system levels, ABC allows modeling at different degrees of details, facilitating (technical, syntactical and semantic) interoperability, autonomicity and distributed intelligence, and validating multiple design choices, before their actual deployment. In addition, agent-based methodologies can be extended and then reused for systematically driving the complete development process, also supporting integration with other paradigms (e.g., cloud computing, business process management) and reducing the probability of failure and time-to-market. However, before blindly adopting ABC in the IoT (as well as in any other development context), three main pragmatic aspects need to be considered, otherwise the overheads of dealing with agents could outweigh any benefits of an agent-based solution. First is related to relative immaturity of agent technology (born and raise mainly in the academia more than in the industry) and small number of available agent-based commercial platforms [15], which made no significantly progresses in the last decades [16], especially with respect to standardization and semantic interoperability (while, as discussed, such requirements are fundamental for the IoT). Second is related to the investment (in terms of both time and resources) needed for implementing agent-based IoT solutions, which are typically more costly than conventional centralized and service-oriented ones [47] (widely reused, for example, in the Web of Things [53]). Last one is related to old, but still common, misapplications and misconceptions, such as:

- *everything can be profitably agentified*: agents are intrinsically autonomous multi-thread problem solvers, thus an agent-based solution may be inappropriate for systems requiring only a single thread of control [15] (e.g., simple IoT monitoring applications) or unsustainable in the case of constrained IoT resources and devices [16] (unable, for example, to implement mechanisms for automatically handling conflicts among policies, synchronizing accesses to shared resources, developing distributed intelligence);
- *agents are a universal solution*: not all classes of IoT applications are suitable for agent-based techniques (for example, just 30% of control tasks and 60% of diagnostic tasks in the industrial scenario [47]).

Ultimately, as highlighted across this survey, we want to remark that the adoption of ABC paradigm needs to be carefully assessed but, as proved by the several contributions presented in this work, it represents to date the most suitable choices for effectively developing the majority of advanced (current and future) IoT systems.

Acknowledgements This work has been carried out under the framework of INTER-IoT, Research and Innovation action - Horizon 2020 European Project, Grant Agreement #687283, financed by the European Union. It was supported in part by PAS-RAS bilateral project "Semantic foundation of the Internet of Things", as well as a collaboration agreement between University of Novi Sad, University of Craiova, SRIPAS and Warsaw University of Technology.

References

1. Atzori, L., Iera, A., Morabito, G.: The Internet of Things: a survey. Comp. Networks **54**, 2787–2805 (2010)
2. Mattern, F., Floerkemeier, C.: From the internet of computers to the Internet of Things. In: From Active Data Management to Event-Based Systems and More. Springer, pp. 242–259 (2010)
3. Patel, P., Cassou, D.: Enabling high-level application development for the Internet of Things. J. Syst. Softw. **103**, 62–84 (2015)
4. Luck, M., McBurney, P., Preist, C.: A manifesto for agent technology: towards next generation computing. Auton. Agents Multi-Agent Syst. 203–252 (2004)
5. Fortino, G., Guerrieri, A., Russo, W., Savaglio, C.: Integration of agent-based and cloud Computing for the smart objects-oriented IoT. In: Computer Supported Cooperative Work in Design (CSCWD), Proceedings of the 2014 IEEE 18th International Conference on IEEE, pp. 493–498
6. Savaglio, C., Fortino, G., Zhou, M.: Towards interoperable, cognitive and autonomic IoT systems: an agent-based approach. In: 2016 IEEE 3rd World Forum on Internet of Things (WF-IoT). IEEE, pp. 58–63 (2016)
7. Jennings, N.R.: Agent-Based Computing: Promise and perils (1999)

8. Fortino, G., Rovella, A., Russo, W., Savaglio, C.: Towards cyberphysical digital libraries: integrating IoT smart objects into digital libraries. In: Management of Cyber Physical Objects in the Future Internet of Things. Springer, pp. 135–156 (2016)
9. Fortino, G., Guerrieri, A., Russo, W., Savaglio, C.: Towards a development methodology for smart object-oriented IoT systems: a metamodel approach. 2015 IEEE International Conference on In: Systems, Man, and Cybernetics, IEEE, pp. 1297–1302 (2015)
10. Ricci, A., Santi, A.: Agent-oriented computing: agents as a paradigm for computer programming and software development. In: Proceedings of the 3rd International Conference on Future Computational Technology and Applications, Wilmington: Xpert Publishing Services. Citeseer, pp. 42–51 (2011)
11. Poslad, S.: Specifying protocols for multi-agent systems interaction. ACM Trans. Autonom. Adapt. Syst. (TAAS) **2**, 15 (2007)
12. Fipa, A.: Fipa acl message structure specification. In: Foundation for Intelligent Physical Agents. http://www.fipa.org/specs/fipa00061/SC00061G.html (2002). Last accessed 30 June 2004
13. Macal, C.M., North, M.J.: Tutorial on agent-based modeling and simulation. In: Simulation Conference, 2005 Proceedings of the Winter. IEEE, p. 14 (2005)
14. Bergenti, F., Gleizes, M.-P., Zambonelli, F.: Methodologies and Software Engineering for Agent Systems: The Agent-Oriented Software Engineering Handbook. Springer Science & Business Media (2006)
15. Wooldridge, M.J., Jennings, N.R.: Software engineering with agents: pitfalls and pratfalls. IEEE Internet Comput. **3**, 2027 (1999)
16. Nwana, H.S., Ndumu, D.T.: A perspective on software agents research. Knowl. Eng. Rev. **14**, 125142 (1999)
17. Manzalini, A., Zambonelli, F.: Towards autonomic and situation-aware communication services: the cascadas vision, in: Distributed Intelligent Systems: Collective Intelligence and Its Applications, 2006. DIS 2006. IEEE Workshop on. IEEE, pp. 383–388 (2006)
18. Spanoudakis, N., Moraitis, P.: Engineering ambient intelligence systems using agent technology. IEEE Intell. Syst. **30**, 60–67 (2015)
19. Katasonov, A., Kaykova, O., Khriyenko, O., Nikitin, S., Terziyan, V.Y.: Smart semantic middleware for the Internet of Things. ICINCO-ICSO **8**, 169–178 (2008)
20. Ruta, M., Scioscia, F., Loseto, G., Di Sciascio, E.: Semantic-based resource discovery and orchestration in home and building automation: a multi-agent approach. IEEE Trans. Indust. Informat. **10**, 730–741 (2014)
21. Vlacheas, P., Giaffreda, R., Stavroulaki, V., Kelaidonis, D., Foteinos, V., Poulios, G., Demestichas, P., Somov, A., Biswas, A.R., Moessner, K.: Enabling smart cities through a cognitive management framework for the Internet of Things. IEEE Commun. Magazine **51**, 102–111 (2013)
22. Fortino, G., Guerrieri, A., Russo, W.: Agent-oriented smart objects development. In: Computer Supported Cooperative Work in Design (CSCWD), IEEE 16th International Conference on 2012. IEEE, pp. 907–912 (2012)
23. Cicirelli, F., Guerrieri, A., Spezzano, G., Vinci, A., Briante, O., Ruggeri, G.: iSapiens: a platform for social and pervasive smart environments. In: Internet of Things (WF-IoT), IEEE 3rd World Forum on 2016, IEEE, pp. 365–370 (2016)
24. Kato, T., Chiba, R., Takahashi, H., Kinoshita, T.: Agent-oriented cooperation of IoT devices towards advanced logistics. In: Computer Software and Applications Conference, 2015 IEEE 39th Annual. IEEE, pp. 223–227 (2015)
25. Zhang, X., Adhikari, R., Pipattanasomporn, M., Kuzlu, M., Bradley, S.R.: Deploying IoT devices to make buildings smart: Performance evaluation and deployment experience. In: Internet of Things (WF-IoT), IEEE 3rd World Forum on 2016, IEEE, pp. 530–535 (2016)
26. Terziyan, V., Kaykova, O., Zhovtobryukh, D.: Ubiroad: semantic middleware for context-aware smart road environments. In: Internet and Web Applications and Services (Iciw), Fifth International Conference on 2010, IEEE, pp. 295–302 (2010)

27. Mzahm, A.M., Ahmad, M.S., Tang, A.Y.: Agents of Things (AoT): an intelligent operational concept of the Internet of Things (IoT). In: Intelligent Systems Design and Applications (ISDA), 13th International Conference on 2013, IEEE, pp. 159–164 (2013)
28. Ayala, I., Amor, M., Fuentes, L.: The sol agent platform: enabling group communication and interoperability of self-configuring agents in the Internet of Things. J. Amb. Intell. Smart Environ. **7**, 243–269 (2015)
29. Leppänen, T., Riekki, J., Liu, M., Harjula, E., Ojala, T.: Mobile agents-based smart objects for the IoT. In: Internet of Things Based on Smart Objects. Springer, pp. 29–48
30. Pujolle, G.: An autonomic-oriented architecture for the Internet of Things. In: Modern Comput. 2006, IEEE John Vincent Atanasoff, International Symposium on 2006. IEEE, pp. 163–168 (2006)
31. Manate, B., Munteanu, V.I., Fortis, T.-F.: Towards a scalable multi-agent architecture for managing iot data. In: P2P, Parallel, Grid, Cloud and Internet Computing (3PGCIC), 2013 Eighth International Conference on 2013, IEEE, pp. 270–275
32. Wu, Q., Ding, G., Xu, Y., Feng, S., Du, Z., Wang, J., Long, K.: Cognitive Internet of Things: a new paradigm beyond connection. IEEE Inter. Things J. **1**, 129–143 (2014)
33. VICINITY - Open virtual neighbourhood network to connect IoT infra-structures and smart objects. http://vicinity2020.eu/vicinity/
34. Kiljander, J., Delia, A., Morandi, F., Hyttinen, P., Takalo-Mattila, J., Ylisaukko-Oja, A., Soininen, J.-P., Cinotti, T.S.: Semantic interoperability architecture for pervasive computing and Internet of Things. IEEE Access **2**, 856–873 (2014)
35. Cheng, B., Longo, S., Cirillo, F., Bauer, M., Kovacs, E.: Building a big data platform for smart cities: experience and lessons from santander. In: Big Data (BigData Congress), IEEE International Conference on 2015. IEEE, pp. 592–599 (2015)
36. Ganzha, M., Paprzycki, M., Pawlowski, W., Szmeja, P., Wasielewska, K.: Semantic interoperability in the Internet of Things: An overview from the INTER-IoT perspective. J. Netw. Comp. Appl. **81**, 111–124 (2017)
37. Savaglio, C., Fortino, G.: Autonomic and cognitive architectures for the Internet of Things. In: International Conference on Internet and Distributed Computing Systems, Springer, pp. 39–47 (2015)
38. Mitrović, D., Ivanović, M., Budimac, Z., Vidaković, M.: Radigost: interoperable web-based multi-agent platform. J. Syst. Softw. **90**, 167–178 (2014)
39. Xu, X., Bessis, N., Cao, J.: An autonomic agent trust model for IoT systems. Proc. Comp. Sci. **21**, 107113 (2013)
40. Karnouskos, S., De Holanda, T.N.: Simulation of a smart grid city with software agents. In: Computer Modeling and Simulation, 2009. EMS09. Third UKSim European Symposium on. IEEE, pp. 424–429 (2009)
41. D'Angelo, G., Ferretti, S., Ghini, V.: Multi-level simulation of Internet of Things on smart territories. Simul. Modell. Pract. Theory **73**, 3–21 (2017)
42. Fortino, G., Russo, W., Savaglio, C.: Simulation of agent-oriented Internet of Things systems. In: Proceedings 17th Workshop From Objects to Agents. pp. 8–13 (2016)
43. Fortino, G., Russo, W., Savaglio, C.: Agent-oriented modeling and simulation of IoT networks. In: Computer Science and Information Systems (FedCSIS), Federated Conference on 2016. IEEE, pp. 1449–1452 (2016)
44. Zambonelli, F.: Towards a General Software Engineering Methodology for the Internet of Things. http://arxiv.org/abs/1601.05569 (2016)
45. Kasnesis, P., Toumanidis, L., Kogias, D., Patrikakis, C.Z., Venieris, I.S.: ASSIST: an agent-based SIoT simulator. In: Internet of Things (WF-IoT), IEEE 3rd World Forum on 2016. IEEE, pp. 353–358 (2016)
46. Manate, B., Fortis, F., Moore, P.: Applying the prometheus methodology for an Internet of Things architecture. In: Proceedings of the 2014 IEEE/ACM 7th International Conference on Utility and Cloud Computing. IEEE Computer Society, pp. 435–442 (2014)
47. Marik, V., McFarlane, D.: Industrial adoption of agent-based technologies. IEEE Intell. Syst. **20**, 2735 (2005)

48. Razzaque, M.A., Milojevic-Jevric, M., Palade, A., Clarke, S.: Middleware for internet of things: a survey. IEEE Int. Things J. **3**, 70–95 (2016)
49. Fortino, G., Guerrieri, A., Russo, W., Savaglio, C.: Middlewares for smart objects and smart environments: overview and comparison. In: Internet of Things Based on Smart Objects. Springer, pp. 1–27 (2014)
50. Kubler, O., S., Framling, K., Zaslavsky, A., Doukas, C., Olivares, E., Fortino, G., Palau, C. E., Soursos, S., Podnar Åarko, I., Fang, Y., Kro, S., Heinz, C., Grimm, C., Broering, A., Miti, J., Olstedt, K., Vermesan, O.: Digitising the Industry: Internet of Things Connecting the Physical, Digital and Virtual Worlds. River Publishers, Chapter 9, vol. 49 pp. 431–448, (2016)
51. Morris, A., Giorgini, P., Abdel-Naby, S.: Simulating BDI-Based Wireless Sensor Networks. 2009 IEEE/WIC/ACM International Joint Conference on Web Intelligence and Intelligent Agent Technology, vol. 2, IEEE, pp. 78–81 (2009)
52. Bresciani, P., Perini, A., Giorgini, P., Giunchiglia, F., Mylopoulos, J.: Tropos: an agent-oriented software development methodology. Autonom. Agents Multi-Agent Syst. **8**, 203236 (2004)
53. Guinard, D., Trifa, V., Pham, T., Liechti, O.: Towards physical mashups in the web of things. In: Networked Sensing Systems (INSS), Sixth International Conference on 2009. IEEE, p. 14 (2009)

Teaching, Learning and Assessment of Agents and Robotics in a Computer Science Curriculum

Ioanna Stamatopoulou, Konstantinos Dimopoulos and Petros Kefalas

Abstract This paper presents our experience in integrating agents and robotics in our Computer Science Curriculum. We present a series of modules throughout our curriculum that progressively address these themes and other AI related topics, which ends with a specialised final year module central to teaching and learning multi-agent systems and principles of robotics. As part of this module a Robotics Challenge is organised, allowing students to integrate the knowledge they obtained in previously attended modules, and to practically apply knowledge and skills in order to solve a real problem.

Keywords Agents · Multi-agent systems · Robotics · Computer science education

1 Introduction

Agents, Multi-Agent Systems (MAS) and Robotics have gained increasing attention in research, mostly related on how they can be designed and implemented in order to exhibit intelligent behaviour. We believe, however, that topics could be more extensively incorporated in Computer Science Curricula in order to better prepare students towards the design and development of distributed, smart, complex systems, when they face these challenges in their professional careers.

[1]https://www.acm.org/publications/class-2012.

I. Stamatopoulou (✉) · K. Dimopoulos · P. Kefalas
The University of Sheffield International Faculty, CITY College,
3 L. Sofou, 54624 Thessaloniki, Greece
e-mail: istamatopoulou@citycollege.sheffield.eu

K. Dimopoulos
e-mail: k.dimopoulos@citycollege.sheffield.eu

P. Kefalas
e-mail: kefalas@citycollege.sheffield.eu

M. Ivanović et al. (eds.), *Intelligent Distributed Computing XI*,
Studies in Computational Intelligence 737, https://doi.org/10.1007/978-3-319-66379-1_28

The Joint ACM/IEEE Task Force includes *"Agents"* in its 2013 version of Computing Curricula[1] within different Knowledge Areas, such as HCI, Intelligence Systems, and Social Issues and Professional Practice. *"Robotics"* is included in both Intelligent Systems and Platform-Based Development. In the 2012 ACM Computing Classification System,[2] the topics are listed in Computing methodologies, Artificial intelligence, Distributed Artificial Intelligence as well as in Computer systems organization and Embedded and cyber-physical systems, respectively. As such, universities have MAS in their programmes either as a separate module or incorporated within more generic modules, such as the ones listed as Course Examplars[1].

Our aim in this paper is to report our experience from teaching a number of modules with agents, MAS and Robotics as a main theme. We provide an overview of how these topics are covered throughout our curriculum and focus on the design of a final year module, the concepts we introduce on software and robotic agents, and how we assess students. The feedback we have so far is extremely positive and we are prompted to encourage colleagues to borrow and expand our ideas for their own teaching, learning and assessment methods.

The rest of the paper is structured as follows: Sect. 2 briefly presents a series of modules throughout our curriculum that progressively address agents, MAS and Robotics, while Sect. 3 focuses on the module Agents and Robotics that is offered in the last semester of our Bachelor's programme. Section 4 discusses in detail a Robotics Challenge that is organised as part of this module and Sect. 5 evaluates our approach by presenting the results of a questionnaire completed by students. Finally, Sect. 6 concludes the paper.

2 Agents in Computer Science Curriculum

In our curriculum we cover agents, MAS and Robotics from various perspectives throughout all three levels of the Bachelor's degree and at Master's level, aiming towards a gradual enhancement of skills. Table 1 provides a summary of all related modules' aims, along with their European Qualifications Framework (EQF) levels.[3]

In the first year of studies in the Programming Principles and Algorithms module a reactive agent platform is used to introduce basic programming skills, through the use of a tool called Mentor[4]. that facilitates the visual outcome of a programmed reactive agent. The concept is similar to that of RoboMind Academy,[5] with the difference that the programming language of Mentor is Java. The environment provides a two-dimensional space in which an agent/virtual robot nay be programmed to perform simple problem solving tasks by perceiving and affecting the environment (Fig. 1).

---

[1]http://www.acm.org/education/CS2013-final-report.pdf.

[3]http://www.accreditedqualifications.org.uk/european-qualifications-framework-eqf.html.

[4]http://robotseducate.us

[5]https://www.robomindacademy.com.

Table 1 Agent-related modules' aims and EQF levels

Module	EQF level	Overall aims
Programming principles and algorithms	5	To introduce problem analysis, algorithmic thinking, and design practices, such as incremental code writing
Artificial intelligence techniques	6	To introduce main principles of AI: knowledge representation techniques, reasoning and search algorithms, as well as principles of natural language understanding
Intelligent systems	6	To expand to more specific areas of AI, such as neural networks, fuzzy systems, planning, and machine learning, as well as their main applications
Agents and robotics	6	To introduce fundamentals of intelligent agents, multi-agent system design, principles of robotics, through hands-on implementation of robotic agents
Knowledge technologies for innovation	7	To provide an overview of knowledge technologies, accompanied by a series of case studies, demonstrating their applicability on smart systems and their potential for business innovation

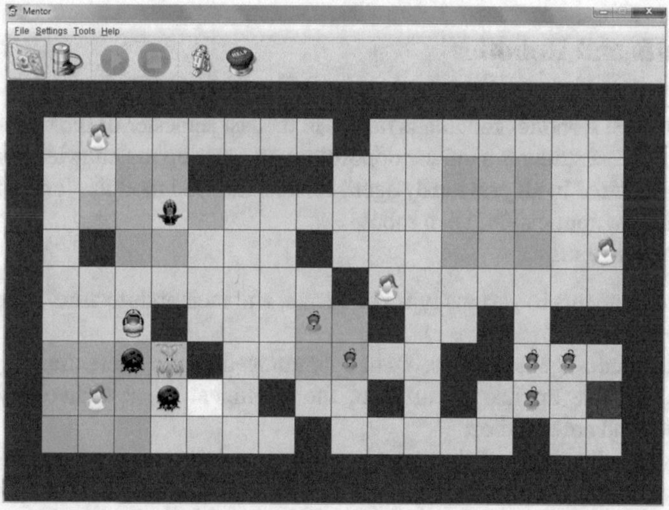

Fig. 1 The tool used in our Programming Principles and Algorithms module that allows students to program a virtual robot

In the second year of studies the Artificial Intelligence Techniques module teaches students structured knowledge representation techniques, and search and constraint satisfaction algorithms for problem solving. Although these are fundamental principles of the broader AI field, we take the opportunity to discuss how the world and a problem domain could be represented to form an agent's knowledge and beliefs, and what kind of techniques agents may employ to reason about their state and the world.

In the third year of studies the module Intelligent Systems comes as a sequel of AI Techniques and expands on a variety of AI areas, aiming at breadth instead of depth. Students are exposed to how neural networks and stochastic systems, fuzzy reasoning, planning, and learning can facilitate the reasoning of intelligent agents. Also in their third year students attend the Agents and Robotics module which is extensively discussed in Sect. 3 of this paper.

Finally, postgraduate students, irrespectively of their background, attend a module entitled Knowledge Technologies for Innovation. The module discusses how knowledge technologies may be exploited so as to increase the performance of classic enterprise systems and facilitate quality decision-making aiming towards product/service innovation. The module is an opportunity to demonstrate a good number of applications that stand at the frontiers of innovative smart business management products. The interested reader may find more information about this module in [7].

3 Agents and Robotics

The Agents and Robotics module is taught in the last semester of the undergraduate studies and offers students a unique opportunity to wrap-up the knowledge and skills they have acquired in all previously agent-related attended modules, and apply them in one coherent application, i.e. a robot.

This module aims to:

- introduce students to various types of agents, and their architecture, strengths and limitations;
- introduce multi-agent systems, agent communication and interaction;
- discuss possible application areas of the intelligent agent technology through examples and case studies;
- discuss the advantages of the agent-based approach to engineering complex software systems;
- introduce students to mobile robots, the related issues involved, and their applications;
- investigate robotic technologies relevant to sensing, perception, action and reaction;
- discuss the evolution of robotics in the immediate future, and determine innovative applications;

- underline the similarities and differences between software agents and mobile robots.

We use Bloom's taxonomy [2] to describe the cognitive level of the learning objectives, and we expect that by the end of the module students are able to:

LO1 *explain* the basic notions of agent systems and the difference between agents and other programs;
LO2 *describe* the fundamental agent architectures and sensibly *design* reactive and BDI agents;
LO3 *discuss* the issues involved in designing multi-agent systems, particularly with respect to communication and interaction, and *apply* techniques for addressing them;
LO4 *demonstrate* an overall understanding of biology inspired agents;
LO5 *argue* that the agent paradigm is an alternative point of view to software engineering and *realise* the related agent-based software engineering methodologies;
LO6 appropriately *taxonomise* robots;
LO7 *explain* the problems involved in designing new robots regarding sensing and perceiving the environment, controlling the movement, and decision making;
LO8 *design* and *construct* simple robotic automata capable of performing simple behaviours.

Contents of the agent part of the module include: definition of the notion of agency, agents types and architectures (primarily reactive and BDI), multi-agent systems, agent communication and interaction, and biology inspired agents.

In the robotics part students are taught basic concepts, types and classification of robots, sensor types, robot movement and actuation, Kinematics of mobile robots, controlling motors and servos.

We use a variety of teaching methods ranging from lectures to workshops and training laboratories. For hands-on practice we use Netlogo and Lego Mindstorms as the tools with which students will practice their knowledge and skills. Netlogo [9] is a cross-platform multi-agent programmable modelling environment extremely suitable for MAS simulation. In NetLogo, the environment consists of a grid of patches and is inhabited by turtles: entities that operate in it, interact with it and among them. The effectiveness of using Netlogo in teaching agents is discussed extensively elsewhere [6, 8] in which the interested reader may also find the reasons why other fully fledged tools for agent development are not preferred.

Lego Mindstorms[6] is a versatile modular robotics platform, developed by Lego, aiming at a commercial and educational audience. We decided to use Lego Mindstorms for various reasons:

- **Modularity**: The Lego platform is inherently modular, allowing for a wide range of robotic constructs. Furthermore, the platform comes with a number of easy to use sensors that allow for a wide range of intelligent behaviours to be developed,

[6]https://www.lego.com/en-us/mindstorms.

while third party sensors can also be used to further expand the versatility of the platform.

- **Ease of constructing custom robots**: Most students are familiar with Lego-type toys, and this helps them kick-start building a robot. In addition, many instructional sources are available online, to guide students towards building a variety of robotic chassis, or even inventing their own.
- **Ease of programming**: There is a variety of programming languages available for programming the platform, many of which are direct imports of known languages. LeJOS (Lego Java Operating System) is a Java import that is flexible, compatible with existing Java libraries, and comes with a number of libraries of its own that support agent-based concepts. The fact that our students our taught programming in the Java language makes LeJOS an appropriate choice.
- **Low cost**: The Lego Mindstorms platform costs a little over 300 euros per set, and includes all the basics needed. A single set may be used by a team of four to five students, making the total cost for a cohort of 20 students about 1,200 euros. There are not many educational robots that cost less that 1,200 per piece having the same flexibility.

The above reasons have made the Lego Mindstorms platform very popular in secondary and higher education [1, 3, 4]. The platform itself has been introduced in the 90s and has since been updated several times. So far, we have been using the NXT 2.0, as we have the sets available for the past 6 years, but we have lately acquired the new version of Lego EV3.

Both main themes of the module, agents and robotics, are assessed through two independent coursework assignments. The first involves the development of a MAS simulation in NetLogo (accomplishing learning outcomes LO1–LO5). This includes the design and implementation of independent agents that collaborate or compete to accomplish a certain task, such as carrying passengers to the airport (Fig. 2), rescue injured people in a disaster situation (Fig. 3), emergency evacuation of a building (Fig. 4), etc. Students are given libraries for BDI archirecture, FIPA exchange of messages and Contract Net protocol implemented in NetLogo [5]. The second coursework assignment involves the completion of a challenge on robot design and task fulfillment (accomplishing learning outcomes LO6–LO8), which is the focus of the following section.

4 Case Study: The Robotics Challenge

The Robotics Challenge event takes the form of a celebration in the Department of Computer Science and the Faculty as a whole. Many students and academic staff from other Departments are watching the setup, preparation and experimentation until the final demonstration. The challenge takes place over two full days with the final challenge taking place at the end of the second day.

Students are divided to teams of 4–5 and have already been introduced to Lego NXT 2.0 and its programming through a series of lab sessions. The number of teams

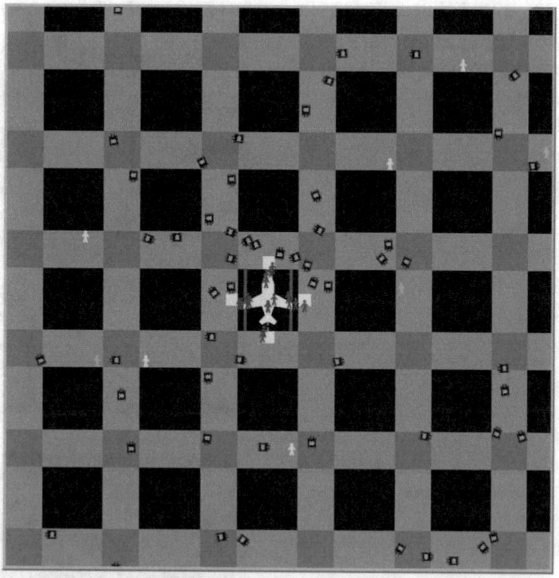

Fig. 2 Taxis negotiate and coordinate in order to carry passengers from any part of a city to the airport

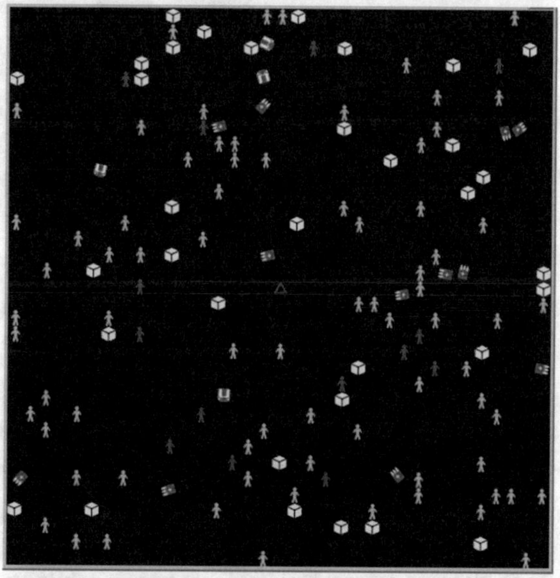

Fig. 3 Rescue units provide first aid to injured people they find in a disaster area. Ambulances collaborate in order to transfer the rescued people to the hospital

depends on the available resources and the number of students in the cohort (usually in the range of 20–25).

As an example consider the following challenge. The terrain the robots operate in is an enclosed area of 2 × 2 m, surrounded by a short wall (Fig. 5). Two patches (A4 papers of different colours) are placed at specific parts of the terrain, representing the

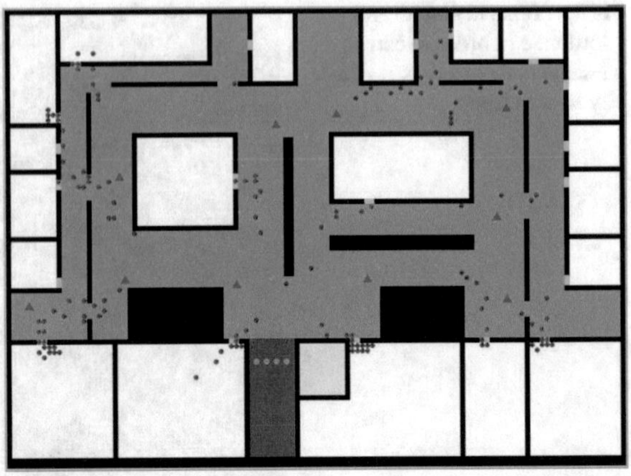

Fig. 4 People evacuate a building upon hearing a fire alarm, by following an exit plan located in each room

Fig. 5 The terrain setup and the robotic agents operating in it

robots' nests. At random places in the terrain objects are placed representing food that can be picked up by the robots. These objects are cylinders of different colours with a diameter of 4.5 cm and a height of 10.1 cm.

The aim of the challenge is to create robots that will explore the terrain foraging for food. To guide students towards completing the challenge, it is broken into smaller ones so that the overall problem can be solved incrementally:

1. Move randomly inside the terrain avoiding obstacles (initially other robots, walls and food cylinders);
2. Explore the terrain looking for the nests, while avoiding obstacles (other robots, walls and food cylinders).
3. Differentiate between food cylinders and other obstacles (other robots and walls), and identify the cylinder's colour.
4. Explore the terrain looking for food cylinders and pick up them up (one at a time), avoiding other robots and walls.
5. Explore the terrain looking food cylinders and take them back to the nests, keeping the location of the nests and of the food cylinders it cannot pickup in memory.

6. Communicate the location of discovered nests and food cylinders that cannot be picked up to other robots.
7. Integrate all the above in creating a robot that explores the terrain avoiding other robots and walls, looking for nests and cylinders, reporting the location of those that cannot be picked up to other robots, and transporting food cylinders to the closest nest. Food cylinders are placed dynamically in random positions inside the terrain and are removed manually when they are placed at a nest.

To complete the challenge students design and develop their own team robot (Fig. 6) and experiment with each of the aforementioned sub-challenges for two days (Fig. 7). Sub-challenges are first discussed with all the teams in a brainstorming session with the instructor, whose role is to coordinate the discussions, advise, assist with clarifications or feedback, and act as an expert reviewer.

Each sub-challenge is implemented as a new behaviour of the robot or as an enhancement on an existing behaviour. For example, for sub-challenge 2, which

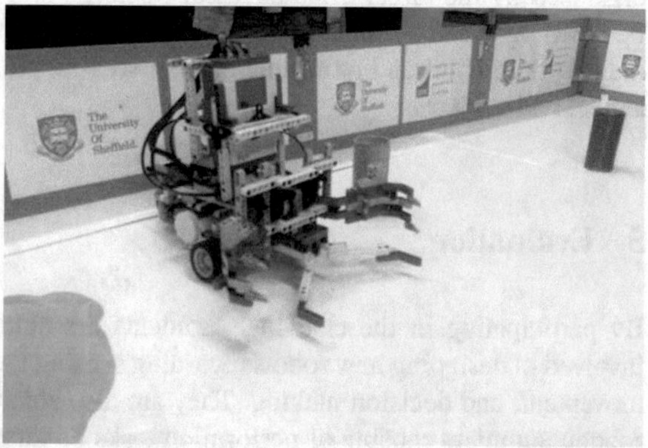

Fig. 6 A final robot that takes place in the competition

Fig. 7 Students working on the robotic challenge

requires the identification of a nest, students need to consider the range of available sensors (camera, colour sensor, light sensor), design and implement the needed behaviour, and incorporate it with the existing obstacle avoidance behaviour of sub-challenge 1, by realising a subsumption agent architecture.

The final robots are ready for demonstration at the end of the second day. After the completion of the challenge, students submit a group report on how they developed their robots, and an individual report about their experience and their contribution to the group.

Throughout the years, we have come up with a number of different challenges. Indicatively, another challenge requires two kinds of robots: rescue-bots and carry-bots. Randomly placed at the grid there are civilians in danger. A rescue-bot locates civilian victims in need and provides them with first aid until the carry-bot arrives to transport the civilians back to the hospital (scenario very similar to the MAS simulation in NetLogo that we ask them to develop in their first assignment).

The most successful robot designs have been demonstrated by the students to the open public in two follow-up occasions: the 7th International Mathematics Week, organised by the Greek Mathematical Society,[7] and the 1st Thessaloniki Science Festival organised by the British Council.[8] Students had the opportunity to present their work to children in a simple manner so as to promote their interest in robotics and STEM subjects.

5 Evaluation

By participating in the challenge, students are able to understand the problems involved in designing new robots regarding sensing the environment, controlling the movement, and decision making. They are also able to design and develop simple robotic automata capable of performing tasks of varying difficulty: from executing simple pre-programmed tasks to learning simple behaviours. The overall experience gives them the opportunity to consider agent architectures in designing a robot, and exercise the skills and knowledge acquired in other modules of the curriculum.

Students who have participated were asked to evaluate various aspects of the challenge as well as self-reflect on what they gained through the process. The questionnaire distributed was electronic and data collected were anonymous. Answers were in a 1–5 Likert scale or in a "Strongly Agree" to "Strongly Disagree" scale, whereas there was an opportunity for free text comments.

We had 100% Agreement in the statements:

- The Robotics Challenge was a positive experience;
- The timing of the Robotics Challenge was good, taking into account my other study obligations in the Department;

[7]http://www.emethes.gr [inGreek].

[8]https://www.britishcouncil.gr/en/events/thessaloniki-science-festival.

- The Lego platform used for the Robotics Challenge was appropriate;
- LeJos used to program the robot for the Robotics Challenge was appropriate.

Students also found that the Robotics Challenge helped them understand the theory (score 4.6/5) and that their programming skills were enhanced (score 3.8/5). 80% of the students believed that the level of difficulty of the Robotics Challenge was just about right, with none of them finding it either too easy or too difficult. Before the Robotics Challenge students did not feel excessively interested in robotics (40% Disinterested or Neutral, 60% Interested), while after the challenge students reported a different level of interest (100% Interested or Extremely Interested). However, student opinions were split 50%-50% on whether there was enough time (2 working days) to complete the task assigned.

The following are some free text comments:

- *"Everything that was taught through all the lectures of the unit, was practically demonstrated in the challenge. It was extremely helpful to understand thoroughly the workings of the sensors, implications when building a robot, different constrains in the map etc."*
- *"The theory helped very much in the overall completion of the challenge"*
- *"It was a very fun process and very helpful"*
- *"A very very positive experience, it was one of the best moments of the semester. Even the mood of the class and the willingness to collaborate and compete at the same time was a very special experience"*
- *"In regards to the time, at the beginning, before starting the challenges it looked as long, but we had such a good time and time flew quickly"*
- *"The things done in the challenge were not of hard complexity, but the outcome of the things learned were of very high importance"*
- *"It was fun and helped us to gain knowledge. Furthermore presenting it to people outside the Department was very interesting"*

6 Conclusions

We have presented our experience in integrating agents and robotics in our Computer Science Curriculum. The concept of intelligent agents is spread throughout the modules at all years of studies. A specialised final year module is central to teaching and learning multi-agent systems and principles of robotics. NetLogo and Lego Mindstorms are used to facilitate hands-on practice with a small learning curve as well as assessment through simulation and realistic tasks, respectively.

We have also presented the Robotics Challenge which allows students to integrate the knowledge they obtained in a number of AI modules, and to practically apply knowledge and skills in order to solve a real problem. By breaking down the challenge into smaller challenges, students are asked to develop robots over a short period of two days without limiting the their ingenuity and inventiveness. The challenge is very well received by students since they have the opportunity to demonstrate their robots outside the University and thus promote Computer Science in general, and agents and robotics in particular.

References

1. Álvarez, A., Larrañaga, M.: Experiences incorporating lego mindstorms robots in the basic programming syllabus: lessons learned. J. Intell. Robot. Syst. **81**(1), 117 (2016)
2. Bloom, B., Krathwohl, D., Masia, B.: Bloom Taxonomy of Educational Objectives. MA. Pearson Education, Allyn and Bacon, Boston (1984)
3. Danahy, E., Wang, E., Brockman, J., Carberry, A., Shapiro, B., Rogers, C.B.: Lego-based robotics in higher education: 15 years of student creativity. Int. J. Advanc. Robot. Syst. **11**(2), 27 (2014)
4. Estrada, F.: Practical robotics in computer science using the Lego NXT, an experience report. In: Proceedings of the 22nd Annual Conference on Innovation and Technology in Computer Science Education, Bologna, Italy (2017)
5. Sakellariou, I., Kefalas, P., Stamatopoulou, I.: Enhancing NetLogo to simulate BDI communicating agents. In: Hellenic Conference on Artificial Intelligence, pp. 263–275. Springer (2008)
6. Sakellariou, I., Kefalas, P., Stamatopoulou, I.: Teaching intelligent agents using NetLogo. ACM-IFIP IEEIII, pp. 209–221 (2008)
7. Stamatopoulou, I., Fasli, M., Kefalas, P.: Introducing AI and IA into a non computer science graduate programme. In: Multi-Agent Systems for Education and Interactive Entertainment: Design, Use and Experience: Design, Use and Experience, pp. 89–100 (2010)
8. Wiens, J., Monett, D.: Using BDI-extended NetLogo agents in undergraduate CS research and teaching. In: Proceedings of the International Conference on Frontiers in Education: Computer Science and Computer Engineering (FECS). The Steering Committee of The World Congress in Computer Science, Computer Engineering and Applied Computing (WorldComp) (2013)
9. Wilensky, U., Rand, W.: An Introduction to Agent-Based Modeling: Modeling Natural, Social, and Engineered Complex Systems with NetLogo. MIT Press (2015)

Author Index

M. Ivanović et al. (eds.), *Intelligent Distributed Computing XI*,
Studies in Computational Intelligence 737, DOI 10.1007/978-3-319-66379-1

Printed in the United States
By Bookmasters

Printed in the United States
By Bookmasters